W9-CKH-401

Membranes in Ground Engineering

Scanning electron photomicrograph of a thin permeable membrane demonstrating melded structure and variable pore sizes. ($\times 22$ and $\times 88$) (Courtesy ICI Fibres.)

Membranes in Ground Engineering

P.R. RANKILOR
Manstock Geotechnical Consultancy Services Ltd.

A Wiley-Interscience Publication

JOHN WILEY & SONS
Chichester — New York — Brisbane — Toronto

Library of Congress Cataloging in Publication Data:

Main entry under title:

Rankilor, P.R.
Membranes in ground engineering.
1. Membranes (Technology)
I. Title
624'.177 TA660.M/ 80-40504

ISBN 0 471 27808 4

Typeset by Photo-Graphics, Yarcombe, Nr. Honiton, Devon.
Printed by The Pitman Press, Bath, Avon.

Contents

Acknowledgements

I would like to thank all the manufacturers of Membranes who have corresponded with me and who have supplied me with the technical information upon which this book is based. In particular I would like to thank ICI Fibres for their support, and for the practical information supplied by their Technical Development Department on the subject of laboratory equipment and membrane testing. Finally, I would like to thank my wife and family for their support during the writing of the book and all of the many people who have been associated with the artwork and preparation of the text.

P.R. RANKILOR

Preface

The objects of this book are threefold; first to outline the author's view of current design procedures in the installation and use of permeable membranes in Civil Engineering Ground Structures. Secondly, to provide a concise list and comparative reference of as many of the commercially-available membranes as possible at the time of writing. Thirdly, the book is intended to show how to install the membrane in the ground, and how to set up adequate testing facilities both in the field and laboratory to ensure that the correct product is being received on site, and that the specified quality is being maintained.

The author has collected and examined over 130 membranes from 32 different manufacturers all over the world, and has photographed each one and shown each one at a uniform scale with the standardised description, alongside the manufacturer's published technical data.

It is hoped that this book will form a useful guide and standard reference. However, owing to the fact that the author has tried to avoid theoretical comment and ambiguity, by referring in the text to particular membranes, the book will no doubt need periodic up-dating. In this regard, the author would be pleased to hear from any manufacturer, designer, or specifier on any subject contained within this book in order that future editions may be up to date, more accurate, and more useful.

Section 1

General aspects of membranes

1

Historical development of membrane utilisation

Although not necessarily true of all scientific subjects, it is singularly apparent that in the field of Civil Engineering, many of today's techniques and 'discoveries' have already been preceded by Nature in the simple accomplishments of animals constructing their homes or modifying their habitats. Indeed, the modern engineer is often surprised to find that Early Man possessed a surprising flare for conceptualising engineering principles, whilst only lacking the numeracy to quantify his ideas, and the technical proficiency to extend them. Within the limits of such raw materials as were available, Palaeolithic Man had already learned to exercise some control over his environment, and had also begun to adapt natural materials to perform new tasks. The use of clothing is a typical example of the adaptation of a natural material to a previously non-existent purpose, allowing Homo Sapiens to, in effect, modify his micro-environment and thus extend his geographical habitat to eventually cover the whole of the globe.

It should be recognised that from Man's first awakening through millenia, to within only a decade of the present day, virtually all technical and engineering development has been based upon the direct control and application of isolated principles and materials. Only in the last few years has the concept of 'Composite Materials' been developed, and their potential realised. As recognised earlier, Nature had already pointed the way in the home-building techniques of such simple creatures as House Martins and Swallows who make nests from a combination of mud and straw (see Figure 1.1). The mud is coated onto the straw, with the critical result that the final composite material exhibits properties which would be lacking in each of the component parts. The essential success of a composite is that it should combine singular properties of two or more components in some unlikely combination, producing a result which exhibits substantially beneficial properties for some special purpose. In general engineering terms, one may look at the concepts of GRP (glass reinforced plastic) or carbon-fibre composites, as typical examples where strength, flexibility, corrosion resistance and lightness of weight are all produced as an end product of the specialised combination of filaments and polymer.

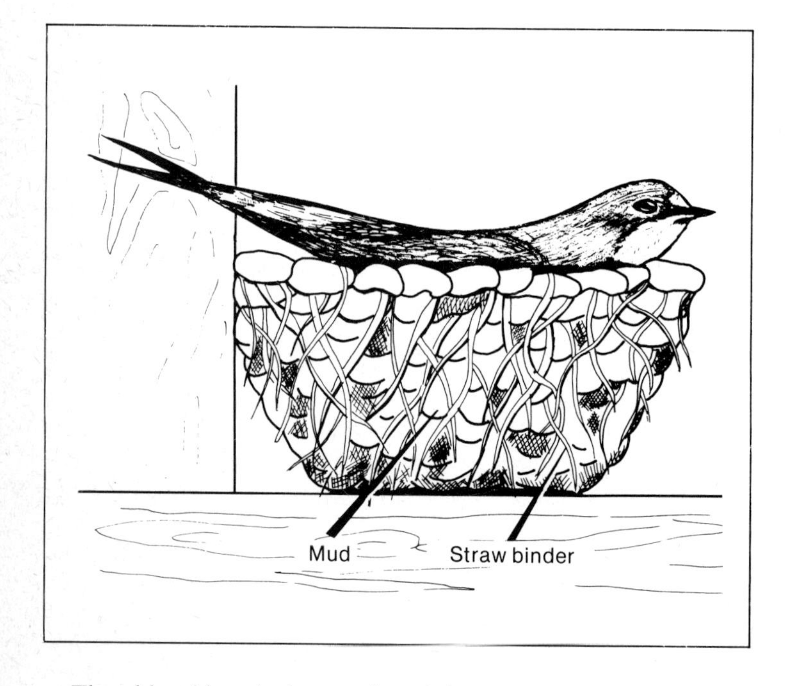

Figure 1.1 The swallow's nest provides a good example of a ''reinforced'' composite structure

The oldest historical examples of the use of 'fabrics' as an aid to road construction over soft ground include the use of woven reed mats by the ancient Romans. In a style remarkably similar to our present-day techniques, they would lay the mats over marshy ground before overlaying with stone. Even before this, pathways were constructed over logs laid in soft ground, as evidenced by archaeological findings in the United Kingdom dating back to 2500 BC. Here, in the Southern English marshes between the Polden Hills and the Mendips, wooden tracks were built — using brushwood mattresses of birch and bundles of hazel branches. Many kilometres of track were built in this way, testifying to the success of the technique. In the Middle and Far East, for millenia it has been the practice to reinforce large earth structures with reeds, rushes or bamboo. Similarly, soil-strengthening by the driving of bamboo or mangrove tree piles at frequent intervals into soft coastal muds still comprises one of the most universally applied techniques for stabilising building foundations in tropical countries.

In a first appreciation of membranes in a Civil Engineering ground environment, it is important to realise that such membranes, whether acting as filters, separators or reinforcing elements, are creating new composite materials hitherto naturally unavailable, and displaying exceptional properties which can be utilised by the Engineer in the construction of virtually any type of Civil Engineering endeavour to the ends of financial saving, increased efficiency and better performance.

The reason for the recent rapid development of composite structures is the ever-accelerating development of technical expertise and materials in engineering generally. In particular, the synthesis of strong, non-degradable synthetic textile materials has permitted the inclusion of modern membranes in long-term Civil Engineering structures, such as coastal defences, dams and major highways.

Coming to more recent times, the industrial development of modern synthetic polymer chemicals such as polyamides and polyester has really taken place subsequent to 1940. Although one cannot be certain, it is believed that the earliest applications of permeable membranes were in filter structures in the United States of America. The Carthage Mills Company claim to have originated the concept of replacing graded filter systems and gravel blankets with plastic filter cloth in 1958. Certainly, it can be said that published data prior to that time would be difficult to find, and it is equally certain that this then-revolutionary concept was slow to be taken up by the conservative world of Civil Engineering. Many eminently referrable books published or revised as late as 1967 still bore no reference to the use or design of membranes in soil structures.

The 1967 revision of the famous textbook *Soil Mechanics in Engineering Practice* by Terzaghi and Peck[1] bore no mention of permeable membranes, and neither did the more specialised publication of the same year — *Seepage, Drainage, and Flow Nets* by Cedergren.[2]

However, despite the lack of general recognition of geo-fabric technology at this time, a few papers were published somewhat sporadically, the earliest known to the author being that by Agerschou in 1961.[3] In this paper Agerschou described the use of woven materials to protect coastal structures from soil migration and eventual collapse. The Japanese were quick to see the potential of this new technology and doubtless, as in the case of America, the existence of a large commercially orientated cloth-producing industry would be no disincentive to the rapid development of the field. By the time of the 1966 Coastal Engineering Conference in Tokyo, still few papers had been published, so the paper on membranes by Barrett[4] proved to be somewhat of a landmark leading the way to an ever-increasing awareness of the uses of membranes in Civil Engineering structures.

It is important to recognise that this earlier work was being conducted primarily with woven fabrics, and that the advent of the 'designed' non-woven membranes was to come much later in the mid-seventies. Investigations by Healey and Long in 1971, and Calhoun[5] in 1972 showed that woven fabrics could be successfully used in drainage applications for lining the periphery of linear drains, as an integral part of prefabricated drains, and as a protective wrapping for collector pipes. The Calhoun investigation also produced design rules which enabled engineers to choose the optimum *woven* fabric to suit the parent soil.

During the early seventies, the Japanese-developed filter fabrics (based upon their available weaving resources) were being used in, and exerting influences on South East Asia designs for coastal works. By the early seventies, woven materials of both American and Japanese manufacture were being used as far south as Singapore. The Dutch and mainland Europeans were also well aware of the technical and commercial potential of this new field, but whereas the American development research was spearheaded by the Corps of Engineers, the Europeans had several eminent academic and government institutions to take up the work.

As can be seen in Figure 1.2, by the mid 1970s, the United Kingdom had started to produce ground membranes, and at this time, firms such as ICI in the UK, Rhone Poulenc in France, Chemie Linz in Austria, and Du Pont in America, started to promote the use of non-woven membranes into a field which had thus far been exclusively the realm of woven materials.

It is possible that the Americans, having such a large internal market, concentrated on home development and sales to the exclusion of the potential world market. On the other hand, the Europeans, and the British, through their long-standing trading history, reached out almost immediately for the vast potential markets of Africa, the Middle East and South East Asia. It is fascinating to observe that membranes have been introduced into different countries of the world through the long-established historical trading links set up between one country and another during the 19th Century. The French influenced West Africa and Indo-China, the British made the early membrane trials in Malaysia and Australia, and the Australians in turn initially promoted non-wovens into Indonesia and wovens back into the UK. British influence in the Middle East has been strong, and much technical innovation has taken place there. Although, as shown in Figure 1.2, some countries are still generally unaware of the design potential of membranes in soil

Figure 1.2 (Opposite) The world development of membrane utilisation

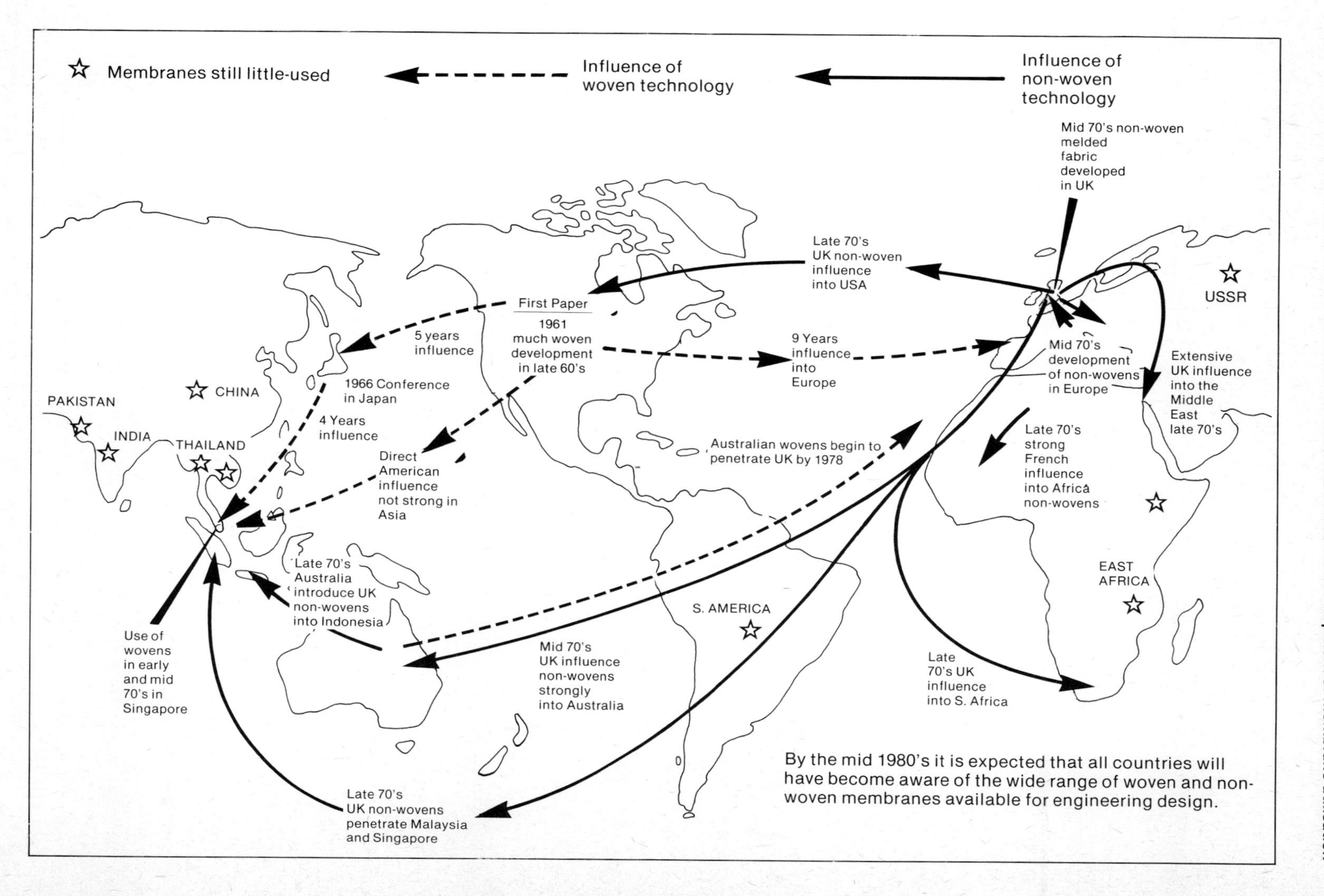
Membranes still little-used
Influence of woven technology
Influence of non-woven technology
Mid 70's non-woven melded fabric developed in UK
Late 70's UK non-woven influence into USA
USSR
First Paper
1961 much woven development in late 60's
5 years influence
9 Years influence into Europe
Mid 70's development of non-wovens in Europe
Extensive UK influence into the Middle East late 70's
PAKISTAN
CHINA
1966 Conference in Japan
4 Years influence
INDIA
THAILAND
Direct American influence not strong in Asia
Australian wovens begin to penetrate UK by 1978
Late 70's strong French influence into Africa non-wovens
Late 70's Australia introduce UK non-wovens into Indonesia
EAST AFRICA
S. AMERICA
Use of wovens in early and mid 70's in Singapore
Mid 70's UK influence non-wovens strongly into Australia
Late 70's UK influence into S. Africa
By the mid 1980's it is expected that all countries will have become aware of the wide range of woven and non-woven membranes available for engineering design.
Late 70's UK non-wovens penetrate Malaysia and Singapore

structures, where membranes are known the large tenders for major industrial projects ensure that a rapid dissemination of information follows on all the different makes and types of membranes. In terms of information, there are now increasing numbers of publications on the special subject of Membranes in Ground Engineering. From the first publication in 1961, the number of papers published annually must closely resemble the qualitative graph in Figure 1.3.

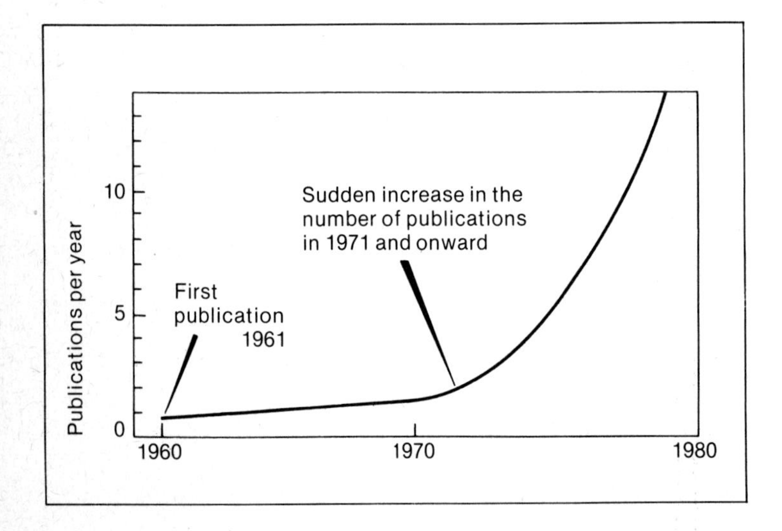

Figure 1.3 Growth in the number of publications on the subject of Civil Engineering Membranes

One unfortunate historical legacy resulting from earlier development work having been primarily undertaken with woven fabrics is that a number of authoritative papers were published comparing woven membranes with earlier inefficiently thick non-woven membranes. Because of the impartiality of the papers, brand names were not mentioned, and yet these papers generally concluded that 'woven materials were superior to non-wovens'. When read today, these papers can lead to a most serious misunderstanding in view of the many new purpose-designed two-dimensional non-woven membranes available today. Recent research works by separate independent sources such as List[6] and the Delft Hydraulics Laboratories[7] show that modern non-woven fabrics can perform as well as or better than woven materials, for example, when subjected to dynamic filtration conditions. This has critical relevance to all coastal structures subject to hydraulic-reversal caused by waves and tides. Similarly the high extensibility of modern non-woven membranes could not be taken into account in early comparative works. In view of this the results of any comparison tests undertaken before about 1974 must be viewed with great care.

Of course, a considerable amount of technical material has been issued by the manufacturers of the various commercially available membranes, and much of that information is presented later in this book. However, at the time of writing, no textbook has been published solely on the subject of membranes in soil structures, and to the author's knowledge the *Designing with Terram* manual produced and distributed by ICI was the first publication made with an overall preliminary attempt at defining design criteria, although it only covered the ICI range of non-woven and woven membranes and had itself been preceded by Calhoun's work on drains[5] and Sweetland's unpublished Master's Thesis[8] on non-woven filter screens. It is felt that there has long been a need for a design guide covering the full range of all membranes, and the Author hopes that this book will help to provide for that need.

2

Aspects of soil mechanics relevant to membrane design

Civil Engineering membranes come into contact with basically two types of soil — granular (such as sand, gravel or crushed rock) and cohesive (such as clay). The critical differences between the two soils are the size of the particles and their chemical compositions. Granular soils have large particles with large pore spaces between. Water can flow relatively easily through these spaces either through gravitational force or hydraulic pressure. In cohesive soils the particles and the pores between are so small that the water has great difficulty passing through. Between these two extremes lies a range of soils which are mixtures of the two, containing various proportions of each, and which are called granular/cohesive soils.

The weight of a granular soil, or the loading upon it, is transmitted downwards through the points where the solid particles touch each other. When water is present within a granular soil, it fills the spaces between the solid particles, and its upper surface is called the 'water table' (see Figure 2.1).

Granular soils can be fine (down to 0.6 mm diameter) or coarse (up to 60 mm diameter or larger). They can be well graded or uniform, angular or rounded (see Figure 2.2).

Granular soils can be loosely packed or tightly packed. This state of compaction is critical to their work capability. When granular material is dumped from a vehicle, it adopts a loosely packed structure, and always needs artificial compaction. The best way to compact a granular soil is by vibration at a frequency of around 25 c.p.s. (for sands) to 35 c.p.s. (for crushed rock). Dead weight rolling will have very little effect.

Moisture control of a pure granular soild being vibrated is not necessary, but as with all soils the correct amount of water in the

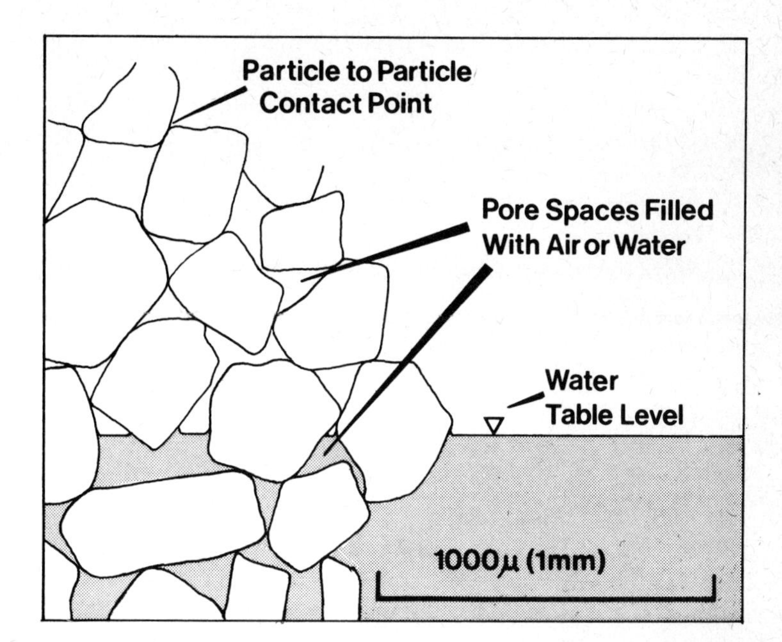

Figure 2.1 Granular soils have particle-to-particle contact

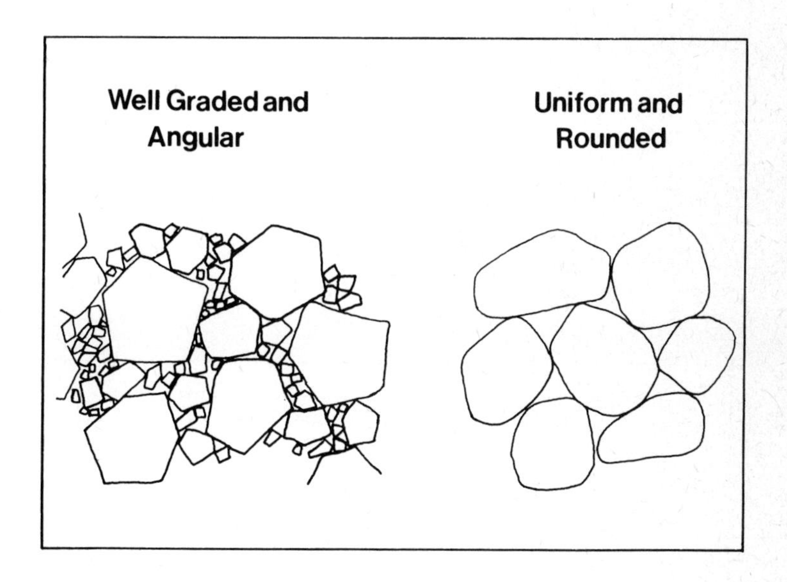

Figure 2.2 Granular soils vary considerably in texture

soil can help to produce optimum compaction. For example vibro-compaction of a completely saturated granular solid is very difficult.

The most important way of defining a granular soil is to pass it through a number of increasingly fine sieves, and plot a graph of the percentage *weight* passing each sieve (Figure 2.3).

The particle distribution diagram defines the soil, and its likely behaviour. The degree of uniformity affects the shape of the curve. The steeper the curve, the more uniform the soil (Figure 2.4.)

A popular method of describing the particle size of any soil is to use the following terms:

D_{60} Being the sieve size passing 60 % of a representative sample by weight.

D_{10} Being the sieve size passing 10 % of a representative sample by weight.

$U = \dfrac{D_{60}}{D_{10}} =$ Uniformity Coefficient. Gives a quantitative description of the steepness of the curve slope.

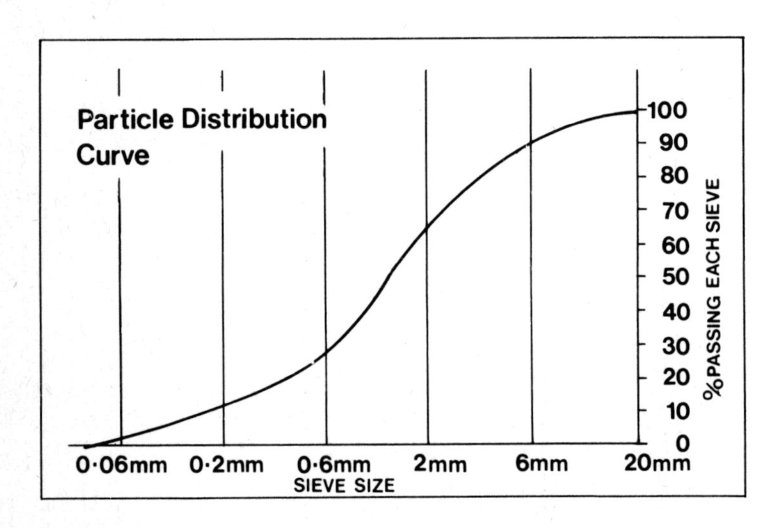

Figure 2.3 Particle distribution diagram describing the variation of particle sizes in a soil

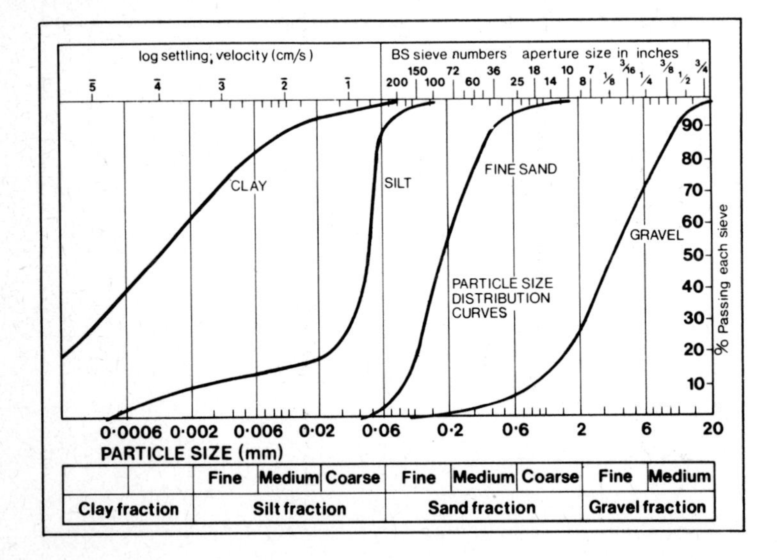

Figure 2.4 Particle size distribution curves for a variety of different soils

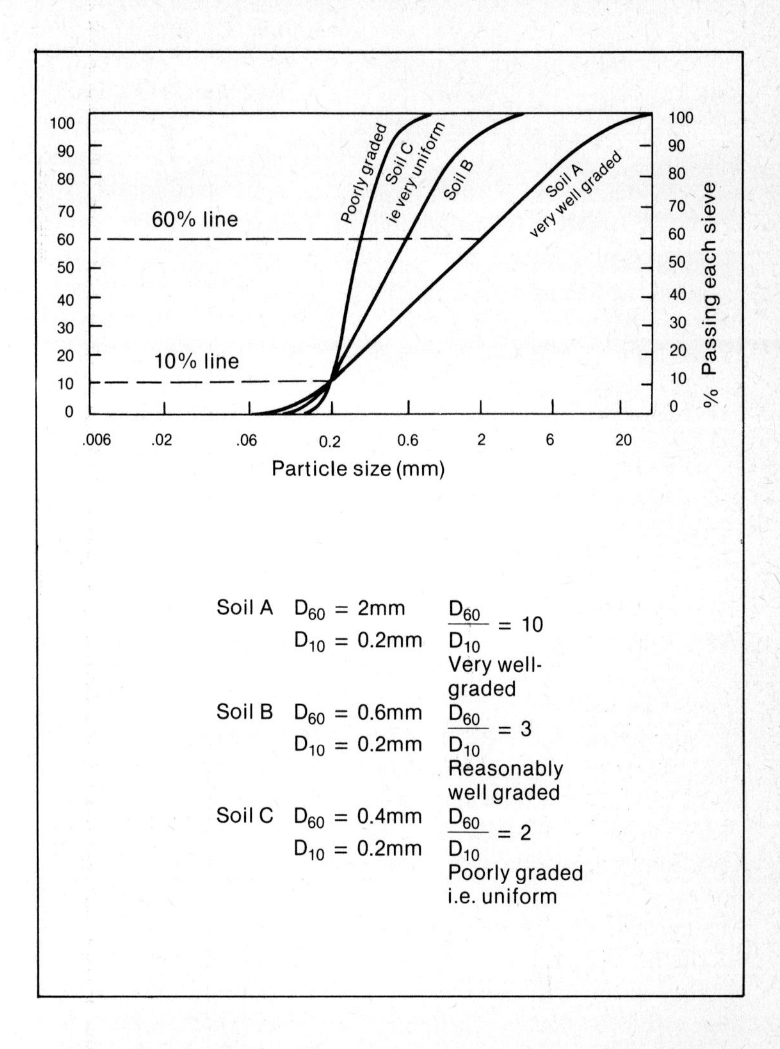

Figure 2.5 The slope of the particle distribution curve describes the uniformity and degree of grading of a soil

As shown in Figure 2.5, the smaller the value of U, the more the soil is comprised of equal-sized particles. The larger the value of U, the better graded (i.e. less uniform) the soil. A gravel with $U = 4$ is classed as 'well graded', but a sand must have a $U = 6$ to be classed as 'well graded'.

Cohesive soils

In a clay structure, there are places where negative and positive charges are adjacent. These attract each other, and resist separation of the particles. These forces give rise to the clay's cohesive strength, and thus its ability to resist shear.

The weight of a cohesive soil, or the loading upon it, is transmitted downwards through the points where the water jackets contact one another. Since these contact points are water-to-

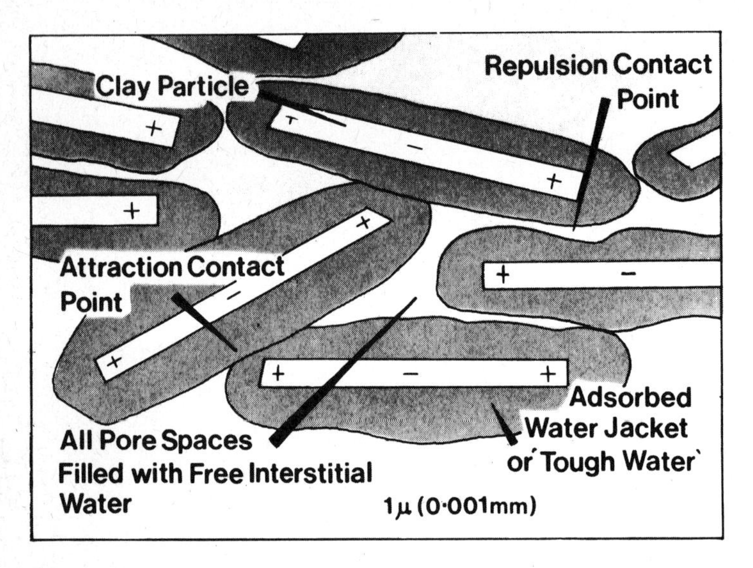

Figure 2.6 The solid particles in a clay are always separated by water

water, the behaviour of cohesive soils is fundamentally different from that of granular soils. For example, under loading, the particle-to-particle geometry can become flattened thus forcing out the interstitial water. With time, as water is squeezed out, the soil decreases in volume, and thus consolidates. This process can take many years to complete. Unlike a granular material, a clay can swell by absorbing more interstitial water if the load is removed or if more water is introduced to it.

It is not correct to try to compact a cohesive soil by vibration. Large pneumatic-tyred rollers, or ideally sheepsfoot rollers should be used on pure clay soils.

The most useful way of classifying a cohesive soil is by the indices known as the atterberg limits. These limits are the percentage by weight of water content at which a particular clay ceases to act as a non-plastic solid, and begins to act as a plastic solid (Plastic Limit, i.e. PL), and the percentage at which the soil begins to act as a liquid (Liquid Limit, i.e. LL). (See Figure 2.7.) A small Plasticity Index (PI) combined with a low LL, means that the soil will quickly turn from a non-plastic solid to a liquid with the addition of only a small amount of water during trafficking.

The tests to arrive at these limits are very simple and are advantageous in that they are performed on disturbed samples, so crude sampling methods will suffice in order to obtain material for testing.

In 1947, Casagrande[9] produced a paper on the use of Atterberg Limits to identify the soils. His Plasticity Chart below shows that a knowledge of the PL and LL can allow the identification of any basic cohesive soil. (Figure 2.8.) A more detailed explanation of Figure 2.8 can be found in such textbooks as Scott.[10]

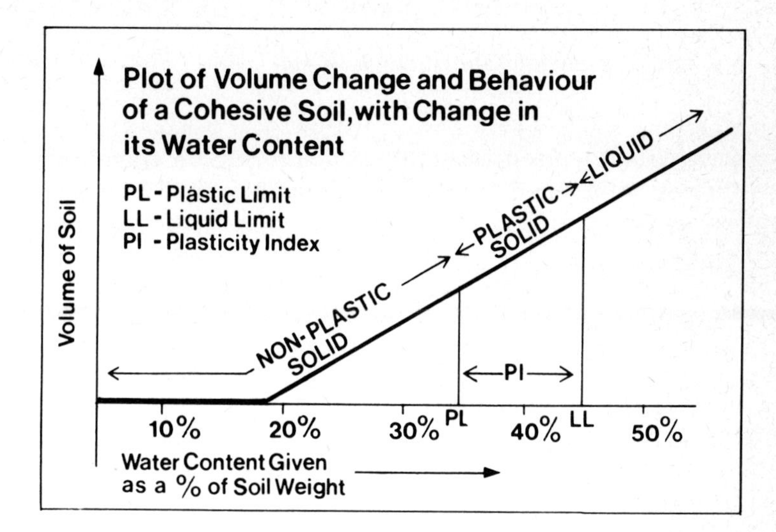

Figure 2.7 Atterberg Limits

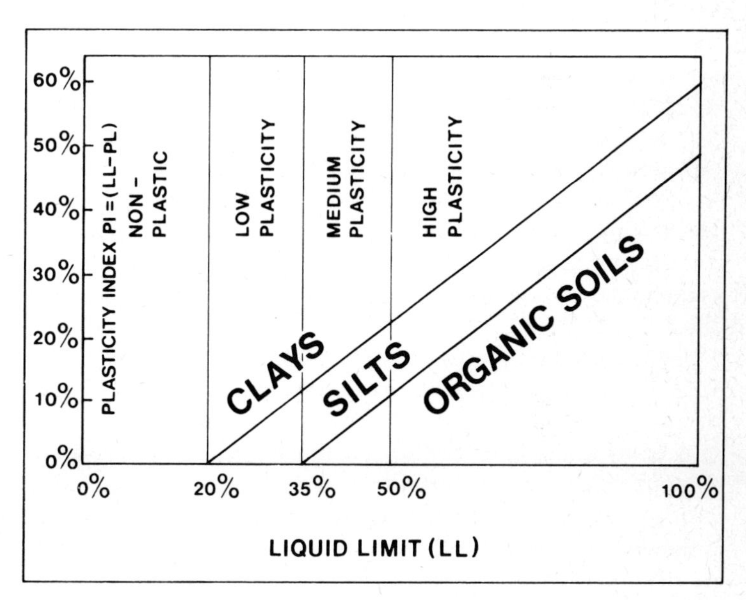

Figure 2.8 Soil Classification Groups

Granular/cohesive soils

Many, if not most natural soils are a mixture of granular and cohesive materials. This is a consequence of natural weathering and sorting processes, and reflects itself in the shape and position of a soil's particle distribution diagram (Figure 2.9). If a soil contains so much clay material that the larger particles are completely separated by it, then the bulk material will behave effectively as a clay. At the other extreme, if so little clay is present that it merely coats the larger grains, then the soil can behave granularly and have a high permeability.

Mechanical behaviour and strength of soil

Whether a soil is a pure clay, or whether it is behaving 'effectively' as a clay, its mechanical reaction to stress is completely different from that of a pure granular or 'effectively granular' soil. However, many soils exhibit mixed properties of both types.

A cohesive soil has an internal strength *without* being confined. It resists deformation by squeezing or shearing. This natural resistance to deformation under stress is called its 'cohesive strength' or 'cohesion' and is denoted in soil mechanics by the letter C.

A pure granular soil has no internal strength, and can only resist deformation if it is confined either within an artificial structure, or by other soil in the ground. Therefore it has no cohesive strength C, but has a different property ϕ (phi).

ϕ is an indication of how strong a granular soil becomes with increasing confining pressure. For example, both a uniform and a well-graded sand will have no strength at zero confining pressure, but if both are confined to the same pressure, the well-graded sand may have acquired considerably more strength than the uniform sand. This would be by virtue of the better interlocking and greater frictional surface area of the graded sand. Figure 2.10 shows that the value ϕ is an angle measured on a graph, and is the angle at which the sand shears internally. It is called the internal angle of friction.

In general terms, therefore, since confining pressure increases with depth in the ground, the strength of a granular soil increases with depth.

In contrast, a *pure* cohesive soil has an internal cohesive strength which does *not* increase with confining pressure. Therefore, a graph of shear strength against confining pressure will be as shown in Figure 2.11. Note that in this case $\phi = 0°$. These soils are therefore known as ϕ zero soils'. In such a soil the strength of the soil does not increase with depth.

As stated earlier, the most common type of soil is one which exhibits both cohesive and granular properties. The cohesive material in the soil gives it an internal cohesive strength, and the granular fraction gives the soil increasing strength with increasing confining pressure. Such a soil therefore has a value of C *and* a value of ϕ, and is known as a 'Cϕ' soil (see Figure 2.12).

The practical significance of the above is that a 'Cϕ' soil will support a larger foundation stress at depth than it will at the surface of the ground. The deeper foundations are placed, the stronger the soil. For design purposes, soil samples can be taken and tested in the laboratory at confining pressures equal to those

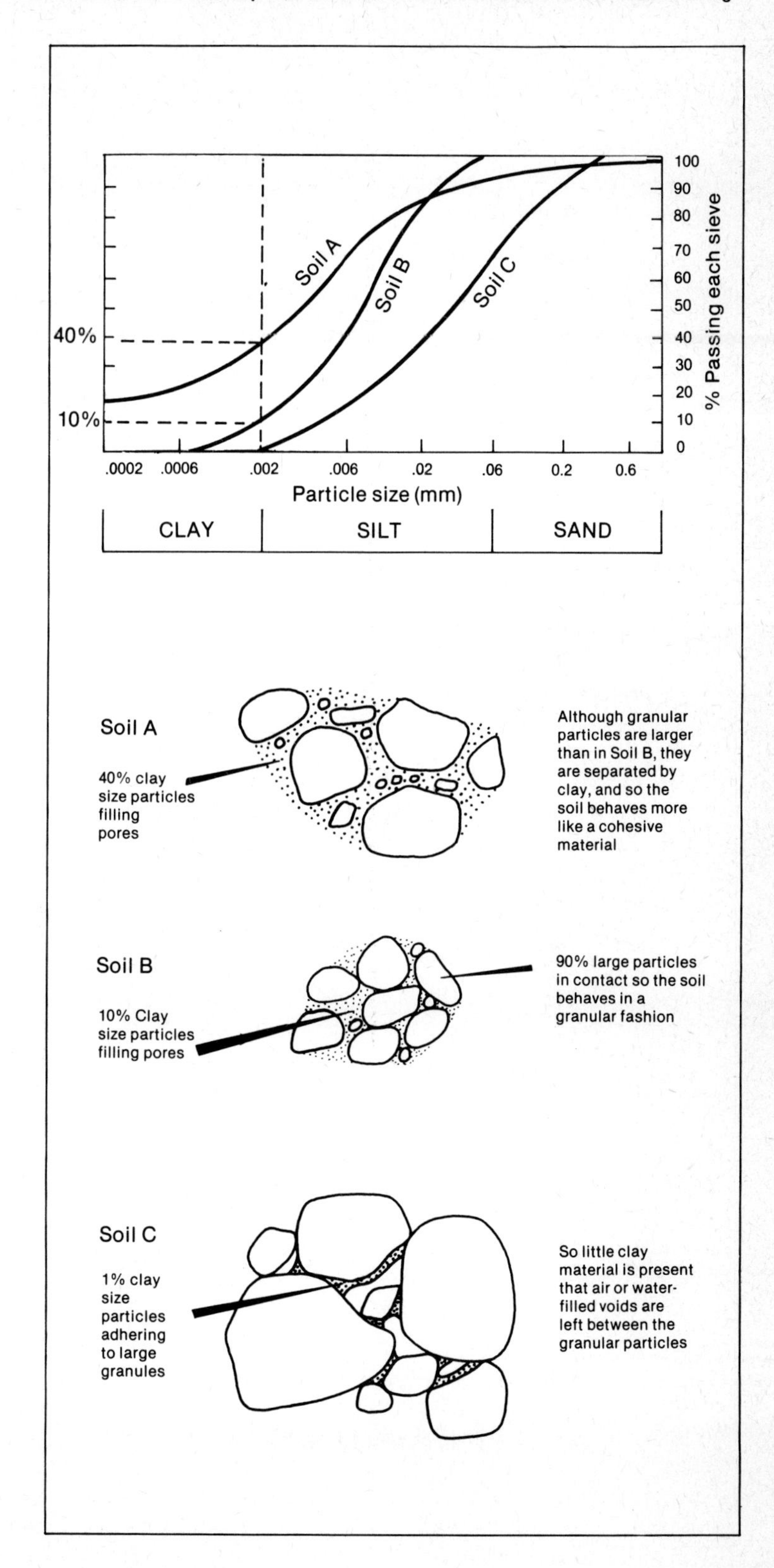

Figure 2.9 The percentage of clay-size particles in a soil governs the behaviour of the soil

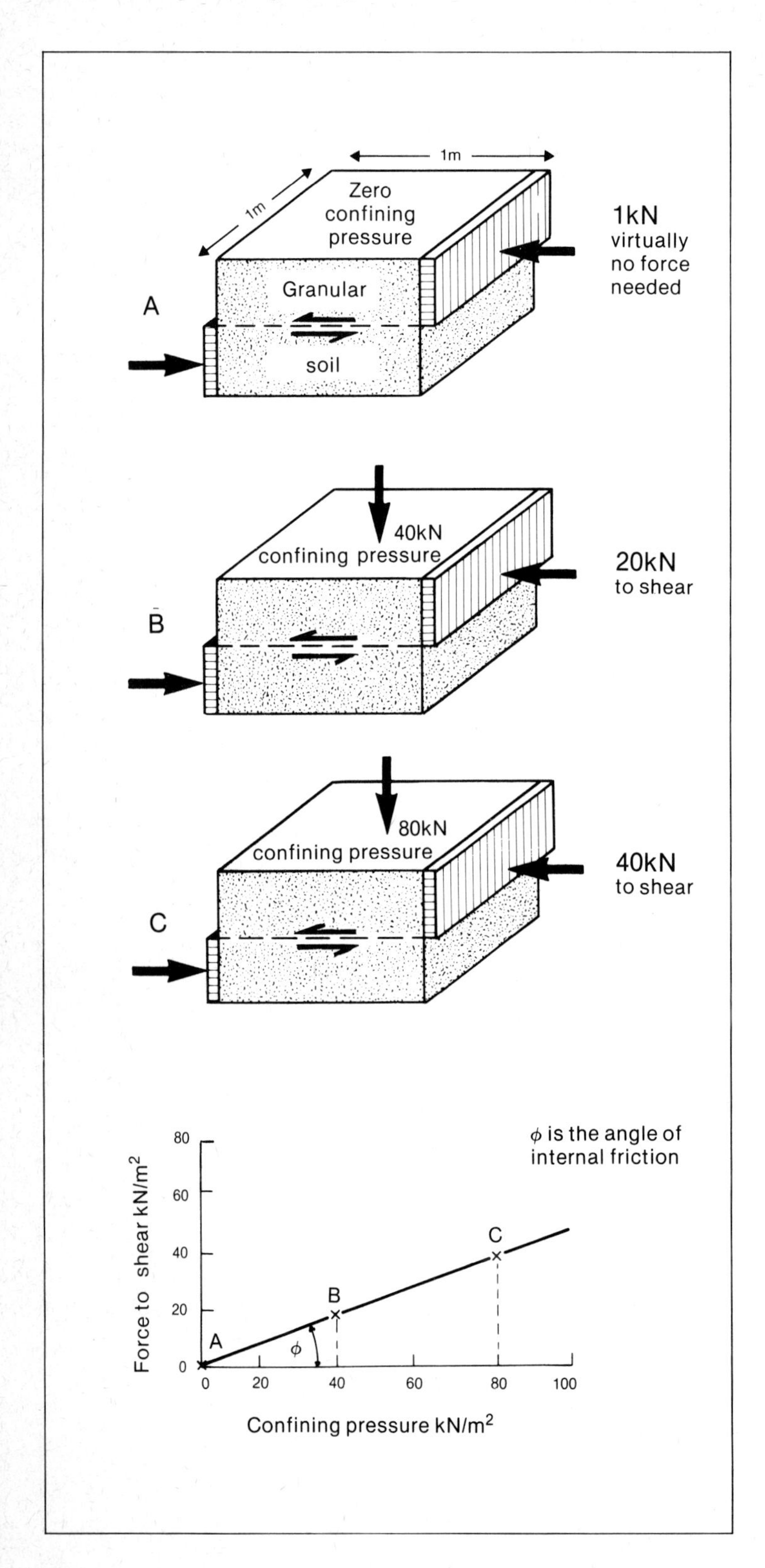

Figure 2.10 The force needed to shear a *pure* granular soil depends upon the confining pressure

Figure 2.11 The force needed to shear a *pure* cohesive soil does not depend upon the confining pressure

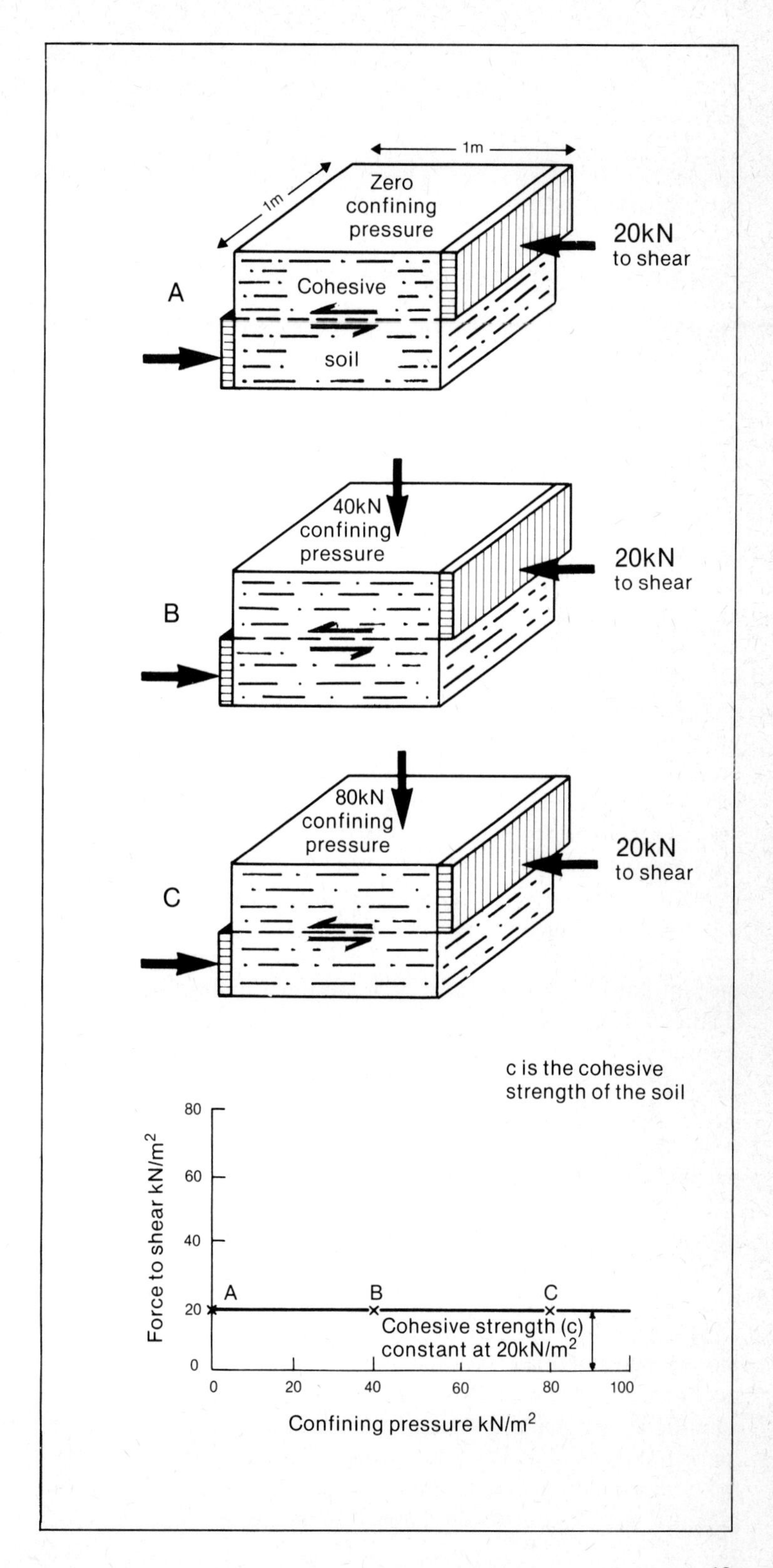

which are to be experienced in the ground below a proposed structure. This gives a realistic picture of the likely soil behaviour both during and after the construction process.

Soil consolidation

When a soil is subjected to loading (as in land reclamation works, or embankment construction), if it is granular, then a small amount of immediate settlement takes place and no more. This is because the load is carried downwards through particle contact points. Even if, as is often the case, the sand is saturated, and the water between the particles is subjected to temporary hydraulic pressure, it does not take part in carrying the foundation load since the permeability of the sand is so high that the water can easily move between the particles and pass out of the area of high pressure. Thus, in a sand any pore water pressures generated by surface loading can be dissipated within seconds or minutes.

The effect of overburden loading on to soils with fine particle constituents such as clay or silt is quite different. In a clay soil, as described earlier, the particle sizes are very small and the water between the soil particles cannot escape easily when pressure is applied. Consequently, although an embankment may take one month to construct, the water from between the particles may take twelve months to drain away from the area of high pressure generated beneath the embankment, into the low pressure areas on either side. Consequently, immediately subsequent to the embankment construction, a large part of the weight of the embankment is being carried by the water between the soil particles in the form of a confined hydraulic pressure (Figure 2.13). Naturally, as the interstitial water is forced out from beneath the embankment, the volume of the soil beneath the embankment decreases and the embankment will settle downwards. This reduction in the volume of the soil, which is a combination of water migration and realignment of the clay particles, is known as 'consolidation'.

When a thick layer of slow-draining soil such as clay is placed under pressure from a building foundation or from land reclamation fill placed on top, then the pore water travels to the nearest dissipation point. This may be the surface, it may be an underlying permeable layer, or it may be intermediate thin layers of more permeable material such as sand. Figure 2.14 shows how pore water pressure dissipates, given different drainage conditions. In general terms, the pore water pressure stays highest in those regions furthest away from any permeable zone. An understanding of the likely location of high pore water pressure zones is important as part of the design capability for

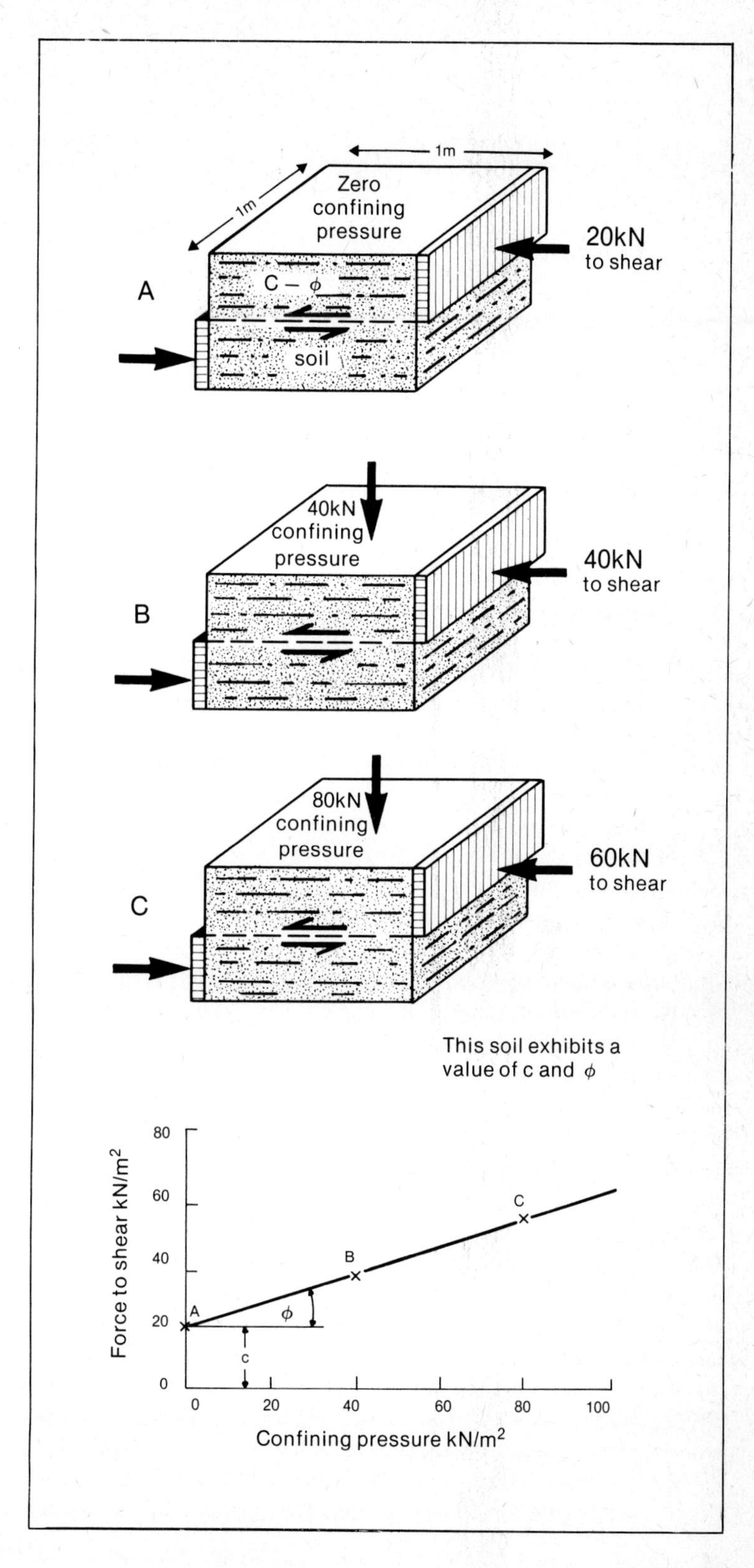

Figure 2.12 The force needed to shear a cohesive/granular c-ϕ soil depends both upon its cohesive strength and the confining pressure

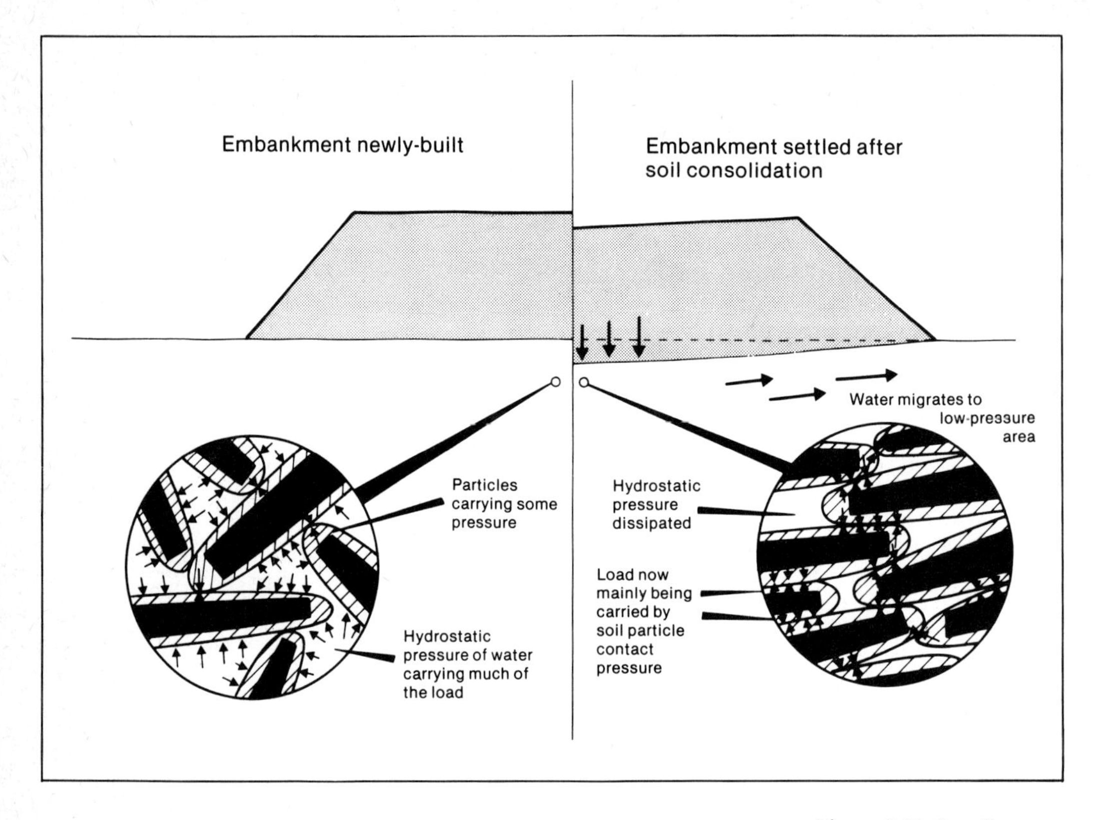

Figure 2.13 Loading causes soil consolidation by driving out pore water and re-arranging the particles

placing membrane drains into the ground. Drains designed for this purpose should penetrate into high pressure areas to allow the migration of water as rapidly as possible.

The reason why pore water pressures are dangerous in soil situations is that they enable loads to be transmitted through the soil by hydraulic pressure instead of by particle-to-particle contact. In practical terms, the more that the soil weight is being carried by hydraulic pressure instead of through the soil, the more the mass of soil is acting like a liquid. Consequently, the greater the pore water pressure, the greater the possibility of the soil failing.

The effect of water content on soil compaction

The presence of too much water in a soil invariably leads to its weakening, and to its inability to perform properly in a working environment. However, during the compaction of a soil for site construction work, both too much and too little water are disadvantageous. It is necessary that for maximum compaction, the soil should be at its optimum moisture content. Figure 2.17 is after Terzaghi and Peck[11] and shows how variations in moisture

Figure 2.14 (Opposite) *Excess* pore water pressure generated by application of a surcharge. Diagrammatic indication of change in vertical distribution with time

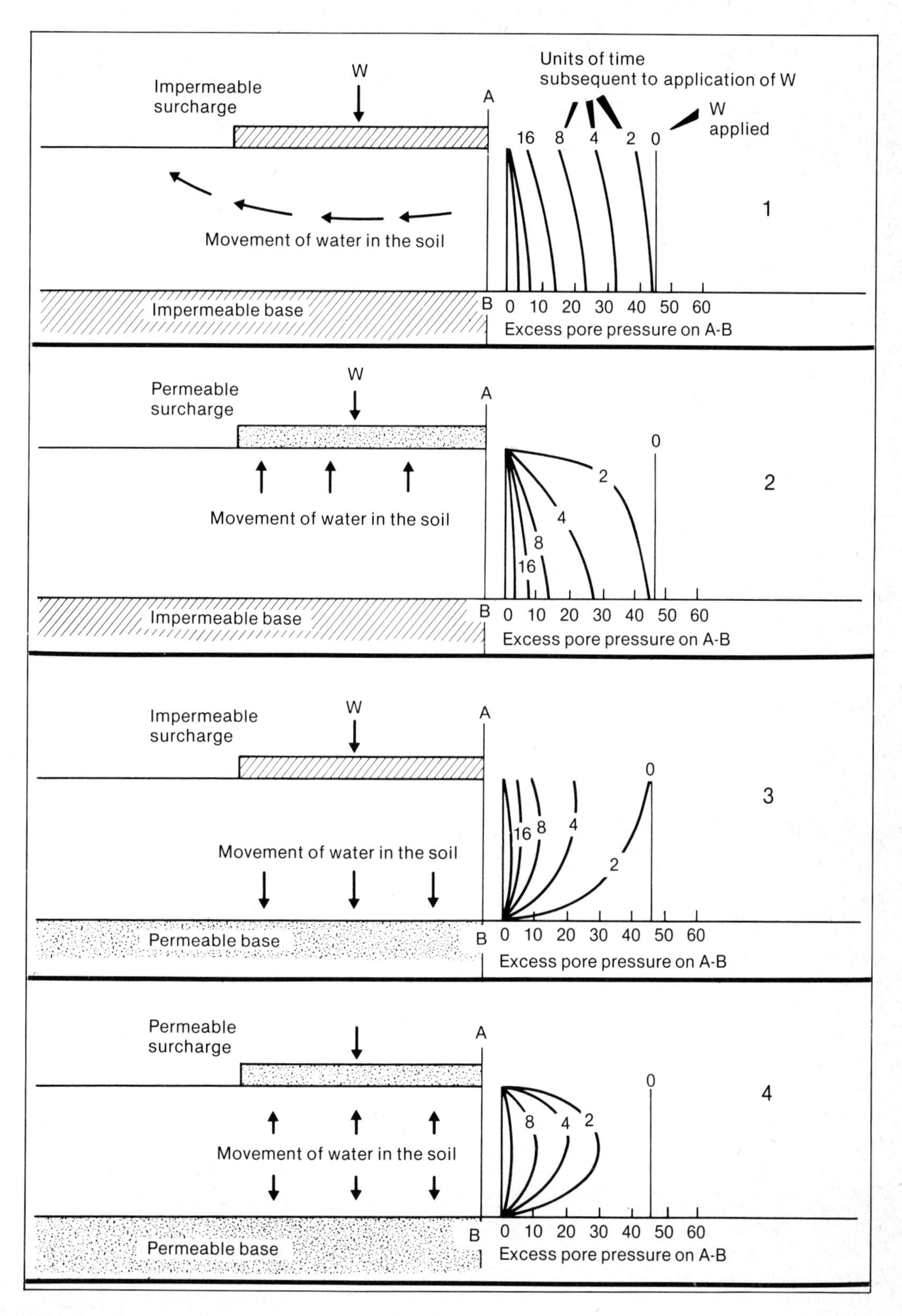
Impermeable surcharge
W
A
Units of time subsequent to application of W
W applied
16 8 4 2 0
1
Movement of water in the soil
Impermeable base
B 0 10 20 30 40 50 60
Excess pore pressure on A-B
Permeable surcharge
W
A
0
2
4
8
16
2
Movement of water in the soil
Impermeable base
B 0 10 20 30 40 50 60
Excess pore pressure on A-B
Impermeable surcharge
W
A
0
16 8 4
2
3
Movement of water in the soil
Permeable base
B 0 10 20 30 40 50 60
Excess pore pressure on A-B
Permeable surcharge
A
0
8 4 2
4
Movement of water in the soil
Permeable base
B 0 10 20 30 40 50 60
Excess pore pressure on A-B

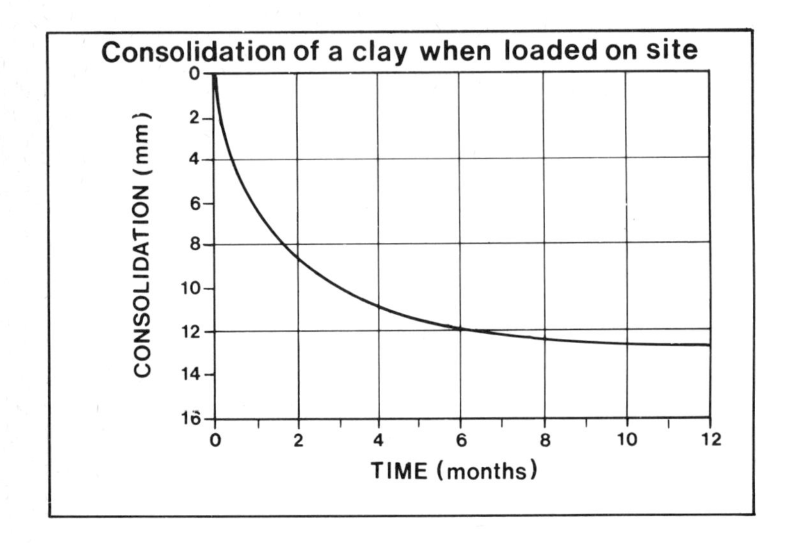

Figure 2.15 Diagrammatic indication of rate of change of consolidation in soil subsequent to loading

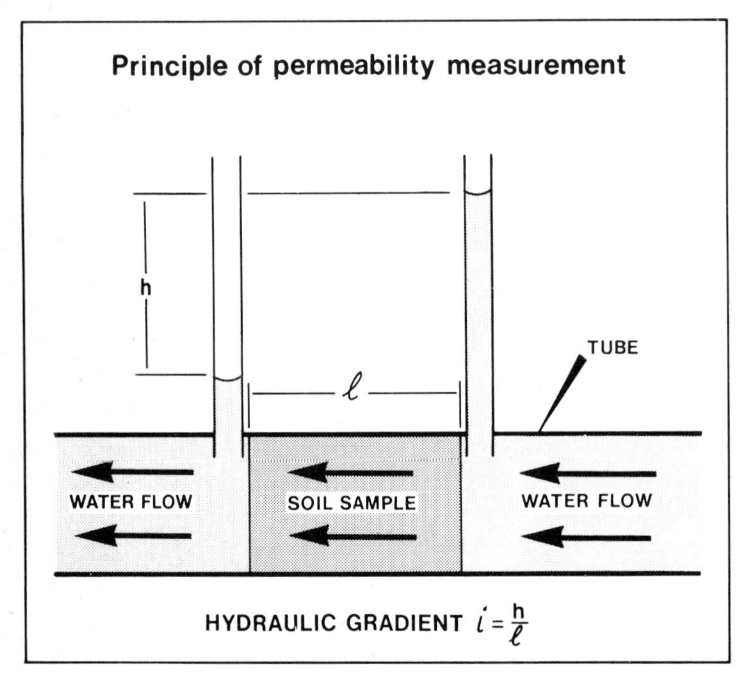

Figure 2.16 Laboratory set-up for measuring soil permeability

content can lead to differences in achieved compaction for a given number of machine passes.

Prior to site work starting therefore, it is good practice to have samples of all materials due to be compacted taken to a laboratory for testing. The compaction test is generally a Standard Proctor, or modified AASHO Test, where the prepared specimen is subjected to repeated blows in a metal cylinder at increasing moisture contents. The curves represent typical values which can be achieved with different soils.

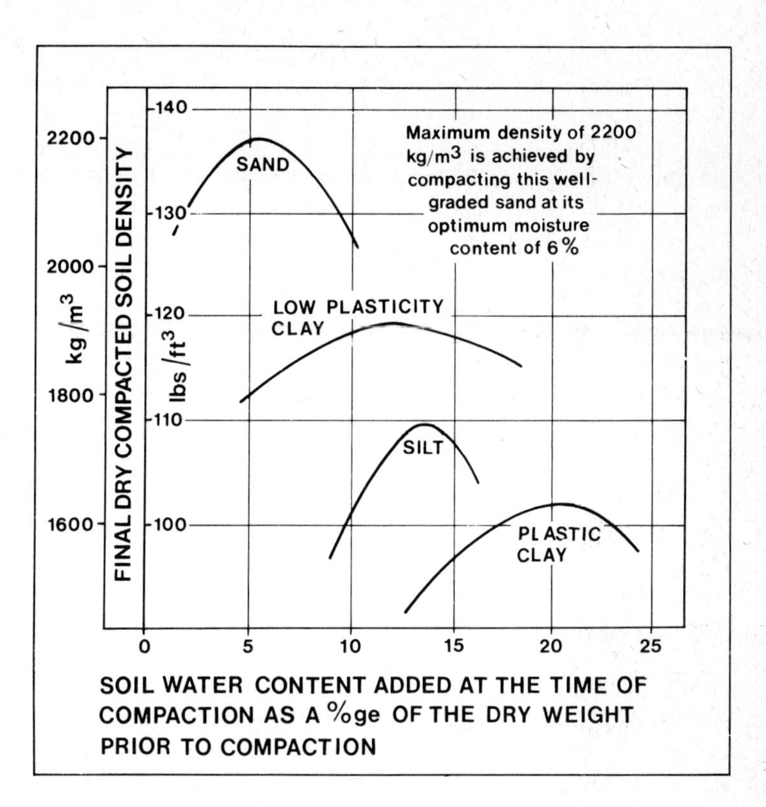

Figure 2.17 Typical densities of different soils at varying moisture content

Other factors also affect the maximum compactable densities. For instance, a well-graded granular material will compact much better than a uniform one (Figure 2.18). This is the reason why well-graded granular material should always be chosen when being laid over membranes for such applications as road construction. A well-graded crushed stone, compacted at its optimum moisture content will ensure that the maximum compacted density is achieved, resulting in maximum working strength. Maximum working strength minimises the stone requirement, thus making for greater economy.

The more often the compacting machine passes over the soil, the more compacted the soil becomes. However, it has been found that the first few passes of the machine achieve most of the possible compaction. Subsequent passes become increasingly less efficient. Figure 2.19 is based on the Transport and Road Research Laboratory research work[12] showing that the efficiency of repeated machine passes falls off very rapidly beyond ten passes. The graphs show that, in general terms 50 % of all compaction possible will be achieved with the first three passes of the plant. A further 40 % is achieved with the next seven passes, and thus 90 % of all possible compaction can be achieved with the first ten passes, if the correct type of roller is used.

Compacting efficiency falls off rapidly with depth, and it is therefore good practice to compact soils in layers of 250 mm thickness (10 inches). Although at optimum conditions, very heavy machines can exert a compaction on soil at greater depths, the above figure of 250 mm is a good guide line to aim for. Therefore, when permeable membranes are to be used in conjunction with road construction works and land reclamation schemes, the above criteria should be borne in mind, and well-graded material should be used at a suitable moisture content, with the correct plant. Since the effects of these criteria are compounded, the benefit derived in terms of structure performance may be several times greater using the right materials with membranes than otherwise.

This is an aspect of the strength of cohesive soils which we did not examine earlier. In our earlier diagrams, such as Figure 2.11, it was assumed that the shear pressure was applied quickly and that there was insufficient time for the water in the soil to migrate out of the soil. This would prevent the soil building up its full strength and would give a rather low figure for the possible strength of the soil. The shear strength of the soil determined quickly in this way, without allowing the water to escape, is known as the undrained shear strength and is denoted in soil mechanics by the letter c. Similarly, any internal angle of friction measured in a $c\phi$ soil is denoted by the term ϕ. Quick shear strength tests therefore give safer results in general since they indicate lower strengths for cohesive soils. However, in major design works it is known that the pressure for large-scale structures will be coming on the soil slowly, and where the pore pressures will therefore have time to dissipate, it is more realistic to undertake slow laboratory tests which themselves allow pore pressure to escape and which therefore allow the soil to build up a higher strength. These types of tests are known as 'drained' tests and the shear strength and internal angle of friction found from such tests are known respectively as c' and ϕ'.

The distance from any point in a clay soil to the nearest permeable layer is known as the length of the drainage path for that point. The object of artificial pore water drainage systems is to provide the shortest drainage paths for the main mass of the soil in the most economical fashion.

Without short drainage paths, it can take many months or even several years for the pore pressures to dissipate within the highly impermeable clay soils. Luckily, as shown in Figure 2.15, most of the consolidation takes place in the early period of time subsequent to loading. The steep part of the curve at the beginning indicates the squeezing out of interstitial pore water, whilst the flatter part of the curve represents the slow driving out of the 'tough' water jackets from around the soil particles them-

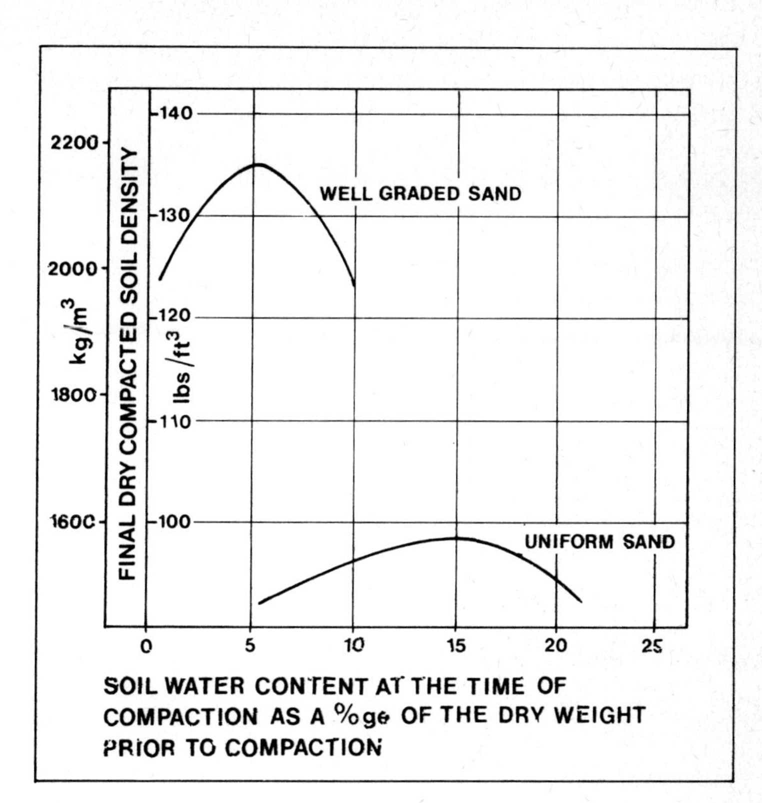

Figure 2.18 Typical variation in density between a well-graded and a uniform sand

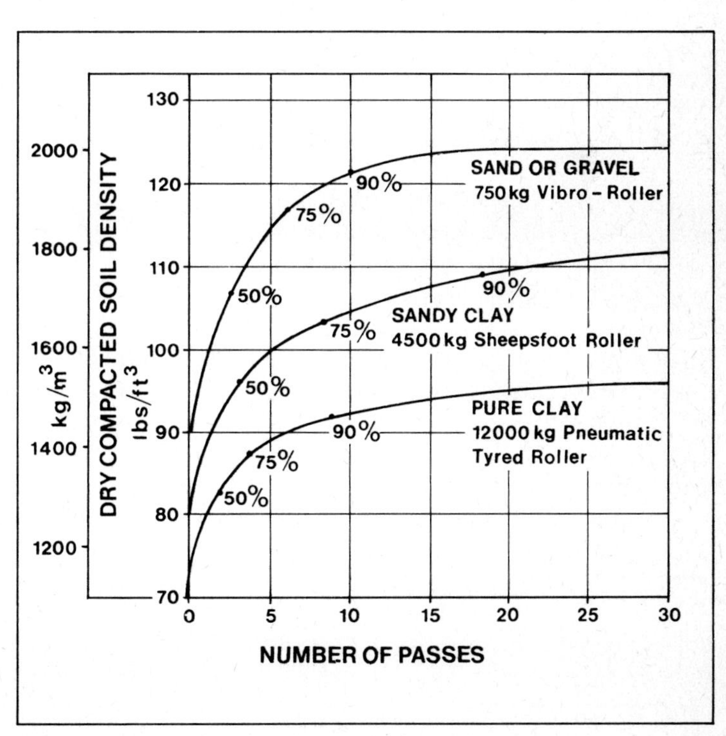

Figure 2.19 Compaction of different soils by rollers

selves. These water films are so thin that they are attracted to the particles of soil by molecular attraction, and are therefore very difficult to move.

Soil permeability

In general terms, the larger the soil particles, the larger the spaces between the particles, and therefore the more permeable the soil is. It follows that a well-graded soil will be less permeable than a uniform one, because in a well-graded soil, smaller particles are present, filling the voids between the larger ones. A simple equation governs the flow of water passing through any unit sq. m. of soil.

$$Q = k \times i$$

where i is the hydraulic gradient, as shown in Figure 2.16 and k is the Coefficient of Permeability (in m/sec), and Q is in m^3/sec.

The permeability (k) of the samples is thus the flow of water through one sq. m. of specimen cross-section divided by the hydraulic gradient (i).

$$k = Q\text{m}^3/\text{sec}/\text{m}^2 \div (h/l)$$

In conducting these types of tests in laboratories, it has been found that typical values of k are:

0.1 m/sec for clean gravel
0.01 m/sec for clean sand
1×10^{-4}m/sec for very fine sand
1×10^{-9}m/sec for clay

For any given situation, the value of k gives a direct comparison of the permeability of one soil with any other, i.e. a direct comparison of the ability of soils to allow water to pass through them or drain out of them. A soil with a k value ten times bigger than that of another, will always allow ten times as much water to flow through it under the same conditions.

This concept also applies to permeable membranes, many of which have permeabilities of k in the region of 0.001 m/sec, which is roughly equivalent to a sand. Therefore, these membranes are more than sufficiently permeable to pass water flows from all types of problem soils such as clays, silts, silty sand or pure very fine sands.

Water in the ground

Drainage systems, apart from those designed to dissipate temporary high pore water pressures beneath structures, are designed to cope with two types of natural groundwater. The first is groundwater resulting from surface saturation as a result of rainfall. The second is groundwater which rises up through the soil towards the surface as a result of hydraulic pressure from elsewhere.

Surface-saturation groundwater

When rain falls onto the surface of the land, a certain fraction evaporates directly back into the air, a further fraction runs off the surface and is carried to the sea as run-off and the remainder is absorbed into the ground. Near the surface, some of this water is taken out of the ground quickly by plant roots and returned through the leaves into the air by a process called transpiration. The rest of the water, which passes down through the soil under the effect of gravity, becomes the groundwater which affects engineering structures and designs. Figure 2.20 shows the general circulation of groundwaters resulting from surface saturation. Note that there is a level in the ground below which the pore spaces of the soil are totally saturated. This level is known as the 'water table'.

Figure 2.21 shows the effect of placing a ground drain below the groundwater table level. The water effectively 'falls' into the drain under its own hydraulic pressure, and consequently the water table surface slopes down close to the base of the drain. This is quite important, since — as shown in Figure 2.22 — several ground drains placed side by side can lower the overall water table over a large area. During a period of heavy rainfall, the water table may rise in level, but will settle back to a lower steady-state condition.

Adjacent to major cuttings and soil slopes, the groundwater flows in much the same way as a result of surface saturation, as it does adjacent to trenches. Figure 2.23 shows that the water table falls down towards the slope, probably meeting the slope near the bottom, where seepage and water outflow will occur. The water in a high slope such as is shown in the diagram will, by virtue of its own weight, exert a hydrostatic pressure on the soil near the base of the slope which could prove to be dangerous in so much as it carries part of the weight of the embankment in hydrostatic loading and this reduces the shear strength of the soil. Ground treatment at the base of slopes is therefore often concerned with removing this water from the ground thus allowing the soil to generate its maximum strength. With only

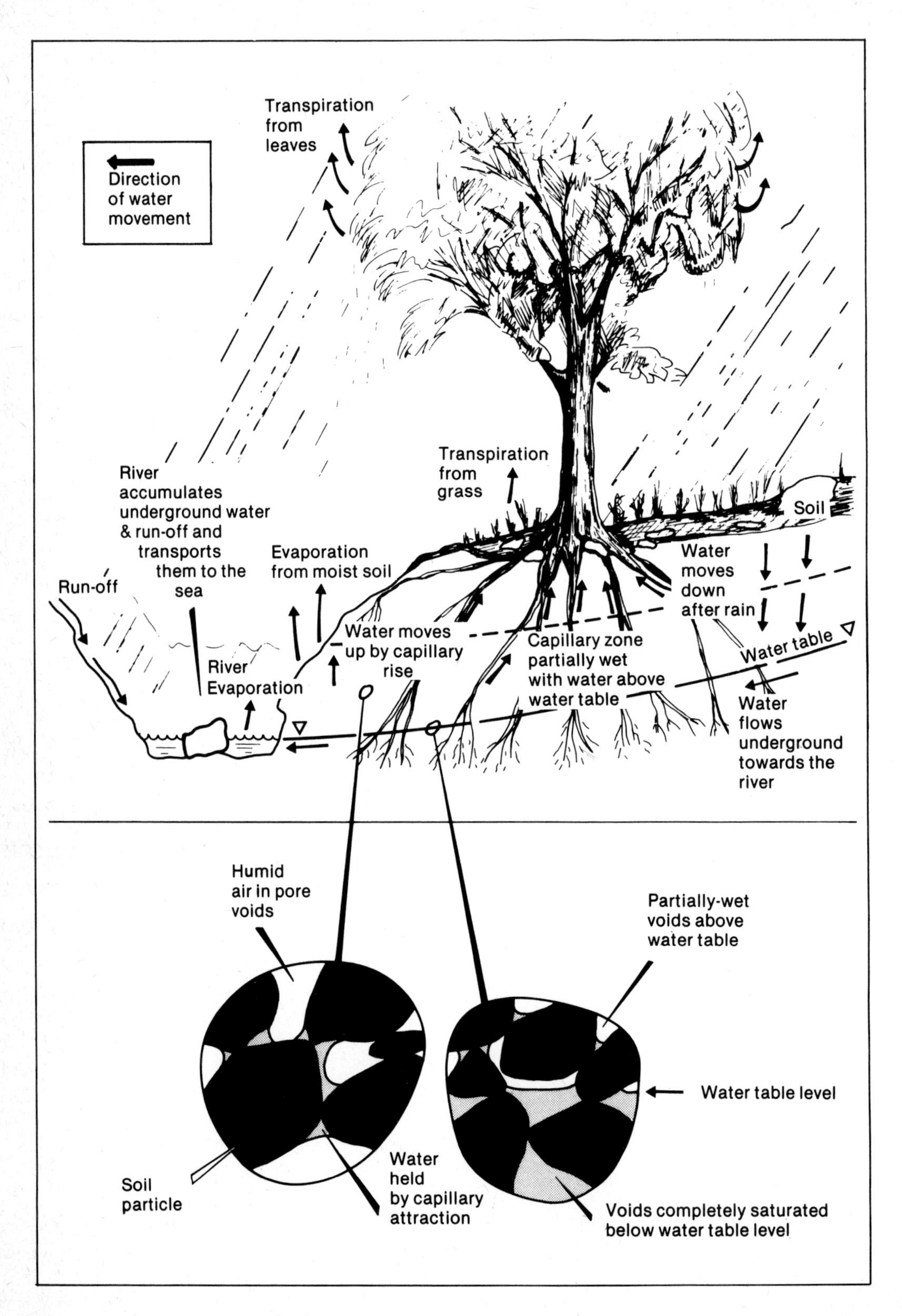

Direction of water movement
Transpiration from leaves
Transpiration from grass
Soil
River accumulates underground water & run-off and transports them to the sea
Evaporation from moist soil
Water moves down after rain
Run-off
Water moves up by capillary rise
Capillary zone partially wet with water above water table
Water table
River
Evaporation
Water flows underground towards the river
Humid air in pore voids
Partially-wet voids above water table
Water table level
Soil particle
Water held by capillary attraction
Voids completely saturated below water table level

Figure 2.20 (Left) Groundwater circulation resulting from surface saturation

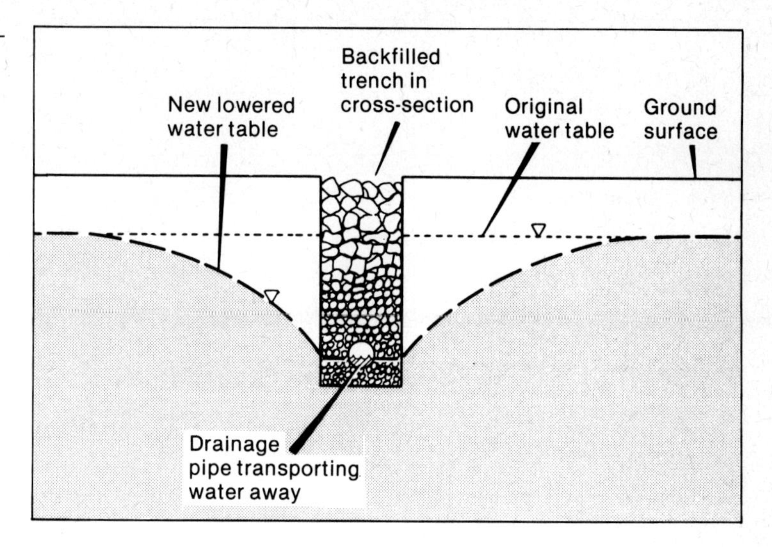

Figure 2.21 The construction of a drainage trench causes lowering of the adjacent ground water level

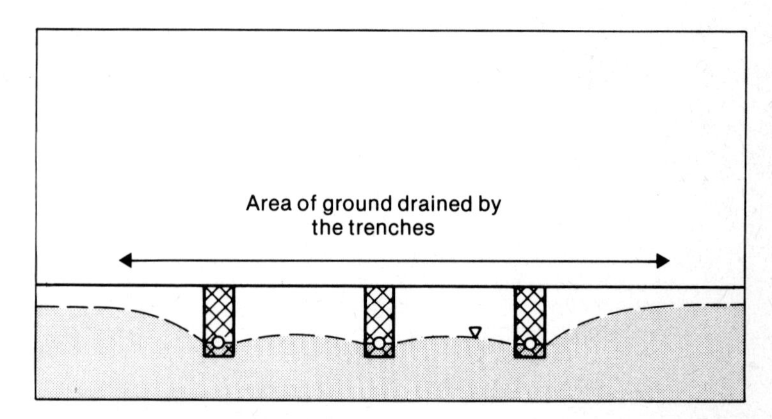

Figure 2.22 Several parallel trenches can lower the groundwater over a large area

minor alterations, Figure 2.23 could represent the downstream toe of a dam where the high water behind the dam might generate pore water pressures at the base of the dam which could cause it to fail. Similarly, the diagram could well represent the upstream face of a dam subsequent to rapid drawdown, where a 'wall' of water would be left within the dam creating high pore pressures in the soils at the upstream toe. These could fail that part of the dam if such a contingency were not designed for.

Rising groundwater as a result of artesian pressure

Fairly often, low lying areas of land are subject to subterranean water pressures which are transmitted through the soil along permeable strata from high level areas. Figure 2.24 gives two examples of how artesian pressures can be generated in soils either naturally or artificially. Drainage systems installed in this kind of environment are usually the 'blanket' type which cover

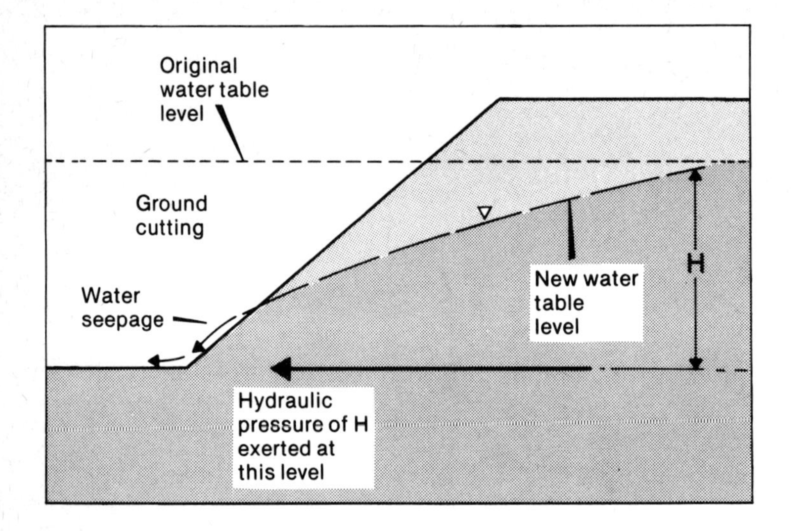

Figure 2.23 Excavating a cutting lowers the groundwater table, and generates hydro-soils problems at the toe of the slope

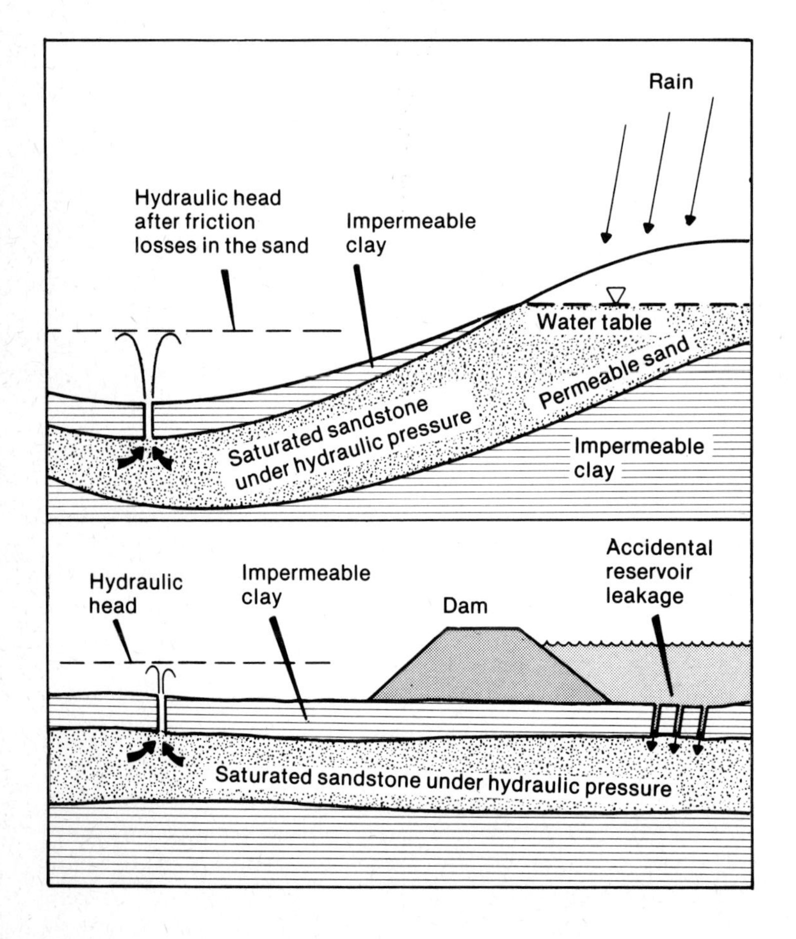

Figure 2.24 The generation of natural and artificial artesian groundwater pressure

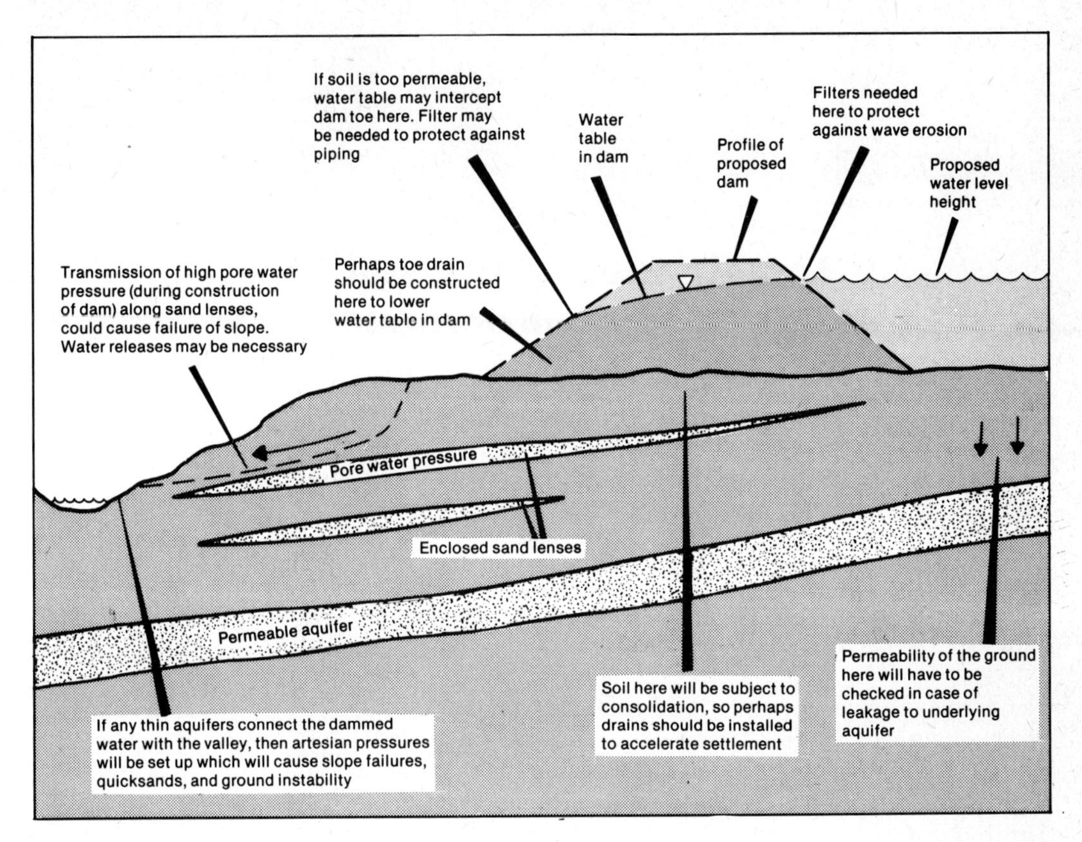

Figure 2.25 Some practical aspects of soil mechanics on a site design

the whole area, thus confining the upward water flow and transporting it away into ground drains. Blanket drains are designed to transport the water horizontally, and on top of the blanket drain soil fill can be placed and construction can follow safely.

Conclusions

All of the complex aspects of soil mechanics come together for consideration at the design stage of a Civil Engineering structure. Figure 2.25 shows some of the aspects requiring investigation for a hypothetical dam project.

3

Theoretical functions of membranes in the soil environment

Permeable membrane functions

The two major functions of a permeable membrane in a soils design are usually either filtration or separation. Rarely does a permeable membrane achieve its desired effect by virtue of its strength — even though at first sight it may appear to do so. This is a common mistake, and engineers should guard against believing that the strength of a membrane is sufficient to support large soil or traffic loads. In most practical cases, separation and filtration are found as co-existing functions, which by virtue of their interacting effects, improve the characteristics of the natural soils, roadstones or filter materials with which they are in contact, and thus provide increased strength and stability in any given fabric/soil system. Naturally, wherever filtration is taking place, the separating element must be present. However, there are occasional conditions where there is no water present (i.e. desert conditions) and in these circumstances membranes can fulfil separation functions without filtration being part of the design.

Separation

It has long been one of the problems of soil mechanics that, although theoretical design techniques have been developed to a high level, practical site work requires rigorous supervision to ensure that design geometry is adhered to and that the high quality materials specified are, in fact, supplied and placed correctly. Worse than this, however, is the fact that once constructed, loading pressures and water movement generally lead to the breakdown of filter or separator design by allowing materials of different quality and specifications to become mixed — even if only marginally. The separating function utilised by inserting a permeable membrane into a design, as shown in Figure 3.1 allows not only material of less critical specifications to be recommended, but also ensures that the design boundaries do not change or blur.

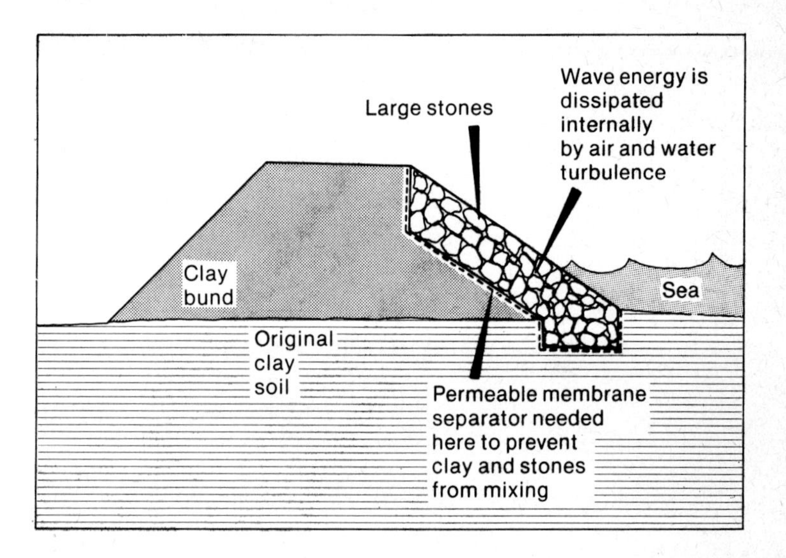

Figure 3.1 Typical use of a permeable membrane as a separator

The photograph in Figure 3.2 shows the penetration of mud from a retaining bund into a large-stone coastal defence in South East Asia. The clay used for the bund construction has more than 70 % passing 200 microns, and the stone placed directly against the bund is of 15 cm diameter average size. The principle of this particular design was that the wall should be flexible, and should be of loose stone construction in order to absorb wave energy by turbulence dissipation internally. If the internal spaces of the stone become filled with mud, (as shown in Figure 3.3) then the structure cannot perform its originally-designed function. This is what is meant by 'design boundaries breaking down'. It is an example where the presence of a permeable membrane is clearly necessary in a primary separation function.

In another kind of design, such as a simple temporary road or haul road, the separation function of a permeable membrane is needed to prevent the upward migration of soft sub-grade material into the crushed stone of the base course. This penetration is caused by the dynamic pumping action of vehicles passing over the road, and as can be seen in Figure 3.4, the soft cohesive sub-grade material separates the granular pieces of stone and effectively reduces the depth of the sub-base layer. If this was designed to have a certain — even if limited — life, then

Figure 3.2 (Left) Coastal defence built without separator, showing mixing of clay and stones

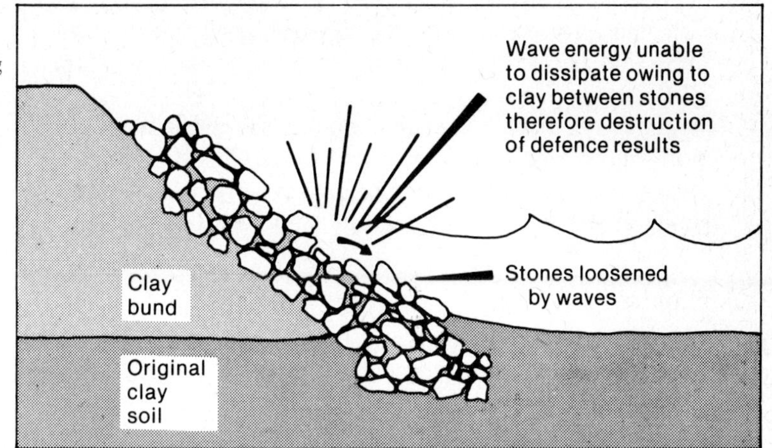

Figure 3.3 Breakdown of coastal defence by clay infilling without permeable membrane separator

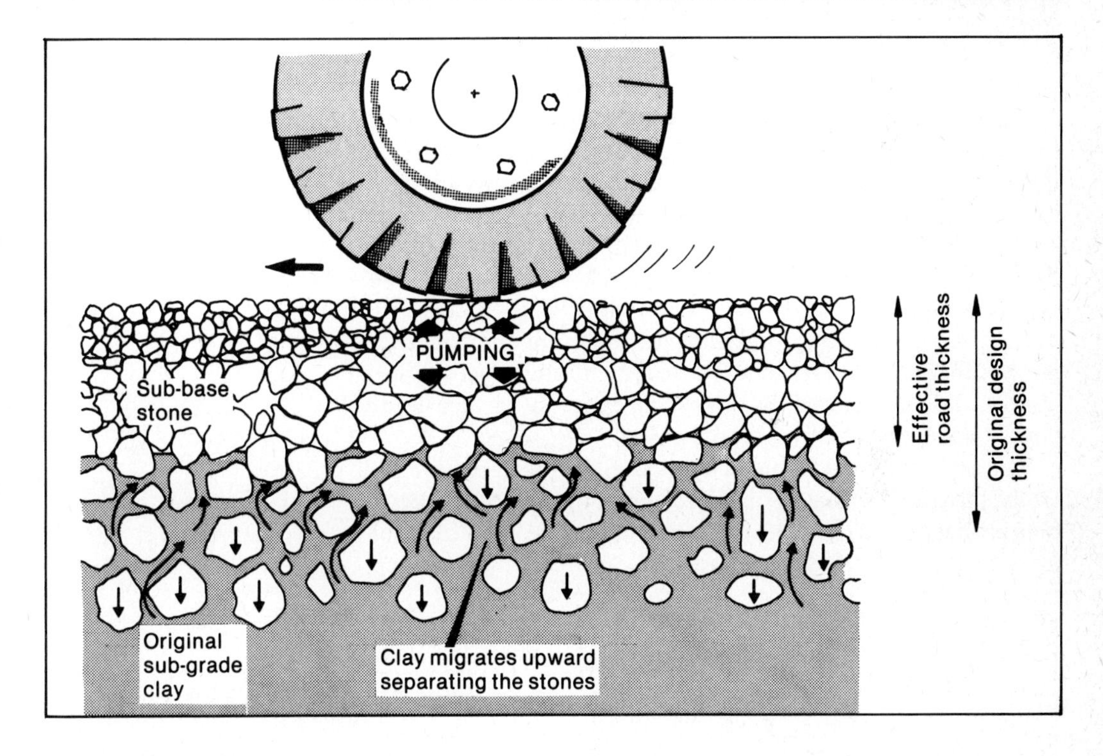

Figure 3.4 Breakdown of a road structure by pumping, because of lack of permeable membrane separator

as the thickness of the sub-base decreases, the potential lifespan of the road rapidly falls and the road becomes quickly destroyed.

The photographs in Figure 3.5 show the remarkable separating effect of a permeable membrane laid on a soft sub-grade soil, overlaid by large stones to form a single base layer, and trafficked by a loaded 8 tonne lorry. The photographs clearly show that the ruts developed by the breakdown of the sub-base /sub-grade interface have reached the depth of the chassis (0.5

m) on the side of the road unprotected by the membrane. Beneath the protected side, the separating action of the membrane has caused the road stone to compact and only minor ruts (0.1 m) have formed.

Therefore, it can be stated that one of the most important functions of a permeable membrane is in its separating ability, preventing the breakdown of design geometry and working mechanics as specified on the drawing board by the engineer.

Figure 3.5 (Opposite) (a) Trial road trafficked without membrane; (b) Trial road trafficked with membrane

Filtration

In considering the separating property of the membrane, it can be considered that the passage of water from a finer soil into a coarser soil across the boundary of the membrane is of a secondary nature. In some situations, however, such as in dams or filter structures, the filtering action of the membrane (or rather the ability of the membrane to develop a soil filter) is critical, since the design is fundamentally a hydraulic one in which soils, stones and membranes are introduced artificially into a structure for some particular engineering purpose. Although different types of drain and filter design are dealt with in detail later, it can be considered that a ground drain is perhaps the most simple and universally understandable example of a filter design (Figure 3.6).

In drainage applications where the controlled extraction of water from soil is being undertaken, filters are constructed to prevent *in situ* soil from being washed into drains. Such washed-in soils cause clogging of the drains and potential surface instability of adjacent land.

The criteria for selection of natural filter materials are well established, and are based on a knowledge of the Particle Distribution Diagram of the *in situ* soil, and that of the filter materials available, as illustrated in Figure 3.7.

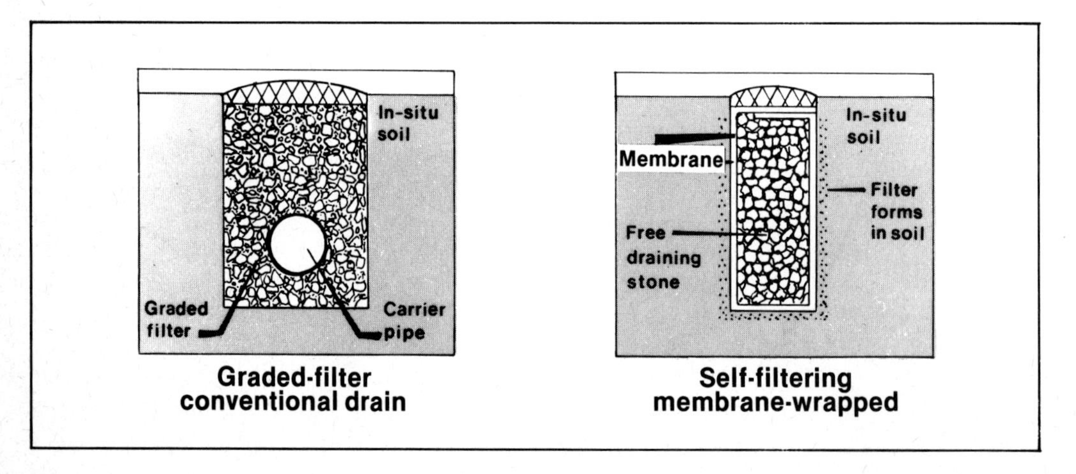

Figure 3.6 Comparison of conventional with membrane-wrapped drain

(a)

(b)

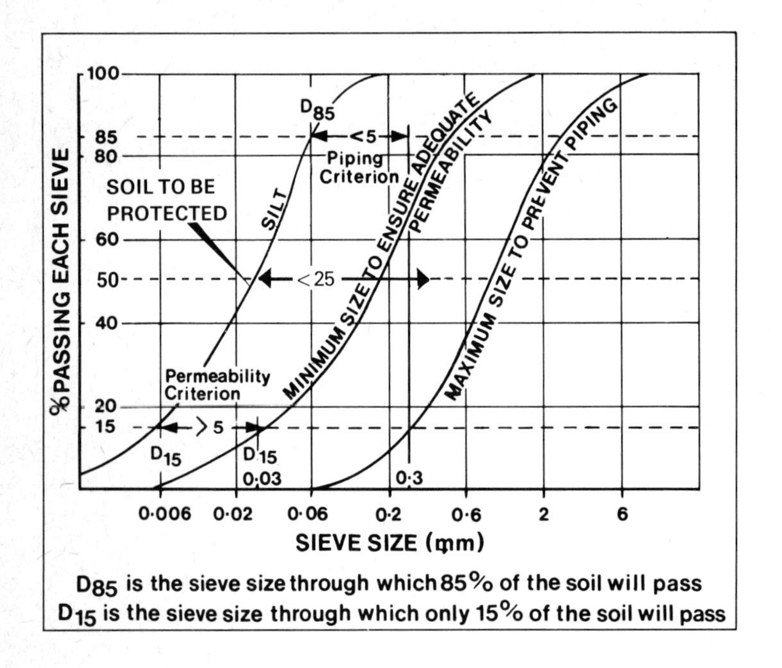

Figure 3.7 Design method for choosing a natural granular filter to protect a soil

If the filter material is freely to filter the soil without allowing the soil to 'pipe', i.e. pass through, then the following criteria must be satisfied:

(1) D_{15} Filter < (5 × D_{85} soil) Piping
(2) D_{15} Filter > (5 × D_{15} soil) Permeability
(3) D_{15} Filter < (25 × D_{50} soil) Uniformity

Criteria specified by Cedergren.[2]

Thus, once the Particle Distribution Diagram of the natural soil is known, it is possible to establish the criteria for a filter or even a series of filters to prevent piping. Multiple filters (or graded filters) simply have a series of curves conforming to the above rules.

A knowledge of the above principles is important since membranes are often used to replace one or more of the costly elements in multi-filter granular systems designed on the above basis. In a granular filter system, the filtering action appears to be achieved by a mixing and grading at the interfaces between filter layers of different particle size, but the development of a filter in a soil adjacent to a permeable membrane is distinctly different.

Research programmes designed to investigate the working principles of filter development in membrane-wrapped drains were instigated at the University of Strathclyde, UK[13] and the University of Tennessee, USA.[14] Both of these research programmes showed that the membrane did not itself filter the soil

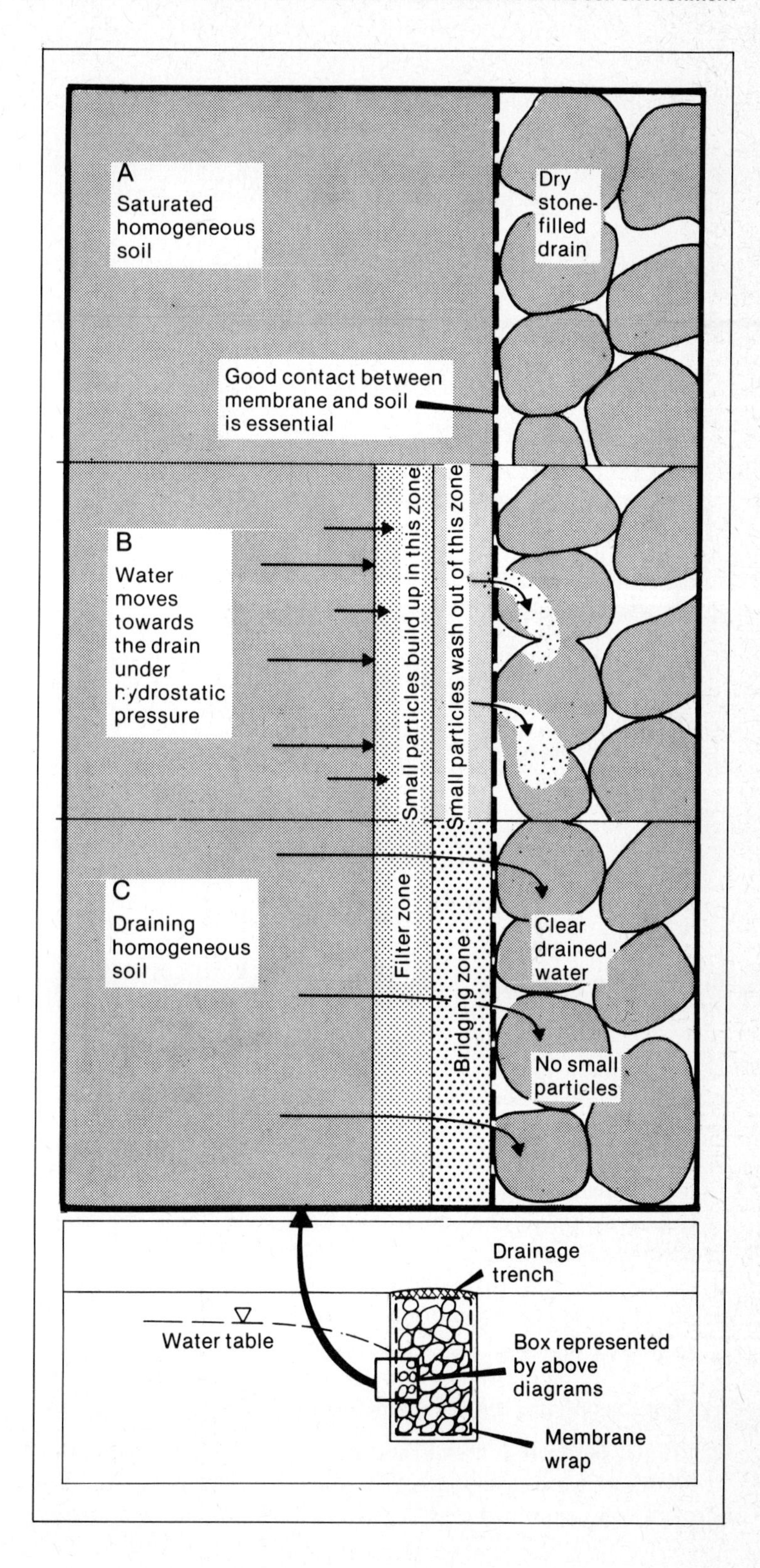

Figure 3.8 Development of an internal soil filter behind a permeable membrane

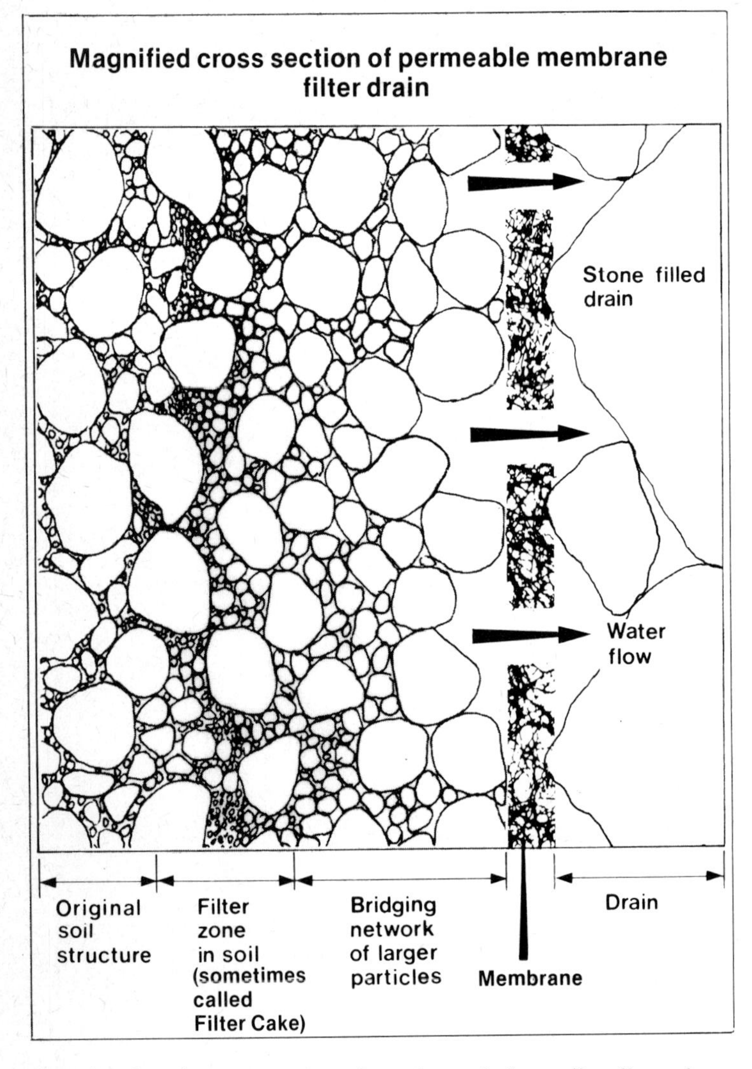

Figure 3.9 Detailed distribution of soil particles behind a permeable filter membrane subsequent to filter cake formation

directly, but by supporting the edge of the soil, allowed an *internal* filter to build up as shown in Figure 3.8. The detailed distribution of particles subsequent to the establishment of a filter is shown in Figure 3.9.

The development of an internal soil filter has three important practical consequences:

1. Since the filter membrane is not itself filtering the soil directly, the *exact* size of the holes in the membrane is not critically important. Membranes with reasonably different pore properties may be used to filter the same soil.

2. If the permeable membrane is removed, the bridging network will break away, followed by the filter cake, and the soil will recommence piping. In this context, it is *important*

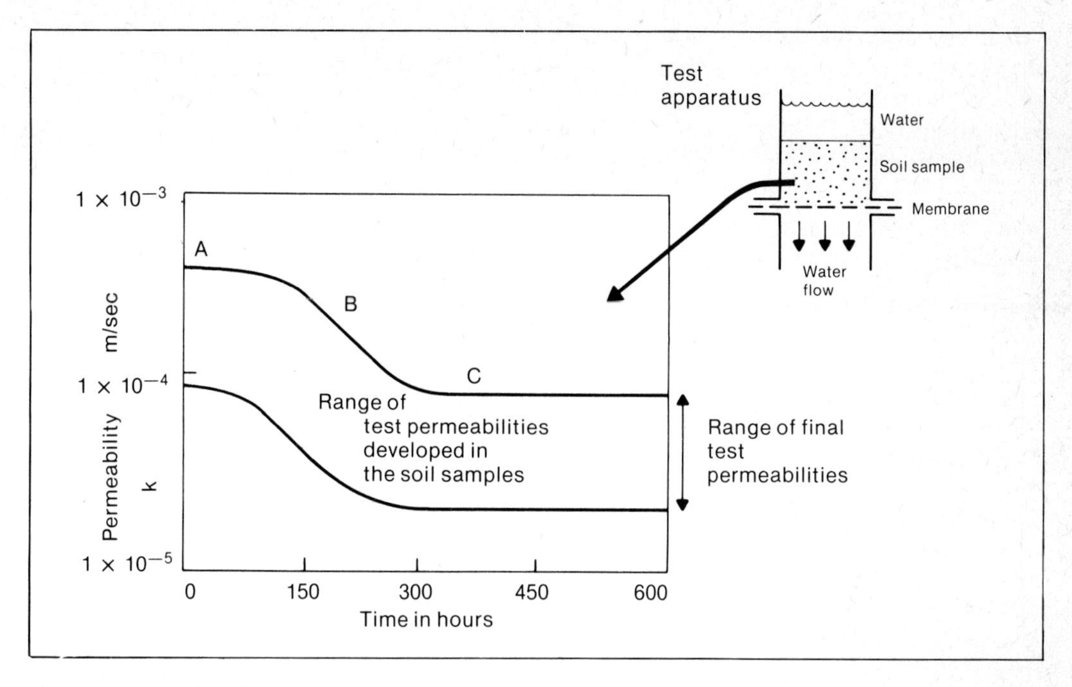

Figure 3.10 Results of permeability tests on soil/fabric samples (Ref 14)

to recognise that good contact between the membrane and the soil is necessary for filtration to work.

3. The final permeability of a simple soil/fabric filter system will be less than that of either the membrane or the soil in its original state, and will actually be the permeability of the filter cake which builds up behind the membrane. For example, in research work conducted by the University of Tennessee,[14] a simple non-reversing hydraulic gradient apparatus was set up as shown in Figure 3.10. At first, piping was observed through the filter membrane, but as the filter cake built up behind the membrane, piping decreased and eventually stopped. Simultaneously, permeability also decreased and stabilised with time at a constant level. In Figure 3.8 the letters A, B and C correspond to the main stages of filter cake formation. This is of critical importance insomuch as in soil filtering applications involving slurries or in settlement lagoons, potential water flow rates through given areas of filter membrane should never be calculated on the basis of the permeability of the membrane alone. Consideration should always be given to the lowest possible permeability of any particular soil layer which could build up on the inside of the membrane as a result of its filtering function.

In most ground drainage situations the hydraulic gradient usually acts in one direction only — although it may vary in value, thus affecting the amount of water being drained out of the ground at any time. However, under certain conditions — such as on the coast within influence of waves or tides — the filter cake can be subjected to reversing hydraulic gradients. For example, at low tide, water may be seeping out from a coastal slope and running into the sea, but at high tide the pressure of water may cause a localised reversal of flow. Similarly, filter structures can be subjected to rapidly alternating pressure surges during the breaking of storm waves. The effect of these flow reversals on a filter cake is to break it up, and permit piping to re-start.

As an example of the combined action of separation and filtration at a membrane/soil interface, it is considered that one of the beneficial aspects of some soil/membrane designs is the reduction in the water content of the original soil, resulting in an increased shear strength. The filtering action of the membrane allows water to escape from the soil and resists the unwanted migration of fine particles. However (as illustrated in Figure 3.11) without the membrane, the surface area of sub-grade soil exposed to the air is very small. It is a natural feature of any such interface that stones become pressed into the sub-grade leaving very little area from which water can escape. On the other hand, with a membrane present at the interface, it has been suggested that a substantially larger area is available for water loss. This means that when the soil is placed under an overburden pressure, pore water can readily escape and the shear strength of the sub-grade can build up.

Reinforcement

Membrane sheets and webbings have now been used for some years in temporary structures such as roads and embankments where the low cost of the installation or perhaps the difficulty of site access has made such designs viable. It is the intention of geo-fabric engineers to bring membranes and webbings increasingly into full-scale permanent structure designs.

The basic principles of incorporating reinforcing membranes into a soil mass are much the same as those utilised in the design of reinforced concrete. The membranes are used to provide tensile strength in the earth mass in locations where calculations show that shear stress would be generated. For example, Figure 3.12 shows that in the case of an embankment made only of granular soil, a failure plane would develop as indicated by the dotted line. However, by inserting membrane sheets or webbing strips, this potential failure plane can be intercepted and the shear resistance of the ground increased locally. A secondary factor is that the insertion of reinforcement modifies the shape of the failure plane to that shown by the dashed line. In practice,

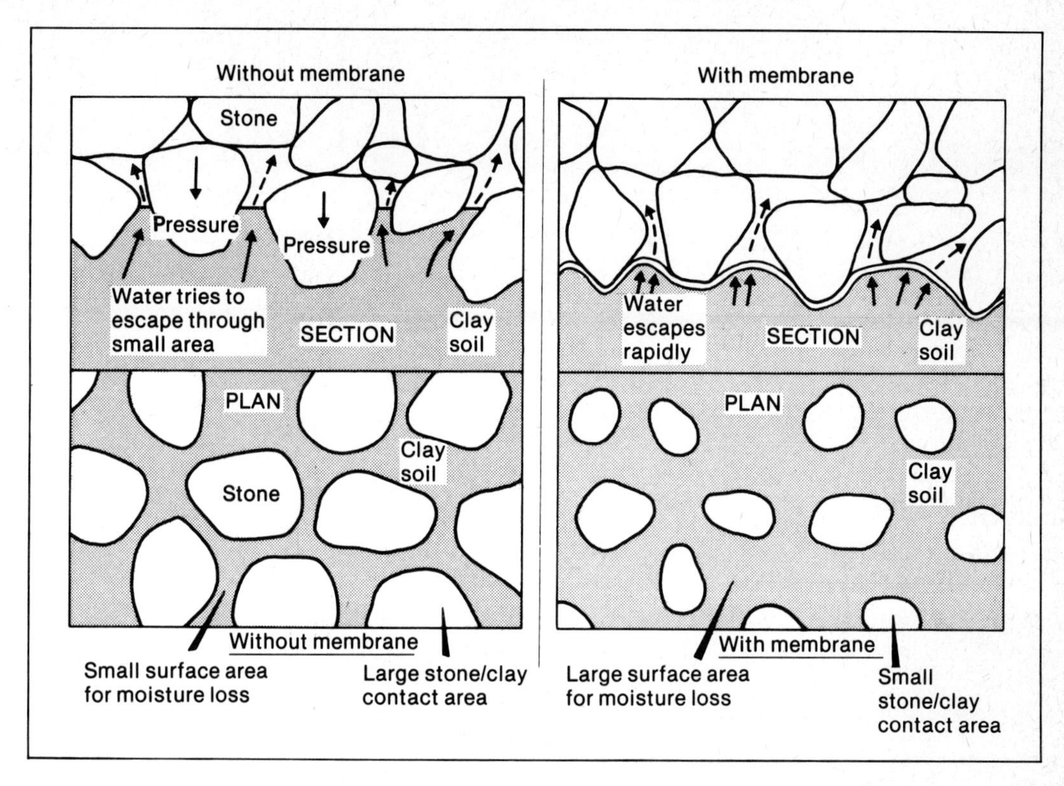

Figure 3.11 Reduction in stone/clay contact area, caused by presence of membrane at sub-base/sub-grade interface

Figure 3.12 Modification of potential failure surface by insertion of reinforcing membranes in fill retaining wall construction

membrane sheets can be laid in the ground and overfilled as illustrated in Figure 3.13. This results in the redistribution of stress internally within each of the membrane-encapsulated layers.

In the case of webbing strips being used in the soil, then these act almost as passive 'tie-back' bars as illustrated in Figure 3.12. Such webbings are usually attached to facing units of some des-

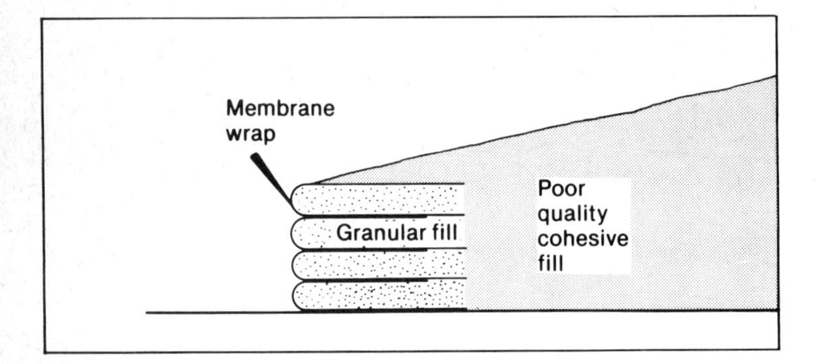

Figure 3.13 Reinforcing fabric sheets wrapped around granular fill to form a 'reinforced earth' wall

cription, or else are wrapped round a close-mesh web containing a filter membrane on the inside. The tensile and shear stress distribution along such a strip placed in the ground can be seen illustrated in Figure 3.14.

Where sheet membranes such as permeable fabrics are used for reinforcement purposes, not only does the distribution of shear stresses on the membrane have to be considered, in an analogous fashion to the previously defined strip stresses, but also, by completely enclosing the earth within membrane 'envelopes', the entire stress distribution pattern within the soil mass is altered substantially. Research work by McGown[15] showed that not only are stress patterns altered, but strain directions are also substantially changed. Figures 3.15 and 3.16 show that for a simple model set up, stress and strain paths are considerably altered by the insertion of a single discontinuity in a granular material. Considering that the granular material has no intrinsic tensile strength, the alteration in the stress/strain paths has resulted in an unexpected increase in resistance to loading by the model footing. Furthermore, considering that a steel sheet was used for comparison with a membrane, the

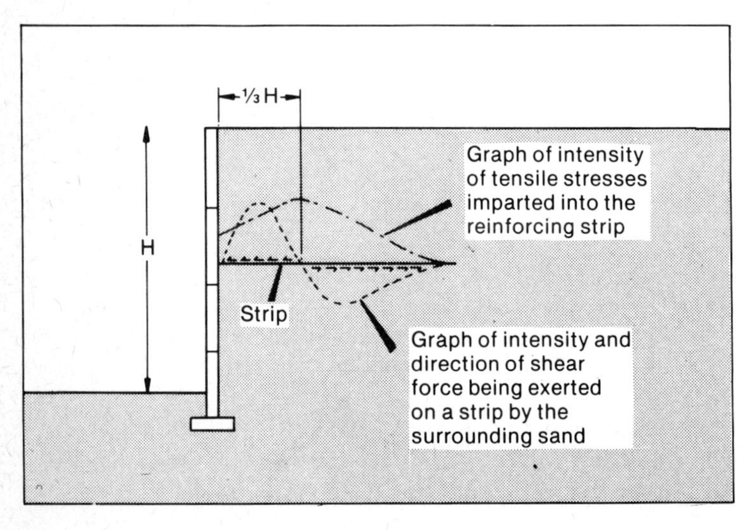

Figure 3.14 Internal and external stresses in and around a typical reinforcing strip in a reinforced earth retaining wall

similarity of results makes it difficult to avoid the conclusion that the discontinuity, and its position, are far more important than the strength and frictional properties of the material used to create the discontinuity.

The fundamental difference between membrane-reinforced designs for earth structures and the more conventional steel reinforced designs, is the larger amount of strain which must be allowed for in the membrane case. Because steel has a high Young's Modulus, it can absorb imposed stresses rapidly, without experiencing much strain. However, in the case of membranes, Young's Modulus is of a much lower value, and there-

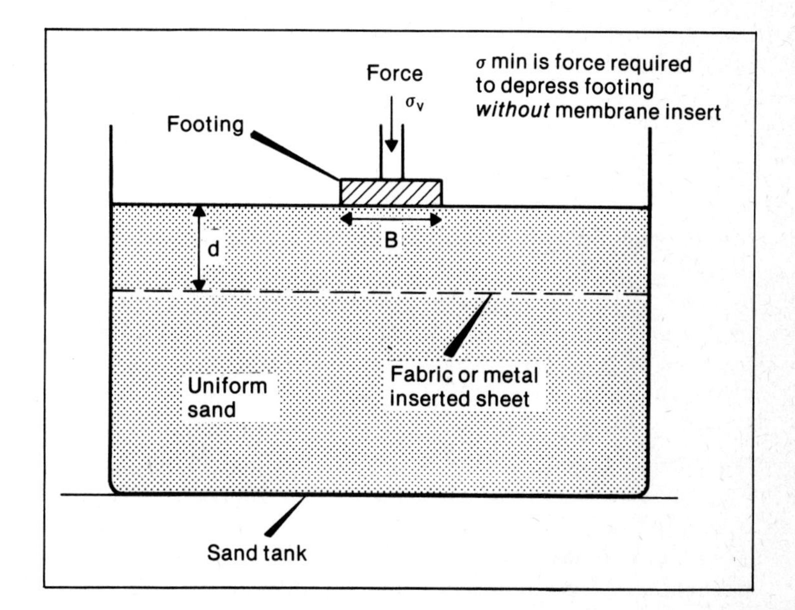

Figure 3.15 Model apparatus for studying improvement in footing bearing pressure caused by inserting a fabric or metal discontinuity in the soil (See Figure 3.16)

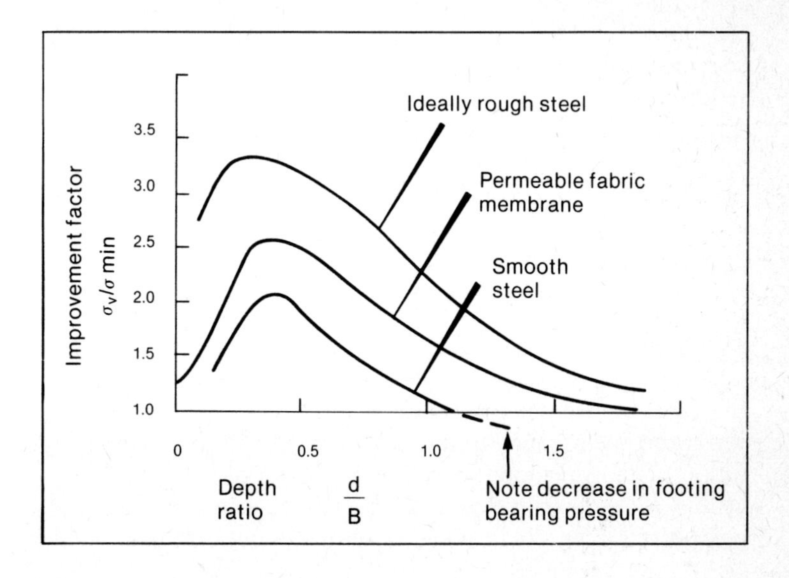

Figure 3.16 Improvement of footing bearing pressure with increasing depth of 'reinforcement' from the surface in the model illustrated in Figure 3.15 (Ref 15). In this case the footing depressed a distance of 0.5B

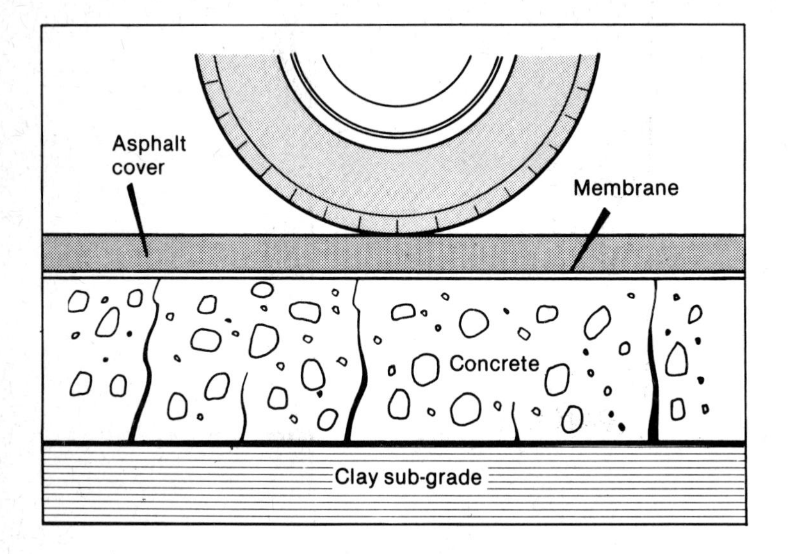

Figure 3.17 The use of a permeable synthetic membrane at the wearing course level of a road to prevent reflective cracking

fore the membrane has to strain possibly several percent before it reaches its designed working load. Therefore it is generally advisable to design with all-membrane components in order that the entire structure should be flexible and capable of self-adjusting internal settlement.

Finally, as may be seen later in the book, membranes are being increasingly used to reinforce the wearing course level of tarmacadam road structures. The objectives behind this use are, for example, to reduce the upward propagation of reflective cracking from broken concrete roads being covered by a flexible coating (see Figure 3.17). The extension of this idea has lead to the concept that membranes can prolong the working life of upper surfaces in new road structures. The theoretical principles here have not been defined, but Figure 3.18 illustrates one possible explanation for the limitation in reflective crack propogation. It can be seen that the membrane, by bridging across the narrow gap between the two rigid concrete blocks, spreads the zone of movement through a wide enough section of the upper layer, that the strain can be absorbed by the elastic properties of that layer, thus reducing fatigue considerably.

Membrane and polymer properties

It is not easy to summarise the overall behaviour patterns of permeable membrane fabrics, but Figure 3.19 can give an indication of the functions utilised in a variety of different applications. This diagram does not however, cover the fundamental properties of the polymer materials themselves. This is highly

Figure 3.18 Possible mechanism for reduction in crack propagation by membrane layer

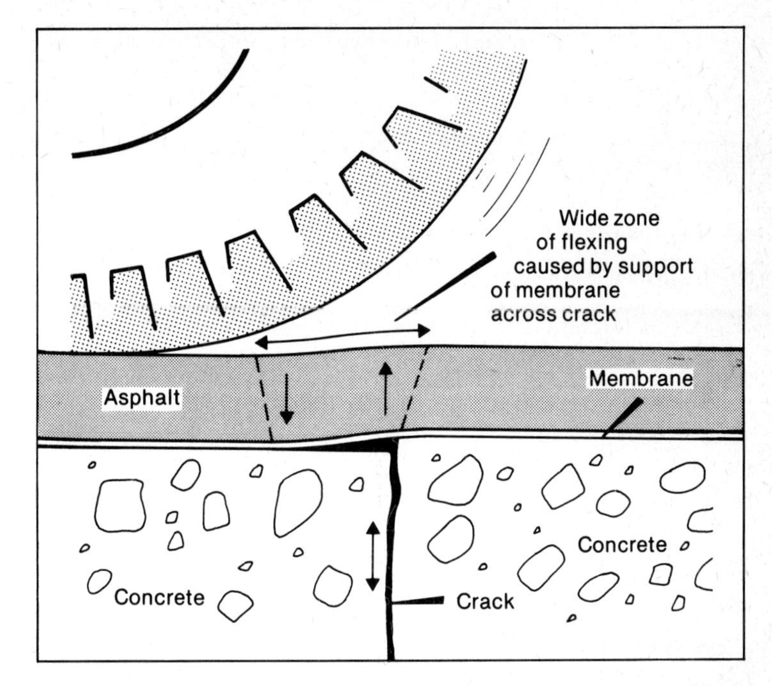

Figure 3.19 Functions of fabrics in different applications

APPLICATION	MEMBRANE FUNCTION			
	SEPARATION	FILTRATION ACROSS MEMBRANE	DRAINAGE ALONG MEMBRANE	REINFORCEMENT
Roads, Railways and Area Sub-grade Stabilisation	[cross-hatched]	[stippled]		[stippled]
Drainage	[stippled]	[cross-hatched]		
Wet-fill Embankments & lagoons	[stippled]	[cross-hatched]	[cross-hatched]	[stippled]
Coastal and River Protection	[cross-hatched]	[cross-hatched]		[stippled]
Land reclamation	[stippled]	[cross-hatched]		
Asphalt Reinforcement				[cross-hatched]
Soil Reinforcement		[stippled]	[stippled]	[cross-hatched]
Marine Causeways and fill areas	[cross-hatched]			[stippled]

critical and it is most important that these properties are considered before choice is made of a membrane group or type for a particular project.

Possible considerations include:

1. Is high tensile strength important, or is high extensibility more important? (Usually the two are mutually exclusive.)
2. Is ultra-violet resistance important, or not?
3. Is moisture absorption important together with any deleterious property changes which may result?
4. Is a low permeability required or a high one?
5. Is a small pore size required or a large one?
6. Is a small % Open Area necessary, or is a large % Open Area required?
7. Is a special chemical resistance or biodegradation resistance required? (E.g. near chemical plants, adjacent to roads, or in low pH soils.)
8. Does the membrane require to be capable of standing up to difficult handling conditions? (Should it have loops, eyelets, pockets, etc.)
9. Is high impact resistance very important? (Is the membrane going to have large/angular rocks dumped onto it?)
10. Is abrasion resistance specially important? (For example in river or marine defence applications where rolling rocks could abrade the membrane.)
11. Is the creep characteristic of the fibre important? (Is the membrane likely to be subjected to stress over a long period of time?)
12. Is the two-dimensional sheet preferable in a particular design to a three-dimensional felt? (Could clogging in a felt be a problem? Should a felt be chosen because water flow along the membrane is required for a particular design?)
13. Is a non-woven material preferable to a woven one, or is the woven type better for a particular purpose? (For example, a non-woven might produce a better filter with less piping than a woven in marine reversing-flow situations. Or a woven might withstand the dumping of large rocks on the sea bed better than a spun-bonded.)
14. Do the particular *in situ* soil and brought-in soils and stones have any peculiar characteristics that might rule out certain membranes? (An *in situ* soil might be gap-graded, or so

uniform that it matches the pore sizes of a certain woven. In this case, blocking or piping might occur on an unacceptable scale, and the particular membrane should be excluded.)

Figures 3.20 to 3.41 are unaltered extracts from a variety of published sources. The Author does not necessarily support the views expressed therein, but includes them as a reflection of public opinion.

Resistance of Polyethylene to biodegradation

Plastics Division experience is that with formulations very similar to that used in Paraweb application, no deterioration in properties which can be attributed to biodegradation has been observed over long periods of soil burial. Thus, water pipe made from MFI 2 LDPE containing well dispersed carbon black and antioxidant, continues to give satisfactory service, for instance in the Welwyn Garden City area, after approximately 25 years burial. Black telephone cable sheathing, similarly formulated, has also given satisfactory service after burial in soil for a similar period.

The normal, commercially available, properly formulated, polyethylene compositons of the type utilised in Paraweb application are, based on our past experience in other technically demanding applications, not expected to show a deterioration in performance which can be attributed to biodegradation.

ICI Plastics

Figure 3.20 Extracts from ICI Plastics Division technical letter

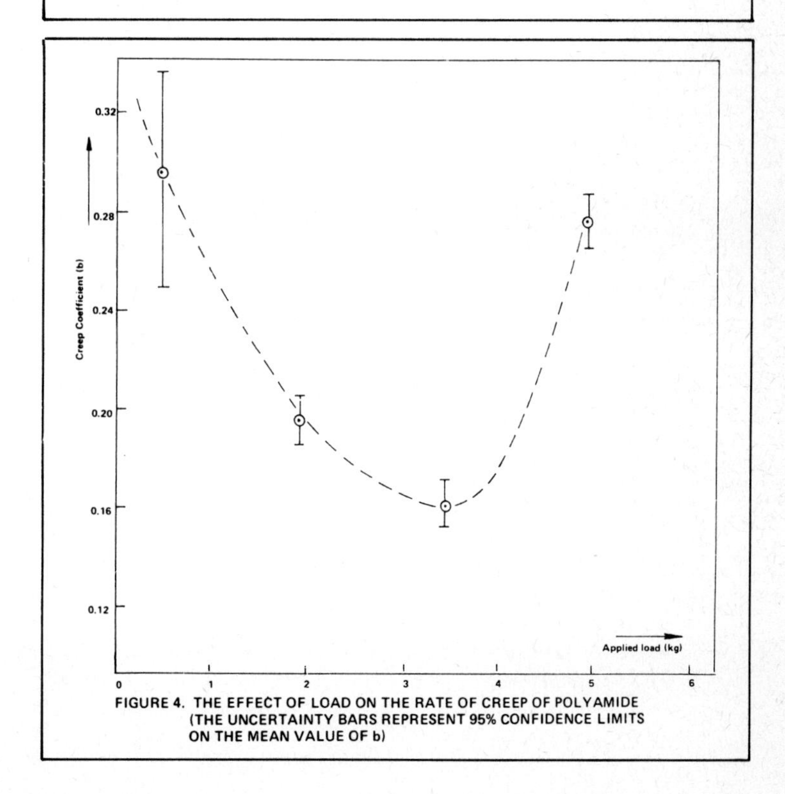

FIGURE 4. THE EFFECT OF LOAD ON THE RATE OF CREEP OF POLYAMIDE (THE UNCERTAINTY BARS REPRESENT 95% CONFIDENCE LIMITS ON THE MEAN VALUE OF b)

Figure 3.21 Extract from a paper by J.A. Finnigan presented at a Conference For The Use Of Fabrics In Geotechnics. Paris 1977

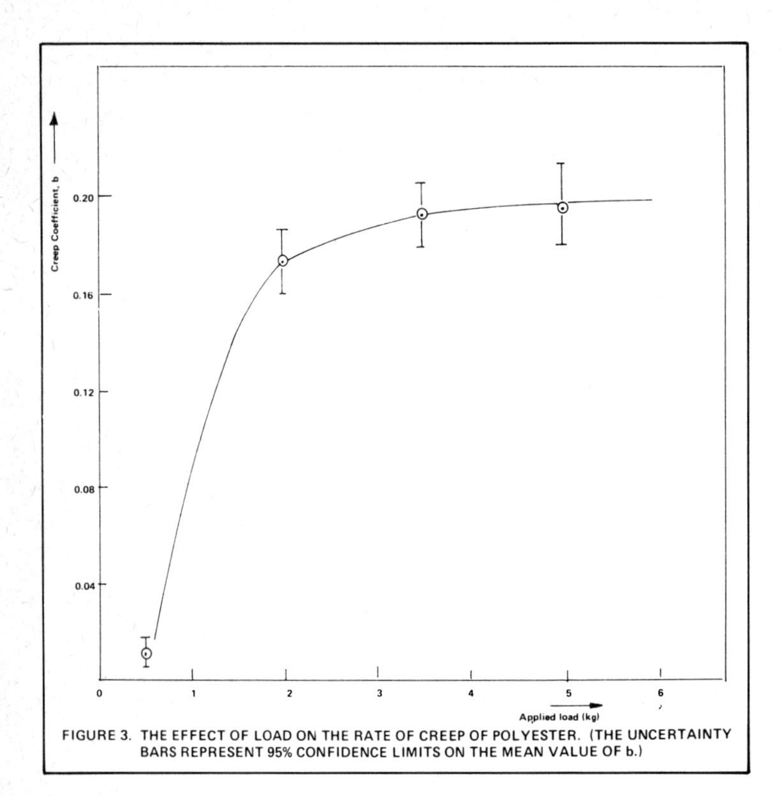

FIGURE 3. THE EFFECT OF LOAD ON THE RATE OF CREEP OF POLYESTER. (THE UNCERTAINTY BARS REPRESENT 95% CONFIDENCE LIMITS ON THE MEAN VALUE OF b.)

Figure 3.22 Extract from paper by J.A. Finnigan presented at a Conference For The Use Of Fabrics In Geotechnics. Paris 1977

Figure 3.23 (Below) Extract from paper by J.A. Finnigan presented at a Conference For The Use Of Fabrics In Geotechnics. Paris 1977

Conclusions

The confirmation that over the period 1 minute to 1000 hours loaded high tenacity polyester and polyamide yarns deform linearly with respect to log (time) and the subsequent derivation of a quick laboratory creep test based on this fact, has permitted a wide appreciation of the levels of creep to be expected from these yarns in various states. Both yarns show low levels of creep and typically one might expect some 1% change in length over 10 years when a yarn is loaded to 20% of its breaking load. For polyamide, as the applied load is increased the creep tendency goes through a minimum before increasing with the applied load, whilst polyester creep increased initially with load and then appears to be limited to a set creep value as the load is further increased.

Piled yarns and fabrics show slightly higher creep values (typically 20 and 50% higher respectively) compared with their constituent yarns. As a first approximation the creep can be lowered by reducing the complexity of the structure. Therefore lower plies, lower twist, straighter warps will all reduce creep in structured items. Where creep may be critical the deformation tendency of both yarns (and fabrics therefrom) can be significantly reduced by hot stretching.

Property	Terylene	Nylon 66	Nylon 6	Nomex	Poly-ethylene	Poly-propylene	Teflon	Polyvinyl Chloride	Viscous Rayons	Jute/Wire
Table 1 – Fibre properties										
Tenacity (grams/denier) appx.	7·8	8	8	5·3	4·5	8	1·4	1·8	2·2	—
Extension at break (%) appx.	9	15	17	22	25	18	15	25	18	—
Specific gravity	1·38	1·14	1·14	1·38	0·94	0·91	2·1	1·69	1·5	1·5
Melting point (°C)	260	250	215	370	120	165	327	—	190	—
Max. Operating temp. (°C) appx.	150	90	Below 65	200	55	90	280	—	Below 150	Below 65
Table 2 – Fabric form										
Woven	Yes	Yes	Yes	Yes	Yes	Yes	Yes	Yes	Yes	Yes
Knitted	Yes	Yes	Yes	Yes	Yes	—	Yes	Yes	Yes	Yes
Melded	Yes	Yes	Yes	—	Yes	Yes	—	Yes	No	No
Spun bonded	—	Yes	Yes	—	—	—	Yes	Yes	—	No
Needlefelt	Yes	Yes	Yes	Yes	Yes	Yes	Yes	Yes	Yes	Yes
Note to Table 2: *Durability, long-term filtration properties, particle retention, water/liquid permeability, thickness and weight are all properties which depend upon the fabric construction process and have to be considered where appropriate.*										
Table 3 – Resistance to:										
Fungus	1	3	3	3	4	3	3	3	3	1
Insects	2	2	2	2	4	2	3	3	2	1
Vermin	2	2	2	2	4	2	3	3	3	1
Mineral acids	3	2	2	2	4	4	4	3	1	1
Alkalis	2	3	3	3	4	4	4	3	4	1
Dry heat	3	2	2	4	2	2	4	2	2	2
Moist heat	2	3	3	3	2	2	4	2	1	2
Oxidising agents	3	2	2	3	1	3	4	—	1	—
Abrasion	4	4	4	3	3	3	3	4	3	3
Ultra-violet light	4	3	3	3	1	3	4	4	2	1
Note to Table 3: Poor – 1; Fair – 2; Good 3; Excellent 4										

Figure 3.24 Extract from a paper by E.W. Cannon, Civil Engineering March 1976

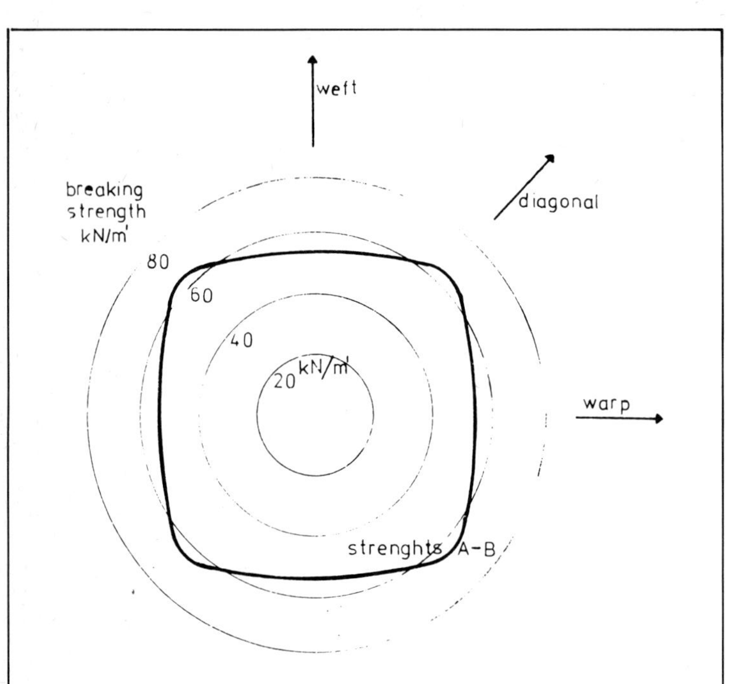

Fig. 4. Breaking strength as a function of the direction of testing for a polyamide (A) and a polyester (B) woven fabric.

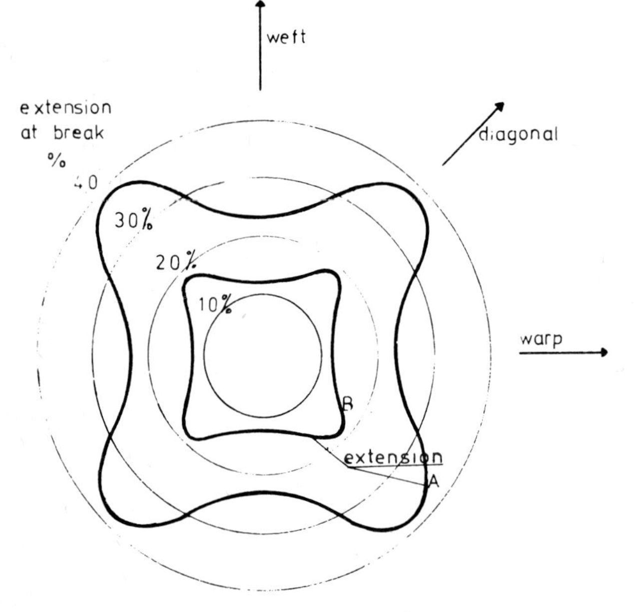

Fig. 5. Elongation at break as a function of the direction of testing for a polyamide (A) and a polyester (B) woven fabric.

Figure 3.25 Extract from a paper by J.H. Van Leeuwen, presented at a Conference on The Use of Fabrics in Geotechnics. Paris 1977

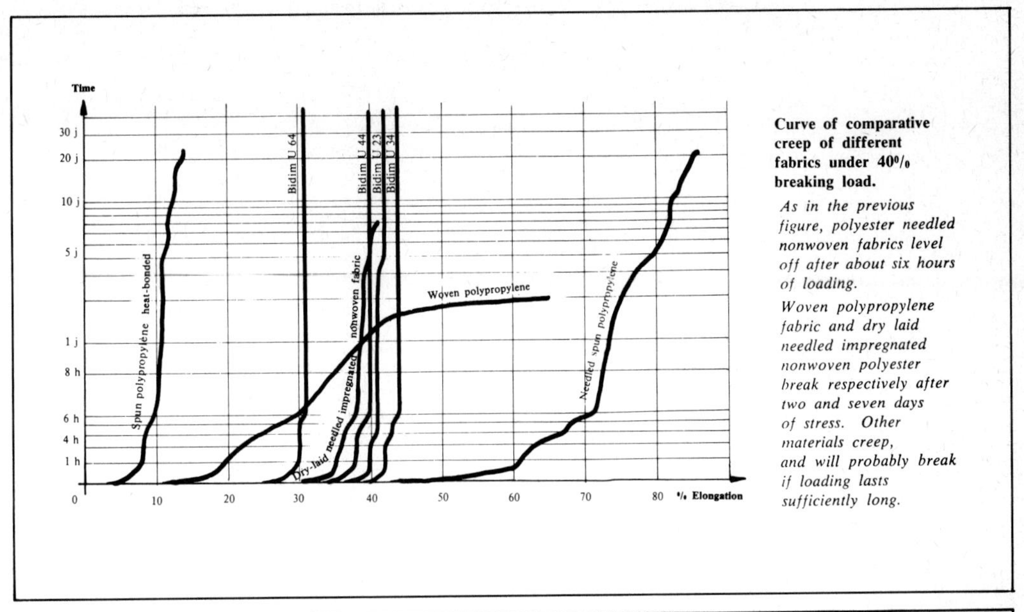

Figure 3.26 Extract from Bidim published literature. Rhone-Poulenc

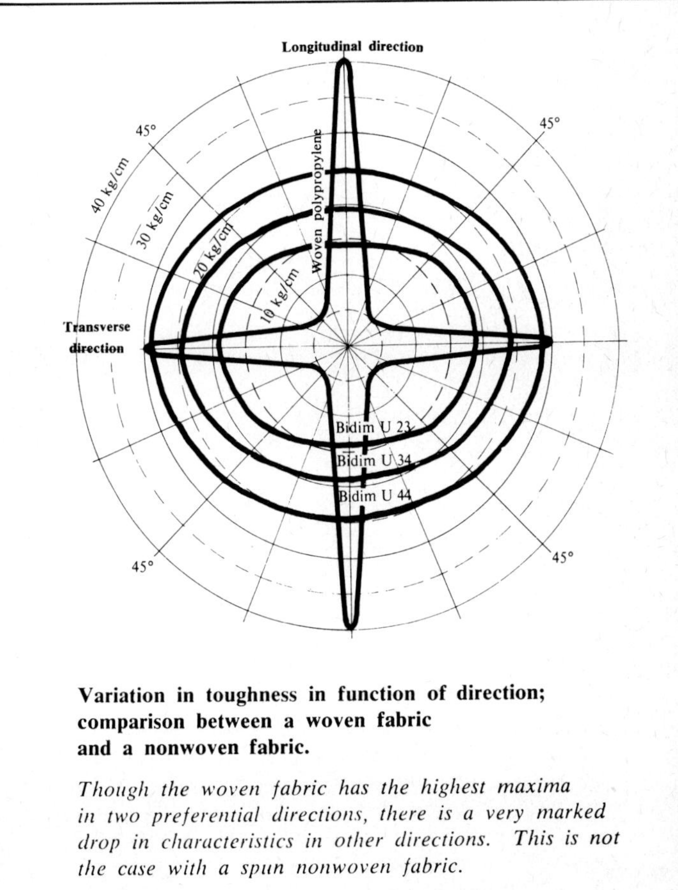

Figure 3.27 Extract from Bidim published literature. Rhone-Poulenc

Permeability:

The coefficient of permeability as measured in the laboratory shows that "Bidim" is a particularly permeable material; in general terms it may be compared to a coarse sand or gravel of 3.10^{-3} m/sec. perpendicular to the plane of the sheet.

The so-called "normal" permeability rapidly decreases as the load increases; it is in the neighbourhood of 4.10^{-4} m/sec. under a pressure of 1 bar, and 3.10^{-4} m/sec. under a pressure of 2 bars (Fig. 26).

Measurement of the coefficient of radial permeability, that is to say in the plane of the sheet, shows that this varies little with the stress applied; 5.10^{-4} m/sec. under 1 bar and 4.10^{-4} m/sec. under 2 bars.

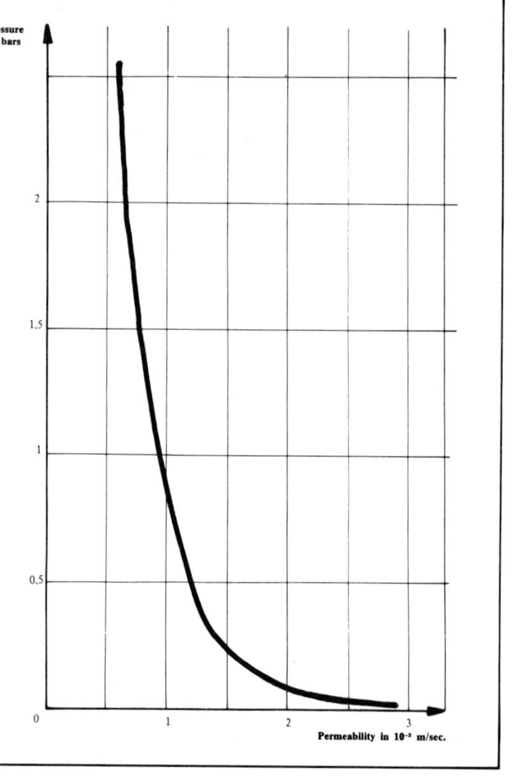

Variation of permeability of a "Bidim" nonwoven fabric, grade U 34, in function of the load applied.

The experimental curve shows that a limit is obtained at about 2 bars.

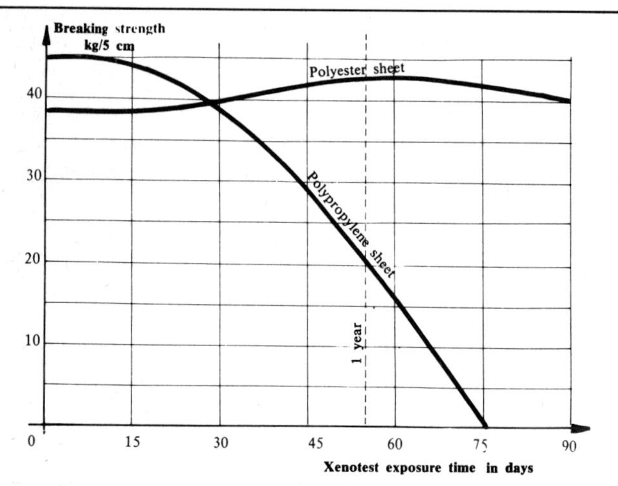

Resistance to ageing (Textile standard AFNOR G-07.001)

The criterion adopted is the textile dynamometric test. We note that two sheets obtained by direct extrusion and whose initial characteristics are close have a markedly different behaviour after a prolonged period in an ageing simulator. The dotted line corresponds approximately to the simulation of one year of exposure to sunshine.

Figure 3.28 Extract from Bidim published literature. Rhone-Poulenc

Figure 3.29 Extract from Bidim published literature. Rhone-Poulenc

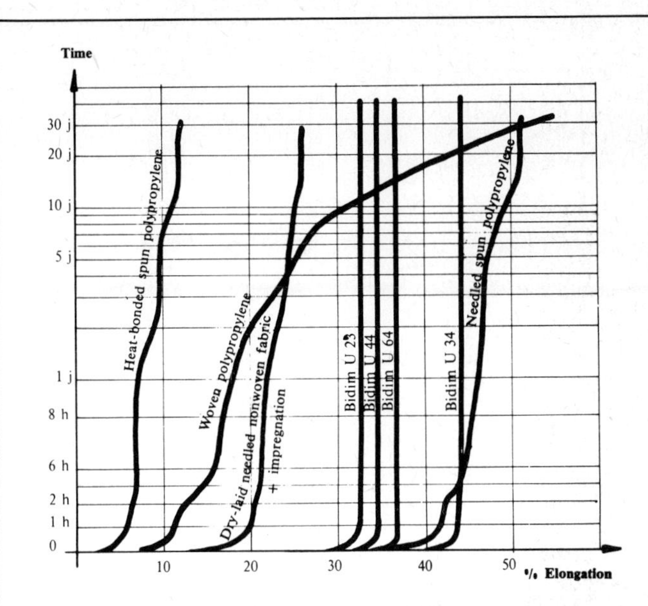

Comparative curve of the creep of different materials under 20% breaking load.

The range of needled polyester nonwoven materials rapidly becomes stable. It is important to note that woven polypropylene behaves in a similar manner to the material from which it is made (see Fig. 5). Deformation occurs steadily over a period of time up to breaking point.

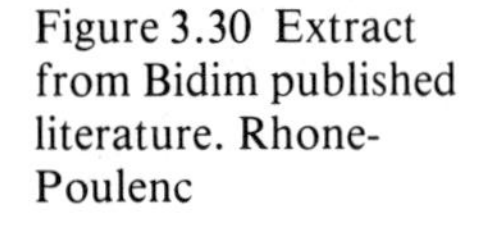

Figure 3.30 Extract from Bidim published literature. Rhone-Poulenc

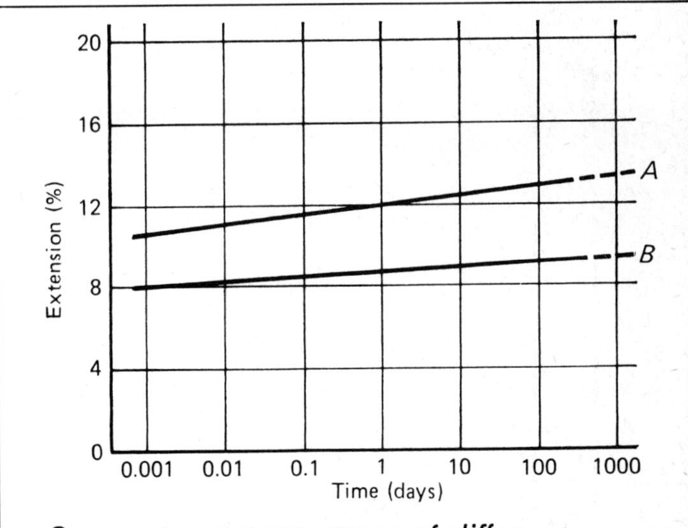

Creep at constant stress of different woven fabrics (Van Leeuwen, 1977)

A. Polyamide woven
B. Polyester woven
Stress level = 50% of standard strength

Figure 3.31 Extract from a paper by J.H. Van Leeuwen, presented at a Conference on The Use of Fabrics in Geotechnics. Paris 1977

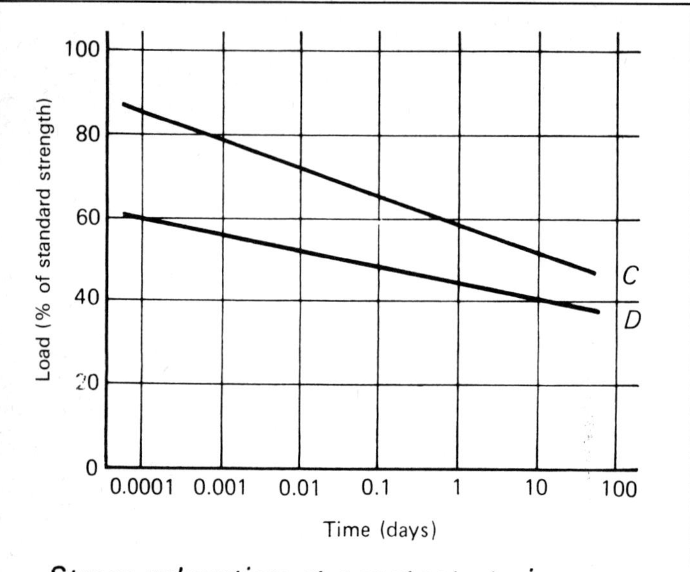

Figure 3.32 Extract from a paper by J.H. Van Leeuwen, presented at a Conference on The Use of Fabrics in Geotechnics. Paris 1977

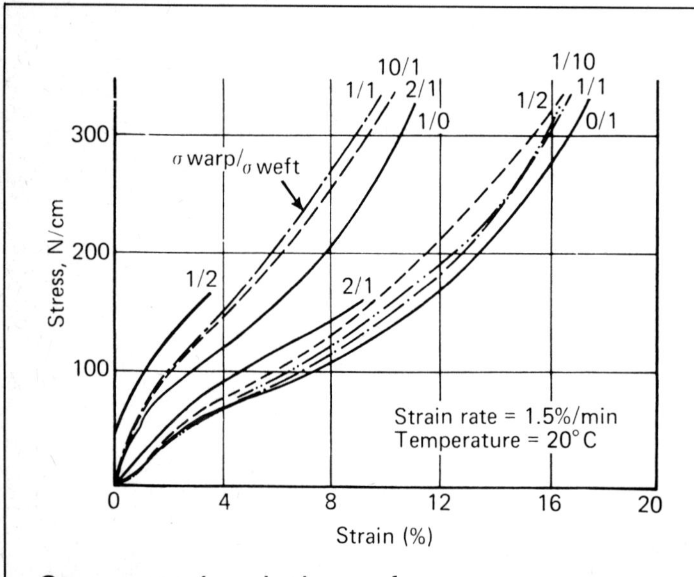

Figure 3.33 Extract from a paper by H.W. Reinhardt on "Experimental Mechanics" Feb. 1976

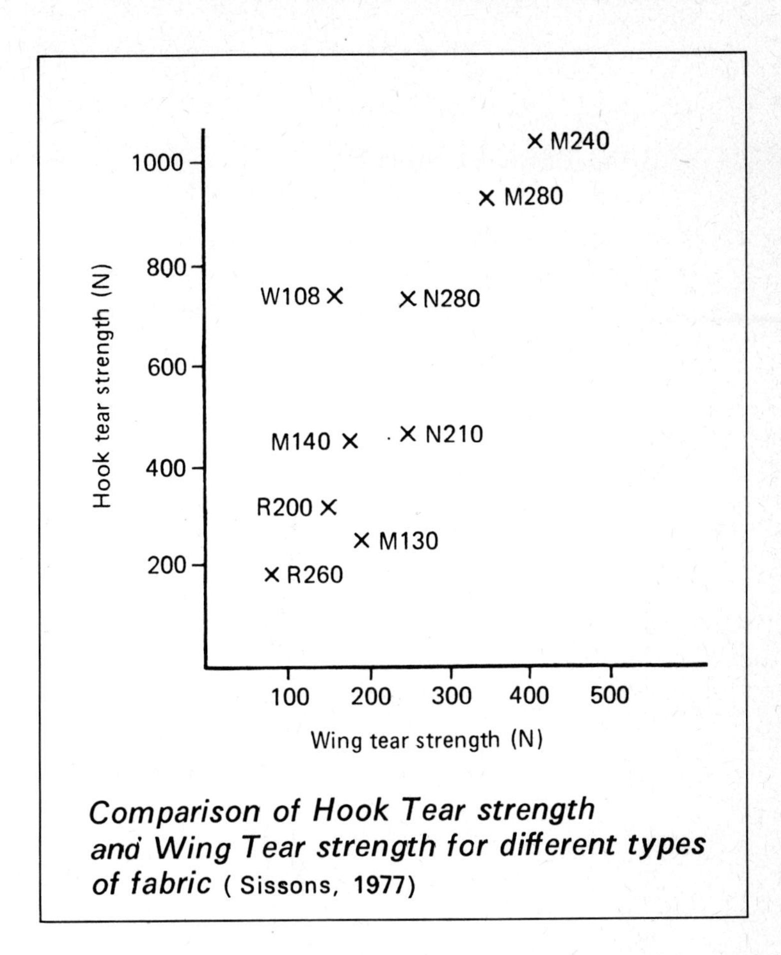

Comparison of Hook Tear strength and Wing Tear strength for different types of fabric (Sissons, 1977)

Figure 3.34 Extract from a paper presented by C.W. Sissons at a Conference on The Use of Fabrics in Geotechnics, Paris 1977

TYPICAL VALUES OF FABRIC PERMEABILITY

(McGown, 1976)

Fabric type	*k* (m/sec)
Woven	Varies greatly
Needle-punched	10^{-3}–10^{-4}
Melt-bonded	10^{-2}–10^{-4}
Resin-bonded	10^{-4}–10^{-5}

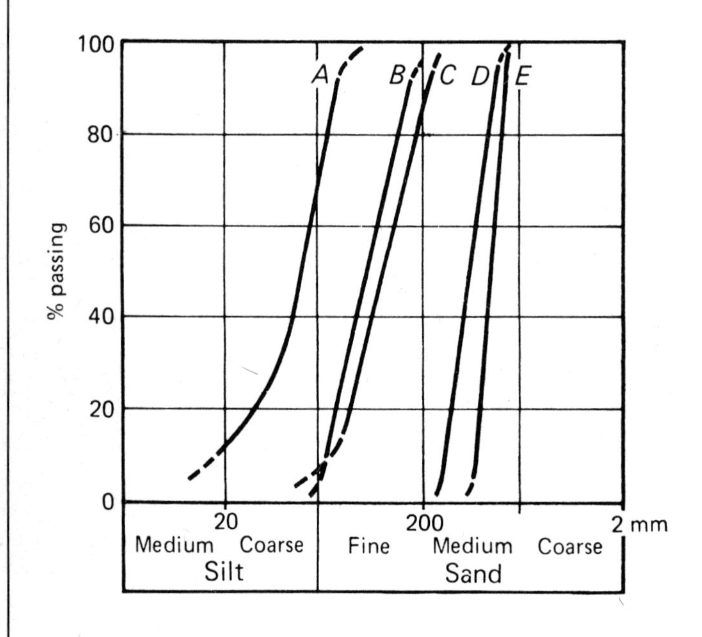

Typical fabric pore size distributions

(McGown, 1976)

A. 250gm/m² resin bonded
B. 300gm/m² needle-punched
C. 140gm/m² melt bonded
D. 380gm/m² hessian woven
E. 185gm/m² woven terylene

Figure 3.35 Extract from a paper presented by A. McGown to a research workshop on materials and methods for low cost road construction. Australia 1976

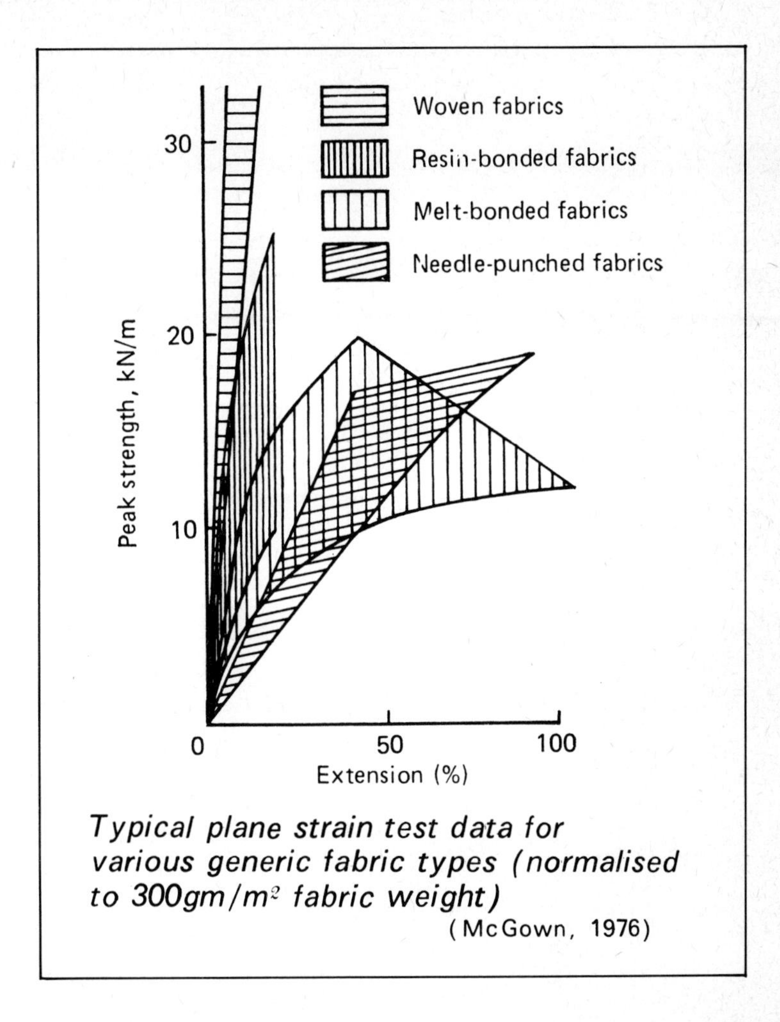

Figure 3.36 Extract from a paper presented by A. McGown to a research workshop on materials and methods for low cost road construction. Australia 1976

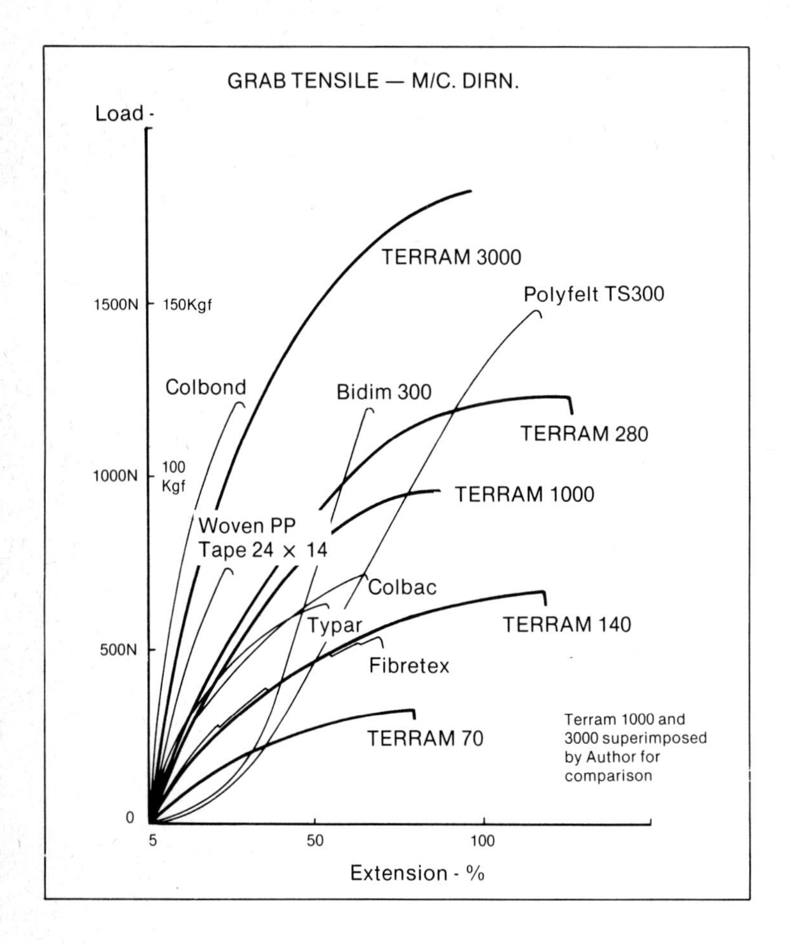

Figure 3.37 Extract from ICI Fibres research test results and publications

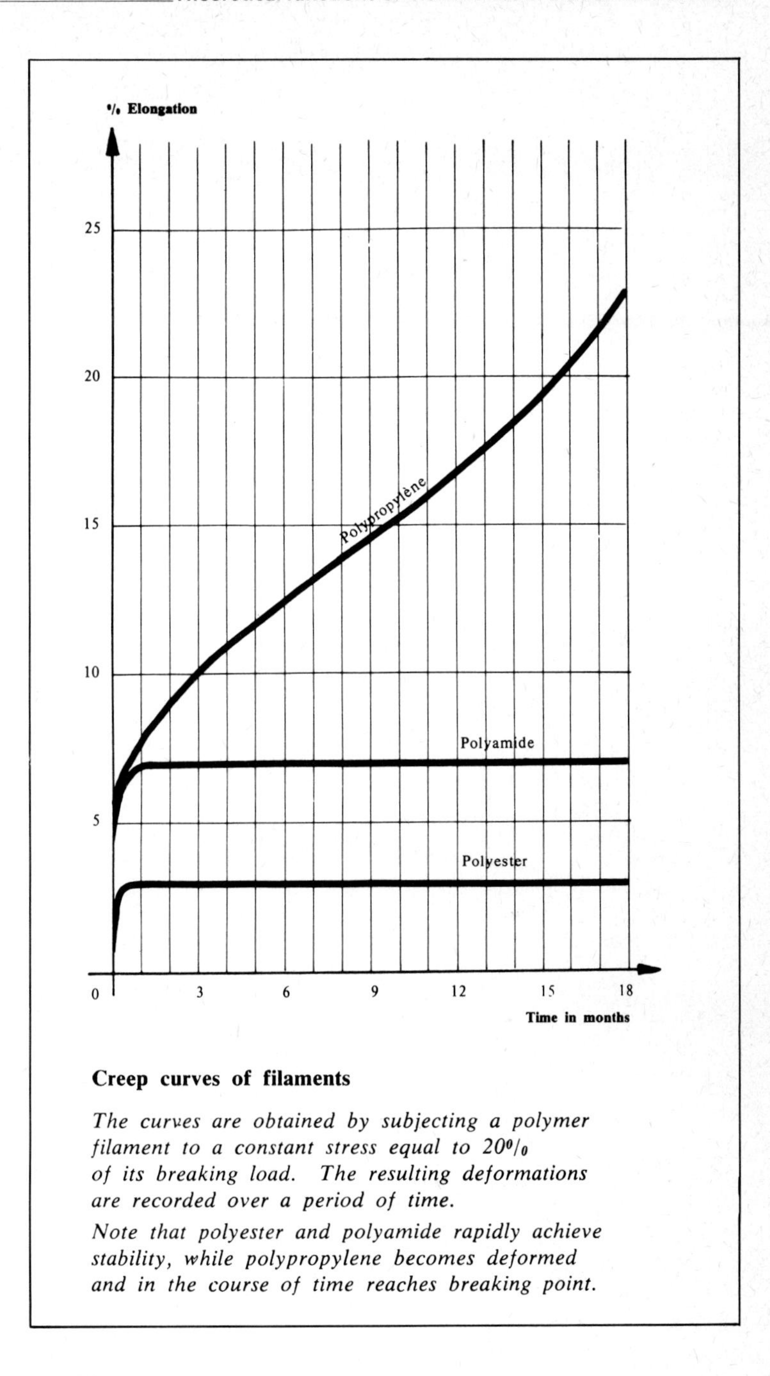

Figure 3.38 Extract from Bidim published literature. Rhone-Poulenc

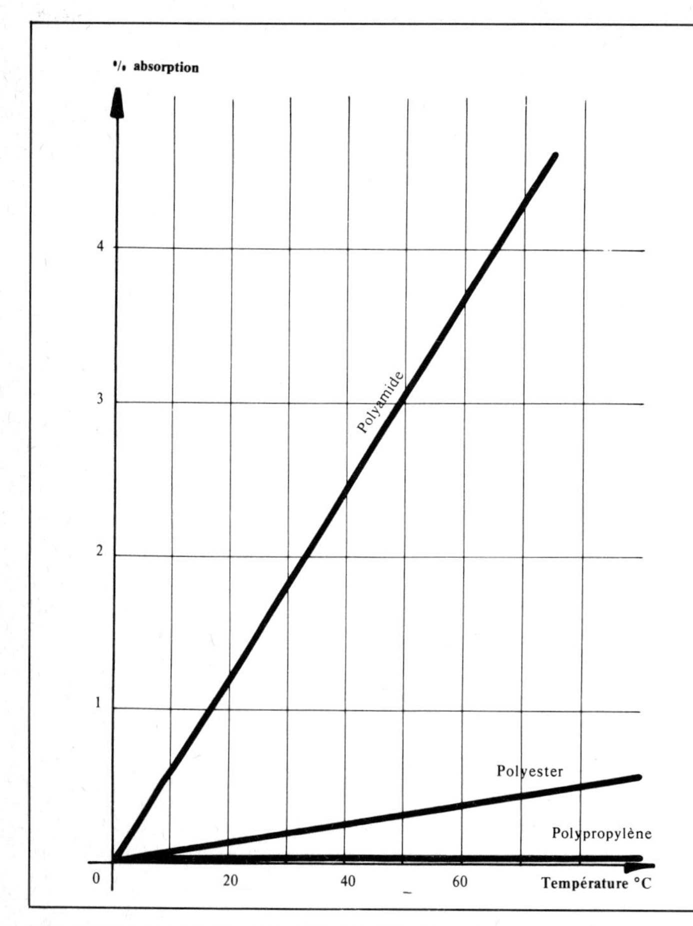

Curve of moisture absorption at different temperatures

Water absorption is a characteristic of polyamides. The dynamometric performances of the filaments are affected, and drop by 20°/₀ to 30°/₀.

Figure 3.39 Extract from Bidim published literature. Rhone-Poulenc

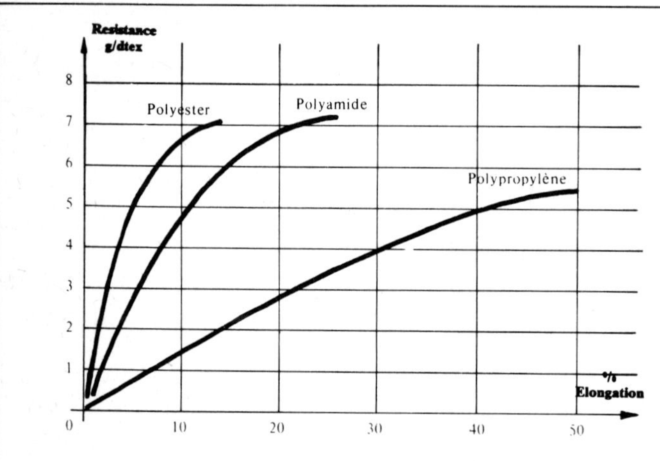

Dynamometric curves of various filaments

Though the three polymers are of comparable strength, the polypropylene curve is seen to have a low and appreciably constant slope up to breaking point.

Figure 3.40 Extract from Bidim published literature. Rhone-Poulenc

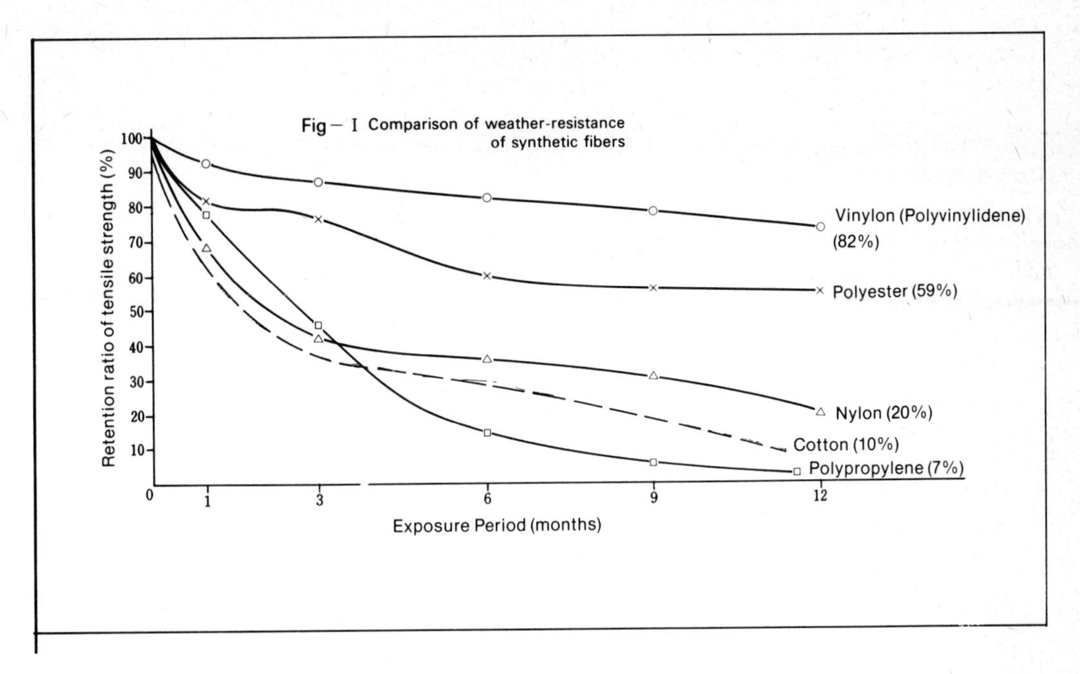

Figure 3.41 Extract from published literature by Kuralon/Taiyo Kogyo Ltd

Soil exposure effects on plastics in civil engineering applications

Soil contains both inorganic and organic chemicals, inorganic materials largely derived from the weathering of rocks and minerals, and organic materials from plants, animals and micro-organisms. In a large majority of the soils, inorganic substances constitute the bulk of the soil material. In addition, the inorganic fraction contains acids and alkalis which may under certain circumstances drastically affect the soil reaction (pH range: 3.0 or less to 10.0 or more). Organic matter normally varies from 1 % to 10 %, occasionally reaching a level of 50 % or more. A great variety of micro-organisms are present, particularly in soils rich in organic matter.

Inorganic chemicals, which are likely to affect buried plastics, include mineral acids, alkalis, salts, gases and water. The intensity of the soil weathering process and the bio-geological environment of the soil determine the distribution and concentration of these chemicals. Knowledge of soil organic compounds is rather limited and qualitative in nature. Flaig reported the observations of Schreiner and Shorey and their associates (work done about 60 years ago), who for the first time isolated several organic compounds from soils.

A search of the literature has not shown any work in which the effect of a specific soil chemical on any buried plastics has been demonstrated. However, laboratory experiments have shown chemical resistance and, in some cases, corrosion of plastic materials.

Corrosion of plastics in soil can result from the action of various chemicals which may be present in substantial amounts in certain aggressive soils such as acid soils (e.g. acid sulphate soils), organic soils (e.g. Histosols containing various organic acids and other compounds), saline-alkali soils (e.g. Solonchak-Solonetz-Solod soils) and calcareous soils derived from calcitic and dolomitic materials. Mineral acids commonly present in soils are sulphuric acid, nitric acid and hydrochloric acid. Organic compounds in soils mostly comprise complex polymers and various aliphatic and aromatic acids. Salts in saline-alkali soils are characterised by sodium chloride, sodium sulphate, sodium nitrate, sodium carbonate/bicarbonate, magnesium chloride, magnesium sulphate, magnesium bicarbonate, calcium chloride, calcium carbonate/bicarbonate, potassium chloride, potassium sulphate, potassium carbonate and ammonium carbonate.

The main process involved in chemical corrosion is oxidation which can cause degradation of a number of plastic products such as nylon, polyesters and polyolefines. Trace metals released in soils through the weathering of rocks and minerals will accelerate oxidation reactions. There are numerous publications on the role of trace elements as catalytic agents in oxidative degradation of plastics. Oxidative processes have reportedly degraded plastic materials in soils, e.g. deterioration of polyvinyl chloride film used in a section of the Dashava-Minsk pipeline in the USSR.

In soil chemistry the following processes are being studied in detail: sulphur transformation (producing sulphur dioxide, hydrogen sulphide, sulphuric acid); ammonification (producing ammonia in gaseous and aqueous state, and ammonium-bearing salts); nitrification and denitrification (producing nitrites, nitrates, nitric acid, nitrogen dioxide, nitrous oxide); ferralliti-sation (producing hydroxides/oxides and ionised forms of iron and aluminium); phosphorus transformation (producing phosphates and phosphoric acid). A literature search has shown that under laboratory conditions several of these chemicals degraded plastic products, e.g.

(a) 100 ppm of NO - NO_2 mixtures resulted in marked changes in tensile strength of polyamides;

(b) 0.2 ppm of SO_2 in air degraded nylon;

(c) Humid air polluted with SO_2 decreased strength of polyester fabric after 30 days' exposure;

(d) Various agricultural corrosive materials like ammonia liquor, manure, Bordeaux mixture, Chernozem soil, etc. degraded a number of synthetic polymers (USSR products), e.g. high/low density polythene, polypropylene, polyoxymethylenes, polyamide P 68, polycaprolactam and polyamide P-AK7; and

(e) In saline and distilled water, ferric chloride decreased the strength of nylon.

Water is absorbed by and diffuses through the vast majority of polymeric solids at ambient temperatures and humidities. A prolonged soil exposure in the tropics could produce deleterious effect on plastics such as polyesters, polyester-based reinforced plastics or plasticised polyvinyl chloride. In plastic composites, there is a possibility that water will diffuse preferentially into the bulk composite along high diffusion rate paths or by capillary action along fissures such as interlaminar planes and particularly channels opened in the region of fibres after mechanical stressing.

The microbiological degradation of plastics, by which is meant the enzymatic degradation by micro-organisms such as fungi and bacteria, has not received the same attention as the photo-degradation of these polymers. However, many papers have been published during the last two decades within the general field of plastics biodegradation, but most of these have dealt with compounded materials such as polymers containing plasticisers and lubricants. This review has been restricted to the deterioration of plastics by micro-organisms which are commonly found in soils.

Soils show a great abundance of micro-organisms, some of which have been found to be detrimental to plastics products. It seems probable that the micro-organisms will be more active in organic soils, with tropical environment inducing a particularly vigorous growth.

The soil microbes may be grouped in four classes, namely: (a) bacteria (weight may exceed 1,000 pounds per acre); (b) fungi (one gram of soil commonly containing 10 to 100 metres of mould filament); (c) actinomyces (numbers varying between 0.1 million and 36 millions per gram of soil); and (d) algae.

Observations on the resistance of plastics to microbiological attack under controlled conditions are reported in Figure 3.42. There is hardly any information available on the biodeterioration of plastics in specific soil environments (e.g. organic soils, waterlogged or strongly acid soils). A search of the literature has only indicated some general trends.

Nylon is reported to be fungus resistant when used as a cable covering, yet fibre and fabric nylon are considered susceptible to microbial attack. Potts *et al.* found a drastic loss of the strength in a caprolactone polyester buried in soil, and suggested that the deteriorating effect could be due to chemical and/or biological agents. A fairly high resistance of glass-reinforced polyester resins to fungal attack was reported by Heap and Morrell. After a tropical exposure, these resins showed only a light surface growth which could easily be removed.

Polythene is considered to be highly resistant to micro-biological attack. Maclachlan *et al.* evaluated the performance of certain synthetic compounds, such as Neoprene, PVC and polythene, used in cable coatings. They reported the occurrence of micro-pores in buried cables; micro-organisms isolated were the following fungi: Spicaria violacea, Metarrjizium anisopliae and species of Fusarium and Stemphyliopsis. Schwartz also found that bacteria may proliferate on a polythene substrate (e.g. water pipes), the polymer matrix remaining unaffected. However, Heap and Morrell observed that polythene cable coatings showed no attack when buried in soils dominated by micro-organisms.

In a soil burial test on plastic sheets, Küster and Azadi-Baksh found some microbial growth in 'Treopaphan' NN 20 sheet (polypropylene as basic polymer), which however was not indicative of any degradation of the polymer itself. In reviewing the micro-biological deterioration of rubbers and plastics in the tropics, Maclachlan *et al.* found polypropylene to have good resistance.

Several investigators showed degradation of plasticised PVC by fungal attack. In a bibliography on the resistance of PVC to attack by various agents, Fox cited a number of papers in which a fairly strong resistance of unplasticised PVC to microbial attack was indicated. Booth and Robb carried out tests on plasticised PVC in an average soil (fine brown loamy soil from Clandon, Surrey, having a pH of 6.9), inoculated with species of Pseudomonas and Brevibacterium, and found degradation within as short a period as eight weeks.

Microbial degradation of styrene copolymers was reported by Callely *et al.* From burial studies (soil characteristics not described), the investigators observed a fair amount of microbial growth, and isolated some forty cultures of bacteria, yeasts and fungi. Belokon *et al.* however, found that the high impact polystyrene was attacked by fungi, but its physico-mechanical properties were not impaired.

A literature search on the effects of soil burial on the mechanical strength of plastics has produced an insignificant amount of information, and in such work that has been done studies indicated only soil effects. Soils used were not properly characterised, and in several cases there were hardly any soil data

Microbiological attack on plastics

Material	Observations on susceptibility
1. Nylon	The material contains polypeptide linkages, and it might be supposed that microbiological attack is highly probable. According to Russian work, nylon suffered from a change of colour and a weakening of the film due to microbiological attack, particularly by Penicillium and Aspergillus.
2. Polyester	Microbial degradation of the material was reported by several investigators. Potts et al found biodegradation of a number of polyesters of varying structure and molecular weight (a group of fungi and bacteria used as test organisms). They found that an epsilon caprolactone polyester of about 40,000 molecular weight, which had no branching, was readily utilised by fungi and bacteria.
3. Polythene	The material shows good resistance to microbiological attack, especially when pigmented with carbon black. The US Naval Department observed that the material of lower molecular weight supported microbial growth, this being in agreement with the work of Jen-Hao and Schwartz. Potts et al found that some of the HDPE & LDPE having molecular weight between 10,000 to 14,000 were appreciably biodegraded. They attributed this effect to the presence of low molecular weight species (< 500 mol wt).
4. Polypropylene	The material shows good resistance to microbiological attack. Potts et al observed biodegradability of a large number of commercial plastics including polypropylene (a group of fungi and bacteria used as test organisms). The microbial growth was thought to be due to the presence of a biodegradable additive in the sample.
5. Polyvinyl chloride	The overall conclusion from several experiments is that the plasticisers of the material, rather than the polymer itself, are directly attacked by microbes, particularly fungi. Hueck reported discolouration of the PVC materials and precipitation of FeS by sulphate reducing bacteria. Schwartz investigated a range of chlorinated lower paraffins of increasing chlorine content, and showed that bacteria could use these easily as a source of carbon up to a chlorine content of 30%, above which growth rate slowed down and became non-existent at 50% (PVC and vinyl chloride monomer have <50% of chlorine).
6. Polystyrene	The material shows a fair amount of resistance to microbial attack. Potts et al found hardly any microbial growth on polystyrenes of molecular weight from 600 to 214,000, and on copolymers of styrene (comonomers included were: acrylic acid, sodium acrylate, dimethyl itaconate, acrylonitrile, ethyl acetate and methacrylonitrile). The chemical structure of polystyrene is basically similar to that of polythene (hydrogen atoms on alternate carbon atoms replaced by phenyl groups). The introduction of the phenyl groups does not render the polymer more bio-inert since aromatic rings themselves are biodegradable.

Figure 3.42 Microbiological attack on plastics

available. As a result, placement of these soils in the world system of soil classification was not possible.

Because of the Bell System emphasis on burying transmission lines, a number of studies were undertaken to determine the behaviour of commercial and experimental plastics in the soil. The investigators used the same two soil types as the burial media, which are located in an 11-acre site in Bainbridge,

Georgia and a 15-acre site in Roswell, New Mexico. These soils have been described by Connolly, and are not considered as severe problem soils for buried plastics. The Bainbridge soil has a fine sandy loam texture with a pH of about 5.2 and the Roswell soil a clay loam texture with a pH of about 8.0. Some of the experiments laid out in these two soils were done on nylon, polyolefines, polyvinyl chloride, polystyrene and glass reinforced laminates.

Nylon

Soil tests on twenty-six thermoplastics were reported by Miner. The materials were commercial grades and contained various types of additives. Because of their proprietary nature, the additives were not disclosed. Changes in the strength properties of polyamides are shown as follows (the effects due to depth and soil type have been averaged):

Type 6 Nylon	*Tensile break strength (% decrease, approx.)*
1 year burial	10
2 year burial	10-12
4 year burial	20
8 year burial	25

These data clearly show that type 6 nylon deteriorated steadily in tensile properties. The material changed from a hard, tough thermoplastic to such a rigid, brittle material that rupture occurred without any appreciable elongation. According to Miner the loss of strength could be attributed to the moisture content of the soil (degradation caused by hydrolysis). It was further observed that type 610 nylon exhibited superior resistance properties than type 6 nylon (610 nylon absorbing 75 % less moisture than 6 nylon).

ICI Dyestuffs Division studied the resistance of nylon (two samples, 'du Pont 48 denier 17 filaments' and 'du Pont 45 denier 15 filaments') and several other yarns to one month's burial in a mixture of soil and well-rotted manure. Nylon showed a fair amount of resistance to soil exposure.

Polyester

The strength characteristics of caprolactone polyester were evaluated by Potts *et al.* The material was moulded into tensile test bars and buried for a period extending up to twelve months (soil characteristics were not described). The results presented below show a drastic reduction in the strength:

Burial time (months)	*Tensile strength (psi)*	*Elongation (%)*	*Weight loss (%)*
0	2610 ± 103	369 ± 59	0
2	1610 ± 180	7 ± 2.0	8
4	520 ± 220	2.6 ± 1.1	16
6	100	Negligible	25
12	Negligible	Negligible	42

Polyolefines

Miner carried out tests on moulded materials made from polythene and polypropylene, and found insignificant change in strength properties of polythene and polypropylene buried on two US soils up to eight years. De Coste's observations on polythene buried as extruded wire coating showed that the materials varied in their retention of tensile properties and depended upon the density of the resins from which they were prepared. Those based on low-density resins showed a relatively small change in tensile properties after up to eight years of burial. The polythene plastics based on high-density resins, in contrast, suffered major losses in elongation, in some cases to the point of embrittlement, and decreased somewhat in strength. De Coste, however, doubted that the embrittlement of the high density polythene was caused by burial, because the shelf ageing for eight years at room temperature produced nearly the same effects on the high-density resins. De Coste further observed that the polypropylene plastics used in wire coating showed small change in strength properties after burial.

Polyvinyl chloride

Changes in the mechanical properties of the semi-rigid polyvinyl chloride compound were recorded by Miner who found loss of mechanical strength due to burial. The PVC compound contained 43 phr (parts per hundred parts resin) of di-n-octyl-n-decylphthalate plasticiser. In addition to the loss of mechanical strength, the semi-rigid PVC had black greasy surface deposits. This material also contained 10 phr of di-basic lead phosphite stabiliser and fatty acid lubricants, which might have migrated to the surface during the soil leaching process. Since changes in the plasticiser content can have a marked effect on mechanical properties, there is no evidence to suggest that the PVC resin itself was degraded.

De Coste performed soil tests on vinyl chloride plastic wire and cable coatings buried for up to eight years. The vinyl plastics ranged from semi-rigid to highly flexible grades, and consisted of both commercial and experimental formulations. The investigator found that the materials varied in their resistance depending on the plasticiser content.

Styrene compounds and GRP laminates

Soil researches in the USA were directed also to burial studies of polystyrene (general purpose, clear), styrene-acrylonitrile copolymer, acrylonitrile-butadiene-styrene (natural) and acrylonitrile-butadiene-styrene (black). The soil exposure up to a period of eight years produced a mild effect on these compounds.

Miner also demonstrated soil exposure effect on the mechanical properties of some glass-reinforced thermoset compounds such as melamine, styrene-polyester and alkyd. He found a significant loss in tensile, flexural and hardness values for all of these compounds. In explaining the causes for mechanical deterioration, Miner suggested that soil moisture has a pronounced effect on the susceptibility of glass-filled materials to burial.

Certain chemicals are likely to dominate particular soil environments such as those exposed to high salt ingress or derived from calcitic ($CaCo_3$) dolomitic ($CaCO_3$, $MgCO_3$) materials. A substantial number of cations (e.g. Na, Mg, Ca, K) and anions (e.g. Cl, CO_3, HCO_3, SO_3) may be present in salt-affected soils. These salts (i.e. the constituents of calcareous or saline-alkali soils), if allowed to act on GRP products particularly in the tropics or arid regions, may cause degradation during prolonged exposure. Ammonium carbonate, commonly produced in the process of soil-N transformation, could have a deleterious effect on certain reinforced laminates. The weathering resistance of plastics composites has been recently reviewed by Scott and Paul. While discussing the question 'How long will GRP last?', the reviewers found virtually no satisfactory answers available from the literature. The reader may consult a recent paper by Klein who reported a detailed investigation on reinforced plastics buried in ordinary soils for several years. (See Appendix II: Further Reading)

Any soil environment may be considered unfavourable for plastics products, because of corrosive properties of the inorganic and organic chemicals and hostility of the macro- and micro-organisms. Figure 3.43 shows the general distribution of the major soil classes (short description in Figure 3.44), of which some are considered to possess a potential danger for buried plastics. Explanatory notes are given in Figure 3.45 to show new soil orders (classification based on modified seventh approximation, Figure 3.44) and their approximate equivalents in 'Great Soil Groups'. For further reference to detailed representation of the great soil groups, the reader may consult the FAO publi-

Figure 3.43 (Opposite) Broad schematic map of the soil orders of the world

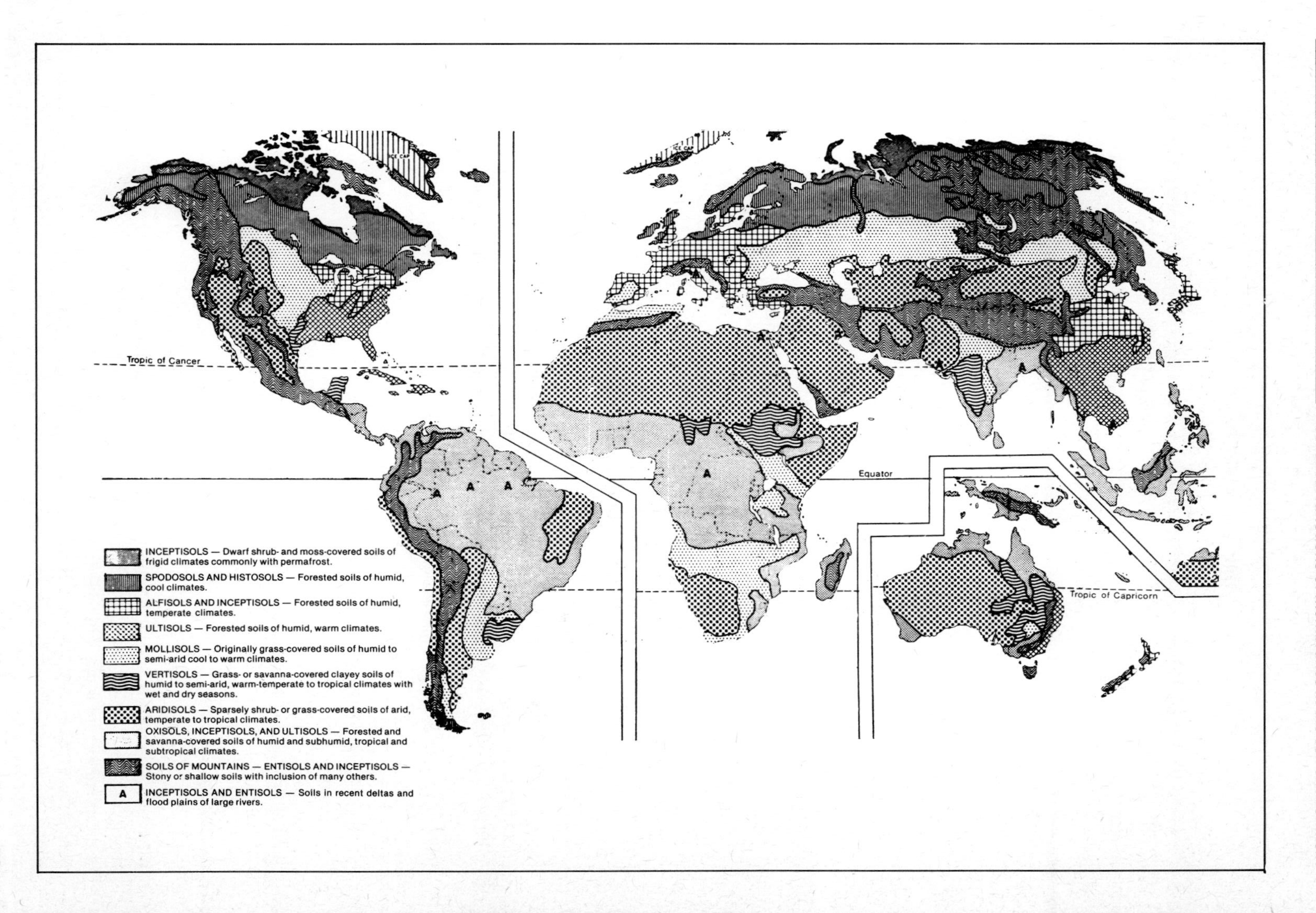
Tropic of Cancer
Equator
Tropic of Capricorn
INCEPTISOLS — Dwarf shrub- and moss-covered soils of frigid climates commonly with permafrost.
SPODOSOLS AND HISTOSOLS — Forested soils of humid, cool climates.
ALFISOLS AND INCEPTISOLS — Forested soils of humid, temperate climates.
ULTISOLS — Forested soils of humid, warm climates.
MOLLISOLS — Originally grass-covered soils of humid to semi-arid cool to warm climates.
VERTISOLS — Grass- or savanna-covered clayey soils of humid to semi-arid, warm-temperate to tropical climates with wet and dry seasons.
ARIDISOLS — Sparsely shrub- or grass-covered soils of arid, temperate to tropical climates.
OXISOLS, INCEPTISOLS, AND ULTISOLS — Forested and savanna-covered soils of humid and subhumid, tropical and subtropical climates.
SOILS OF MOUNTAINS — ENTISOLS AND INCEPTISOLS — Stony or shallow soils with inclusion of many others.
A
INCEPTISOLS AND ENTISOLS — Soils in recent deltas and flood plains of large rivers.

Main soil regions of the world
After Foth and Turk

Soil Order & Area (10^3 sq. miles)	Short Description
1. Alfisols Area: 7,600 % World total : 14.7	Enriched with Al & Fe, subsoil clayey. Agriculturally productive. Most alfisols occur in humid regions where the soil is moist at least part of the year (grey-brown podsolics in this order).
2. Aridisols Area: 9,900 % World total : 19.2	Primarily soils of dry places. Natural vegetation is sparse, consisting of desert shrubs and grasses. Always some carbonate, sometimes accumulation of calcium sulphate (gypsic layer). Organic matter low. Most abundant in the world.
3. Entisols Area: 6,500 % World total : 12.5	Soils of recent origin, and are shallow and without genetic horizon (layer) differentiation. Alluvial soils (⅓ of world's population reportedly obtain food from alluvial soils) belong to this order. Other soils included are lithosols which rest directly on the hard rock and may occur on steep slopes of mountains.
4. Histosols Area: 400 % World total : 0.8	These are "organic soils". Remain saturated with water at least 30 consecutive days a year, and contain more than 20% organic matter. Stability depends on the nature of the plant materials deposited in the water and degree of decomposition.
5. Inceptisols Area: 8,100 % World total : 15.8	Soil profile development at initial stage, but older than Entisols. Two large areas include the Tundra of North America and Europe-Asia. Tundra soils contain large amount of organic matter, mainly because of low soil temperature. Usually have permafrost, are slightly to strongly acid.
6. Mollisols Area: 4,600 % World total : 9.0	Bordering desert regions are areas of adequate rainfall, with a transition from aridisols to mollisols. Well decomposed organic matter (at least 1%), well aggregated soil structure, agriculturally most productive.
7. Oxisols Area: 4,800 % World total : 9.2	Occur only on ancient land surfaces in the humid tropic (earlier latosols and laterites). Subsoil enriched with hydrated oxides of Fe and/or Al and variable amounts of 1 : 1 lattice clay (non-expansible). Stable soil aggregates and erosion resistant.

Figure 3.44 Main soil regions of the world

cations of world soil maps given in the list of references at the end of this book.

Based on available data with regard to the chemical and micro-biological resistance of the plastics materials, the potentially hazardous soils may be placed in the following four categories:

1. Acid Sulphate soils
2. Swelling clay soils
3. Organic soils
4. Other chemically reactive soils

Acid sulphate soils

The micro-biological reduction of sulphate is a common event wherever sulphates and organic matter occur in reduced natural environments. The sulphidic mud, largely composed of pyrites, is the resultant product which is oxidised when exposed to the atmosphere. As a result of oxidation, sulphuric acid is produced which is responsible for the peculiar characteristics of acid

New soil orders and approximate equivalents in great soil groups
After Foth and Turk

Order*	Meaning	Approximate equivalents*
1. Alfisol	Aluminium-Iron enriched soils	Grey-Brown Podsolic, Grey Wooded, Noncalcic Brown, Degraded Chernozem, and associated Planosols and Half-Bog soils.
2. Aridisol	Arid soils	Desert, Reddish Desert, Sierozem, Solonchak, some Brown and Reddish Brown soils — and associated Solonetz.
3. Entisol	Recent soils	Azonal soils and some low Humic Gley soils.
4. Histosol	Tissue (organic) soils	Bog soils (organic)
5. Inceptisol	Inception or young soils	Ando, Sol Brun Acide, some Brown Forest, Low Humic Gley and Humic Gley soils.
6. Mollisol	Soft soils	Chestnut, Chernozem, Brunizem (Prairie), Rendzinas, some Brown, Brown Forest, and associated Solonetz and Humic Gley Soils.
7. Oxisol	Oxide soils	Laterite soils, Latosols.
8. Spodosol	Ashy (podsol) soils	Podsols, Brown Podsolic soils, and Ground-Water Podsols.
9. Ultisol	Ultimate (of leaching)	Red-Yellow Podsolic, Reddish-Brown Lateritic (of U.S.A.), and associated Planosols and Half-Bog soils.
10. Vertisol	Inverted soils	Grumusols

*Equivalent civil engineering terminology non-existent

Figure 3.45 New soil orders and approximate equivalents in great soil groups

sulphate soils. A considerable amount of elemental sulphur can also be produced under special conditions by chemical and micro-biological oxidation of H_2S. Acid sulphate soils occur in all climatic zones of the earth and are usually located in areas that are flat, swampy or marshy, and filled in with peaty material (Figure 3.46).

In industrial areas, soils are being increasingly exposed to sulphuric acid and its precursor sulphur dioxide, both common waste products of smelting, fossil fuel combustion and other related activities. Sulphuric acid also results from natural oxidation of sulphide minerals and organic sulphur in mining areas.

Some characteristics of the acid sulphate soils are presented in Figure 3.47. In a recent symposium held in the Netherlands by the International Institute for Land Reclamation and Improvement, the distribution and chemical and physical properties of the acid sulphate soils were discussed. Such soils do not occur in the United Kingdom, but pyritic soils with a potential for acidity development are present.

With regard to the placement of plastics products which are susceptible to acidity, sulphur compounds and bacterial action, the acid sulphate soils may pose severe problems. These soils do not normally show acidity below pH 3.0, but in the process of chemical reaction the soil pH might fall below 3.0, particularly when the soils are allowed to dry and oxidise. Nylons or nylon-based products, when buried in such soils with strong acidity, are expected to deteriorate rapidly.

Swelling clay soils

Plastics products such as pipes, when buried in soil, are subjected to internal pressure, soil and traffic load, and subsidence. Gilbu and Rolston of the Fiberglass Corporation in Ohio discussed the danger of a buckling effect that may be produced in reinforced plastic pipes buried in soils. A research project on the underground deformation of pipes (made from polythene and polyvinyl chloride) is being carried out at the Hydraulic Experimental Station of the State Rivers and Water Supply Commission of Victoria, Australia. Two soil types are used as the burial media, the investigation aiming at determining the properties of the pipes in buried condition and of the soils in contact with the pipes.

The problem of the underground deformation of pipes is likely to be more aggravated in clayey soils, particularly in swelling clay soils classified as 'Vertisols'. The main characteristics of these soils are high expansion and contraction producing a considerable amount of internal pressure.

Organic soils

Organic soils pedogenetically classified as 'Histosols' include also two large areas of Tundra in North America and Europe-Asia. The distribution of Histosols and Tundras is shown in Figure 3.43 in which Tundras are grouped with 'Inceptisols' (areas indicated by oblique straight lines in Figure 3.43). Figure 3.48 shows the organic matter content of soils in a peat-enriched profile which is located in the Netherlands.

An excellent review of the chemistry of soil organic compounds has been made by Flaig. These observations are qualitative in nature. Flaig concluded that any attempt to quantify the distribution of the organic compounds would lead to contradictions.

Problems of organic soils appear to be threefold, due to organic acids and solvents, occasional water saturation and widespread microbial activity. All species of microbes that are known to grow on plastic compounds are abundantly present in organic soils. A fairly exhaustive discussion of the subject of micro-organisms and organic matter of soils has been made by a number of Russian authors in a publication edited by Kononova. Their observations showed a wide variety of microbes that grow in organic soils, the intensity of the growth

Figure 3.46 (Opposite) World distribution of acid sulphate soils

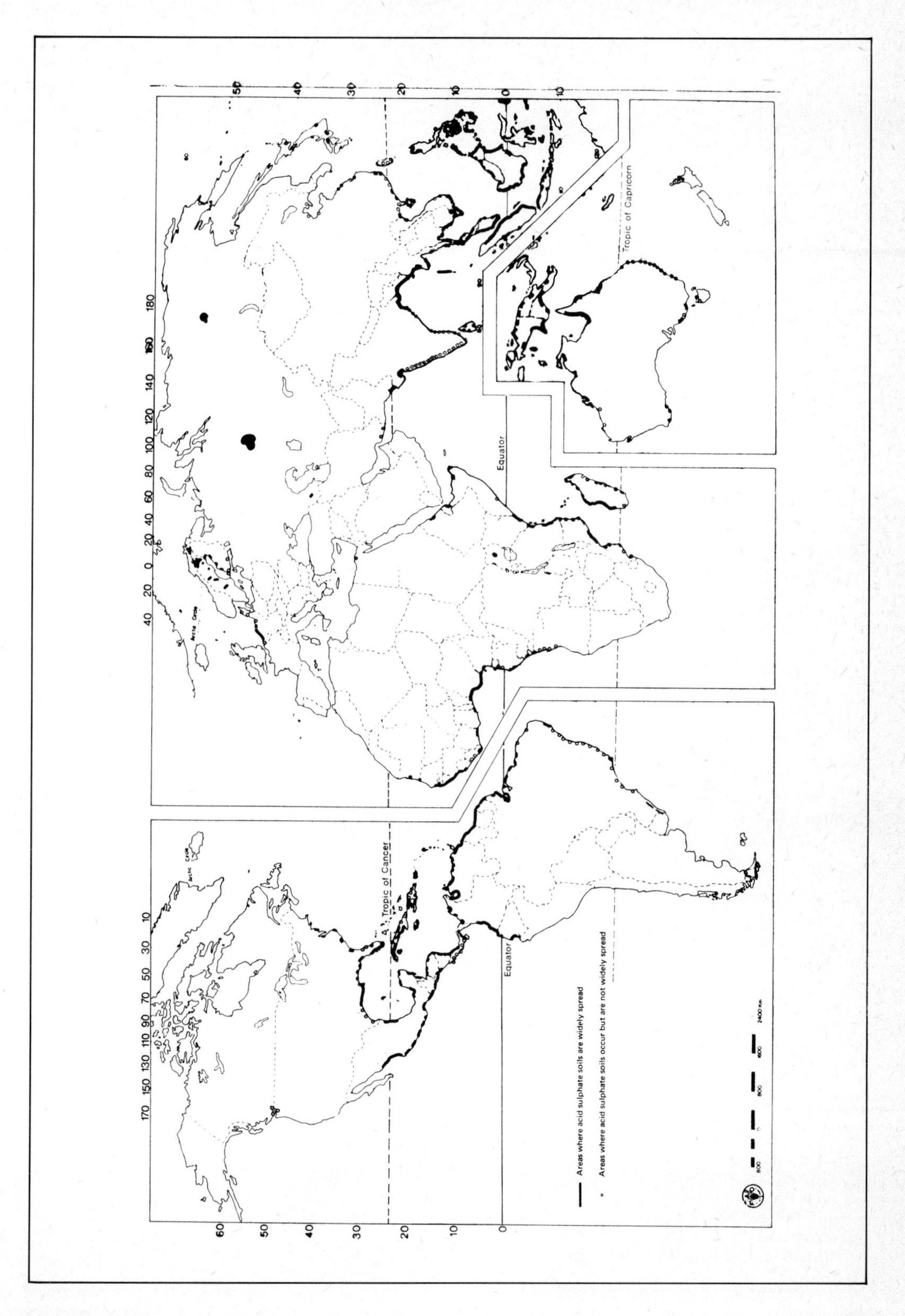
Tropic of Cancer
Equator
Equator
Tropic of Capricorn
Arctic Circle
Areas where acid sulphate soils are widely spread
Areas where acid sulphate soils occur but are not widely spread
800 0 800 1600 2400 Km
170 150 130 110 90 70 50 30 10
40 20 0 20 40 60 80 100 120 140 160 180
60 50 40 30 20 10 0

Characteristics of some acid sulphate soils

(a) Chemical characteristics of some soils of West Malaysia (after Kanapathy)

Depth cm	pH* dry soil	Cl ppm	SO_4 ppm
0-15	3.1	180	1,280
15-30	2.9	200	2,040
30-60	2.4	240	8,560

*pH of acid sulphate soils is normally between 3 and 5, but also lower values have been reported, eg pH 2.3 at a depth of 1 to 2m in Brazil and pH 2.7 of dried pyritic soil from Tubbergen, Holland.

(b) Micro-organisms involved in soil forming process

Bacteria	Resultant products
I Sulphur reducers After Rickard: Desulfotomaculum nigrificans Desulfotomaculum orientis Desulfovibrio desulfuricans Desulfovibrio salexigans Desulfovibrio africanus Desulfovibrio gigas	production of sulphide minerals (which have potential for developing acid sulphate soils) such as pyrite, marcasite, mackinawite and griegite.
II Sulphur oxidisers* After Swaby and Vitolins: (a) Autotrophic thiobacilli Th. thiooxidans Th. thioparus Th. neapolitanus Th. denitrificans Th. ferrooxidans (b) Heterotrophic bacteria Arthrobacter aurescens Bacillus licheniformis Brevibacterium Micrococcus Pseudomonas	Development of acid sulphate soils (role of Thiobacillus group established) with the production of sulphuric acid, jarosite, natrojarosite and also elemental sulphur.

*Some heterotrophic yeasts, such as Debaryomyces and Saccharomyces oxidis, also oxidise sulphur

(c) Frequency of sulphate reducing bacteria in some soils of Senegal (after Jacq)

Soil	Sulphate reducing bacteria (number per g soil)	
	Soil in dry season	Soil in rainy season
Rhizopora mangrove	36,310	16,980
Avicenia mangrove	—	66,070
Bare and saline "tanne"	3,388	1,700
Heliocharis saline "tanne"	—	832
Heliocharis non-saline "tanne"	3,715	871,000
Mangrove paddy soil	20,800	4,074

Figure 3.47 (Left) Characteristics of some acid sulphate soils.
(a) Chemical characteristics of some soils of West Malaysia
(b) Micro-organisms involved in soil forming process
(c) Frequency of sulphate reducing bacteria in some soils of Senegal

Organic matter content of a peaty soil profile* from Tubbergen, Holland
After Van Holst and Westerveld

Depth cm	Description of layer	Organic matter %
0- 60	Fill sand	—
60- 80	Non-calcareous marine deposit	4.1
80-100	Non-calcareous marine deposit	4.9
100-150	Calcareous marine deposit	2.7
150-215	Calcareous marine deposit	5.7
215-240	Detritus	26.3
240-250	Reed-sedge peat	60.5
250-265	Sedge peat	57.2
265-285	Detritus	68.8
285-290	Clay with remnants of reed	16.8
290-310	Non-calcareous marine deposit	5.2
310-400	Calcareous marine deposit	2.1

*Considered as a potential acid sulphate soil

Figure 3.48 Organic matter content of a peaty soil profile from Tubbergen, Holland

being governed by a number of factors such as waterlogging (associated with anaerobic conditions) and nature and extent of peat decomposition.

Other chemically reactive soils

The term 'chemically reactive soils', described in this section, refers to those soils which are saturated with salts or enriched with ferruginous (ferrous-ferric iron) materials.

Salt-affected soils. Salt saturation of the soil can create a potential danger for some of the plastics products through a prolonged contact with salts, but the dilution of the salts in soil solution would lower the amount of risk.

The soluble salts that effectively contribute to soil salinity consist mostly of various proportions of the cations sodium, magnesium and calcium (potassium occurs to a lesser extent) and of the anions chloride, sulphate, bicarbonate and sometimes carbonate. Sodium seldom makes up more than half of the soluble cations. Among anions, bicarbonate and carbonate are usually present in minor amounts as compared to chloride or sulphate. The main source of salt affecting irrigated soils is from surface and ground water. Saline marine deposits that have been uplifted and have led to salinity in soils also occur in various parts of the world.

In Europe, soil names like solonchak, solonetz and solod indicate different stages of salinity and alkalinity. The soil map of Europe, prepared by the FAO, shows the distribution of the salin-alkaline areas. Solonetz and solonetz-like soils represent the most widespread group of salt-affected soils in Europe (covering about 50 million acres), the second most common group of salt-affected soils being solonchaks (covering about one third of the solonetz area). Large areas of salt-affected soils develop in arid regions (Aridisols in Figure 3.43) through irrigation, e.g. current loss of about 1,000 acres/year of arable land in the Indus Valley due to excessive salinity.

Ferruginous soils. Iron is an active constituent in a variety of soils, and its role in oxidation-reduction reactions (involving both chemical and micro-biological processes) has been extensively studied by chemists during the last several decades. Under certain conditions the movement of iron in soil profiles in ferrous state or in a hydrated form as dispersed colloidal particles is followed by precipitation/deposition. This can lead to the development of specific iron-enriched layers in soils, such as 'Spodosols' (formerly podsols) occurring in large areas of the cool-temperate regions and 'Oxisols' (formerly latosols) occurring extensively in the humid tropics (distributions of Spodosols and Oxisols in Figure 3.43).

Ferruginous soils may deteriorate certain plastics materials, as illustrated by a possible degradation of nylon by iron rust, and clogging of pipelines by iron bacteria. Figure 3.49 shows the distribution of iron in a latosolic soil profile located in a lowland Pleistocene terrace in South East Asia. Although results are expressed in oxide form, free iron oxide in colloid and non-colloid fractions represents mostly soluble and mobile form of iron. Like many other constituents, iron may be actively involved in a number of soil forming processes such as the development of acid sulphate soils from the oxidation of iron sulphide minerals (e.g. pyrites) or ferruginous organic soils. The ionic and colloidal distribution of iron in soils can attain a large magnitude as exemplified by the movement of iron in terra rossa (red earth) soils of Southern France and in related red-brown calcareous soils of the British Isles.

Conclusions

Synthetic organic polymers, used in manufacturing plastics for engineering applications, show a fairly strong chemical resistance under laboratory conditions particularly at room temperature. Compounds like polythene, polypropylene and polyvinyl chloride are reportedly bio-inert, but evidence to date does not justify a quantitative measure of their micro-biological resistance.

Distribution of iron in some latosolic (oxisols) soils of the humid tropics*
(after Karim & Khan)

Depth of soil (cm)	Total Fe_2O_3 (%)			Free Fe_2O_3 (%)			Clay Colloid in total soil %
	Soil	Non-Colloid	Colloid	Soil	Non-Colloid	Colloid	
0 - 2.5	4.96	3.57	10.53	2.52	1.60	6.68	20.5
3.8- 6.4	5.24	3.96	9.75	2.74	1.78	6.30	21.5
10.2- 8.9	6.25	4.00	9.45	3.58	1.64	6.40	41.5
25.4-34.3	6.90	4.85	9.45	4.00	2.19	6.12	47.3

*Soil type: Brownish grey fine sandy loam (Brahmaputra-Ganges delta area)

Figure 3.49 Distribution of Iron in some latosolic (oxisols) soils of the humid tropics

Commercial plastics products seem to be more susceptible to chemical and micro-biological attack, depending on the nature of additives used and the amount of monomers present. Soil burial exposure, however, creates an unfavourable situation which needs to be carefully examined.

A considerable number of corrosive chemicals are present in most soils in diluted amounts; but these chemicals may also occur in greater concentrations under severe soil conditions, thus creating a hostile environment. In addition to chemicals, soils contain a great variety of micro-organisms some of which can use buried plastics as growth media and, in certain cases, can actually deteriorate the products.

It is apparent that concentrations of chemicals and microbes vary widely from soil to soil, and even within a single soil profile depend on the seasonal fluctuations and changes in environmental factors such as deforestation of the land, flooding and soil exposure due to excavation. Therefore, data presented for certain soil classes should be regarded as indicative of general trends rather than as absolute parameters.

Under laboratory conditions rigorous chemical (and occasionally micro-biological) tests are generally carried out by plastics manufacturers on a routine basis. Making certain assumptions, the products are recommended for use in underground conditions, and their application might lead to the following sequence:

(a) Average soil conditions (i.e. soils without any severe limiting factors) will affect the products slowly over a long period of time, the main weathering agents being water, carbonic acid and microbes with a moderate growth; and

(b) Severe soil conditions (problem aggravated by one or more factors) will initiate deterioration, its intensity depending on the type of plastics and the nature of soils.

Governing whether an average or severe soil situation exists, are the effects produced due to an interaction of three factors: chemicals, organisms and mechanical treatment. Time adds a significant dimension to the problem. However, relationships between the effects of the specific soil constituents and the deterioration of plastics cannot be quantified from the limited research data available.

A search of the literature has revealed certain severe soil conditions which could cause a drastic reduction in the service life of plastics products. For instance:

(a) In acid sulphate and related sulphidic soils, the formation of sulphuric acid (soil pH may sometimes reach a low level of 2.5) and the pronounced activities of sulphur bacteria are prominent features which can lead to the deterioration of PVC or nylon-based products.

(b) In swelling clay soils internal heavy pressure is created due to the expansion and contraction of the clayey mass as governed by the degree of water saturation. Buried plastic pipes made from materials such as polythene polyvinyl chloride or glass-reinforced polyesters can be ruptured or deformed.

(c) In organic soils, depending on the nature of the decomposable organic matter and the aerobic/anaerobic condition (as determined by the degree of water saturation or waterlogging), micro-organisms can deteriorate those plastics which are made from polyamide, polyester, plasticised PVC, polystyrene and others which support microbial growth.

(d) In salt-affected soils, dominated by alkali and alkaline earth compounds and in extreme cases associated with swelling clays forming solod-vertisol or solonetz-vertisol complexes, chemical corrosion and/or mechanical stress can degrade plastics.

The geographical distribution of these problem soils are shown in world soil maps which have been prepared by the FAO and the US Department of Agriculture. The detailed soil maps are available from regional soil survey organisations in most countries.

There is a clear indication of an increasing need for soil burial tests. This is evident from recent experiments carried out in North America, Europe, Australia. Tests often lack proper characterisation of the soil profiles and of the soil-product interaction processes. However, the need for stringent tests has been emphasised due to an increase in the burial of plastics into hostile soil areas.

It seems evident that there is a need for a better understanding of certain factors such as:

(a) service performance of the products in underground conditions (acelerated weathering tests in modified soil environment);

(b) profile characteristics (chemical, biological, physical and morphological) of the soils used;

(c) characteristics of any backfill materials, and

(d) interaction of the products with soils and backfill materials where used.

(Author's Note: the foregoing section on Soil Exposure has been published by courtesy of ICI. It is based with only minor alteration on an ICI in-house research document written by D.H. Khan. The many references can be found in Appendix II: Further Readings. The Author extends his thanks to ICI, UK, for their permission to publish.)

4

Types of membrane available

The first major sub-division of Civil Engineering membranes is between natural and synthetic materials. Jute and various palm-leaf woven products have been used in tropical countries in the past, and they are mentioned occasionally in this book. However, most of the work in this book is necessarily concerned with the recent developments in modern synthetic membranes and webbings.

There are numerous charts published describing the theoretical properties of chemicals used to manufacture synthetic membranes. Some[16] cover a wide range of materials and even extend into the description of natural materials. However, they can be misleading in places, since they confuse proprietary trade names such as 'Terylene' (ICI's trade name) with the correct chemical terms such as 'polyester'. Also, as another example, some charts give the ultra-violet resistance of polyethylene as 'Poor', which may be true of the pure form, but most industrially produced polyethylene membranes will contain a percentage of 'carbon black' which protects from ultra-violet attack to the extent that many years' exposure can be experienced without degradation. The engineer should be aware of this, since such a performance is really 'excellent', and yet certain charts might dissuade him from choosing a polyethylene product.

Figure 4.1 shows the basic breakdown of the different types of membrane available commercially at present. As might be expected in today's highly competitive environment, some new types and combinations are being developed and tried, but are not included here since they still require field evaluation. The types of membrane mentioned in Figure 4.1 are all commercially proven and acceptable, subject only to their availability, the technical requirements of any particular project, and cost.

As can be seen in Figure 4.1, it is possible to group all the modern materials and the traditional natural ones under the two main headings of 'membranes and webbings'.

Permeable membranes

Woven fabrics

These comprise a series of fabrics made of an extremely wide range of synthetic chemicals, and fundamentally based on traditional weaving patterns — some with later modifications to improve performance in the soils environment. The woven fabric is made on a loom as shown in Figure 4.2. The warp threads run down the length of the loom and are continually lifted and lowered by reeds. In general weaving practice, in order to stand the tension in the machine and the continuous wearing action of the reed, the warp threads are usually stiffer and stronger than the weft. The weft threads have to be thin and flexible since they are generally inserted between the warps from a bobbin or 'pirn' held within a shuttle which is rapidly knocked backwards and forwards through the warp. Figure 4.3 shows the action of the reeds during the weaving process. The number of threads inserted in the weft per centimetre are known as the 'picks per centimetre'. The more picks per centimetre, the tighter the fabric and the stronger the structure. Similarly, the number of warp threads can be set higher or lower at the beginning of the weaving process and thus vary the strength and/or the density of the fabric. It is important to understand that one of the problems of producing woven fabrics for soil purposes is the limitation on variability within the process. For example, there is an obvious maximum number of weft threads which can be packed into every centimetre length. Also, on the weaving loom, the problem of making finer and finer reeds limits the number of warp threads that can be incorporated. At the other end of the scale, if fewer warp threads are inserted and fewer weft picks used per centimetre, then the structure will become loose, as illustrated in Figure 4.4, and some extra method such as heat setting or resin glueing has to be used to prevent the membrane distorting under strenuous site conditions. In general, because of the lack of stability inherent in loose-weave fabrics, the successful range of woven soils membranes tends to comprise the more tightly woven products.

The three most important aspects of woven fabrics are:

(a) the implications of the structure on strength and extensibility, i.e. Young's Modulus;

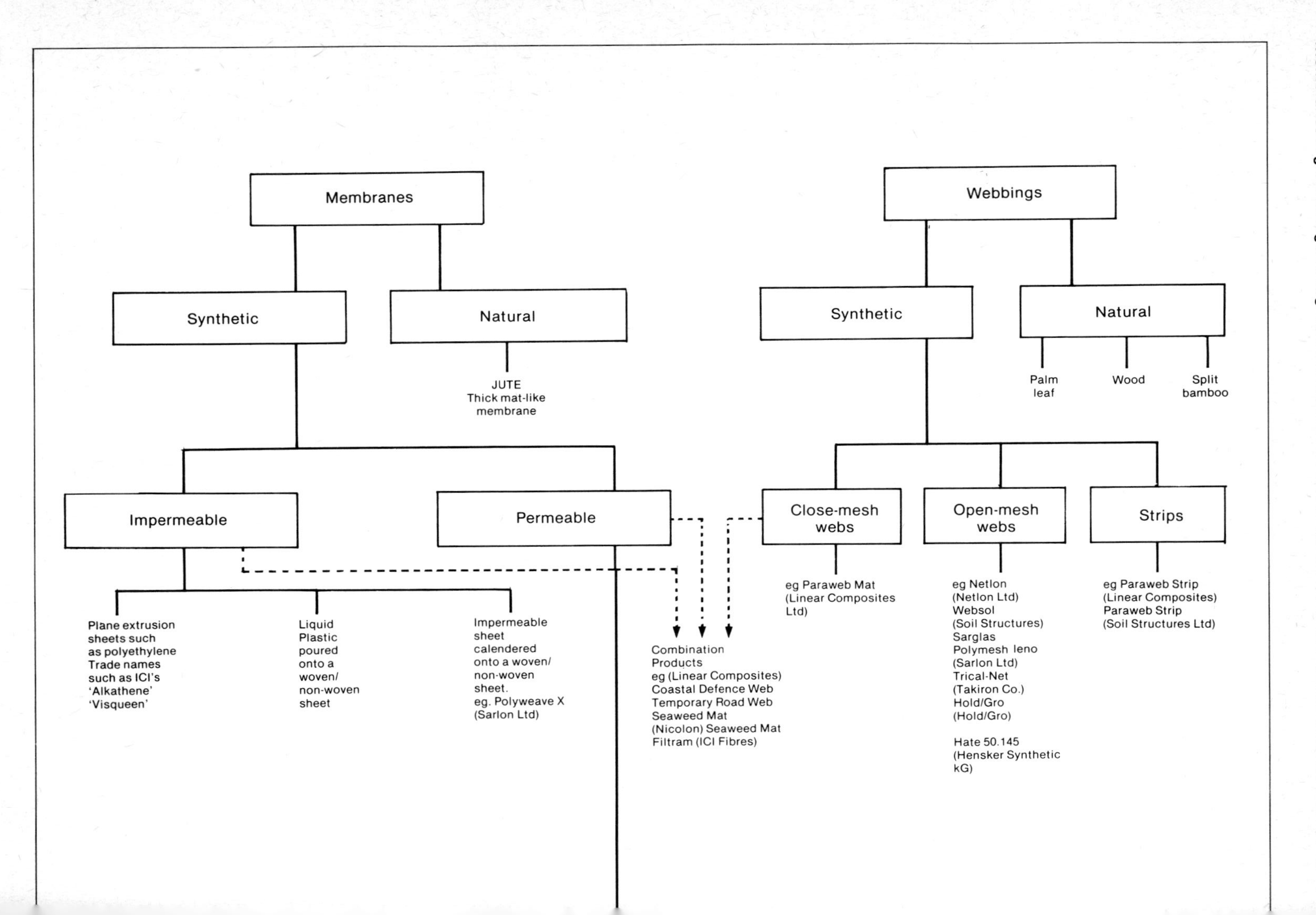
Membranes
Synthetic
Natural
JUTE
Thick mat-like membrane
Impermeable
Permeable
Plane extrusion sheets such as polyethylene Trade names such as ICI's 'Alkathene' 'Visqueen'
Liquid Plastic poured onto a woven/ non-woven sheet
Impermeable sheet calendered onto a woven/ non-woven sheet.
eg. Polyweave X (Sarlon Ltd)
Combination Products
eg (Linear Composites)
Coastal Defence Web
Temporary Road Web
Seaweed Mat
(Nicolon) Seaweed Mat
Filtram (ICI Fibres)
Webbings
Synthetic
Natural
Palm leaf
Wood
Split bamboo
Close-mesh webs
Open-mesh webs
Strips
eg Paraweb Mat (Linear Composites Ltd)
eg Netlon (Netlon Ltd)
Websol (Soil Structures)
Sarglas
Polymesh leno (Sarlon Ltd)
Trical-Net (Takiron Co.)
Hold/Gro (Hold/Gro)
Hate 50.145 (Hensker Synthetic kG)
eg Paraweb Strip (Linear Composites)
Paraweb Strip (Soil Structures Ltd)

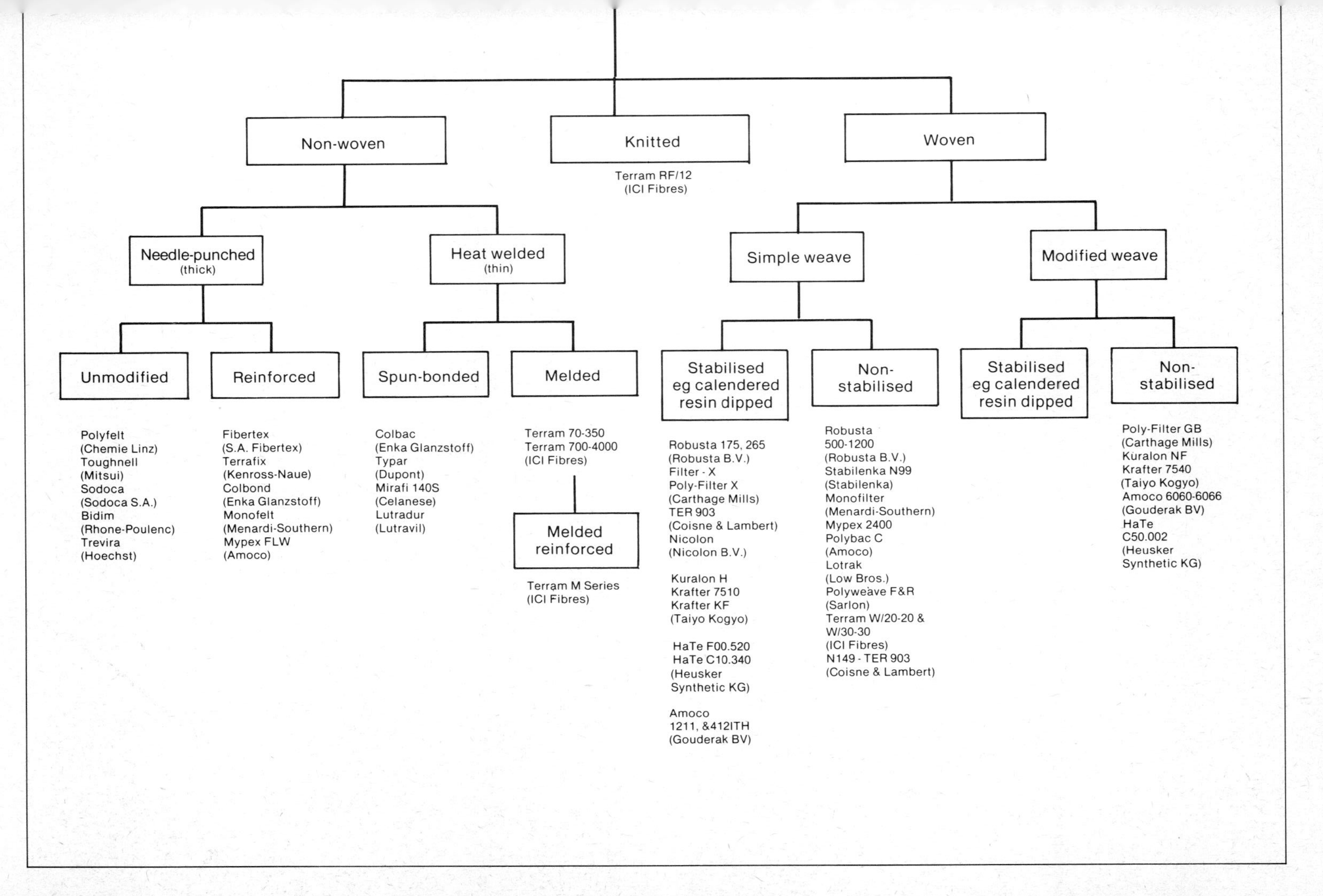

Figure 4.1 Classification of membranes and webbings. A detailed list of manufacturers is contained in the Appendix of Tables/Charts

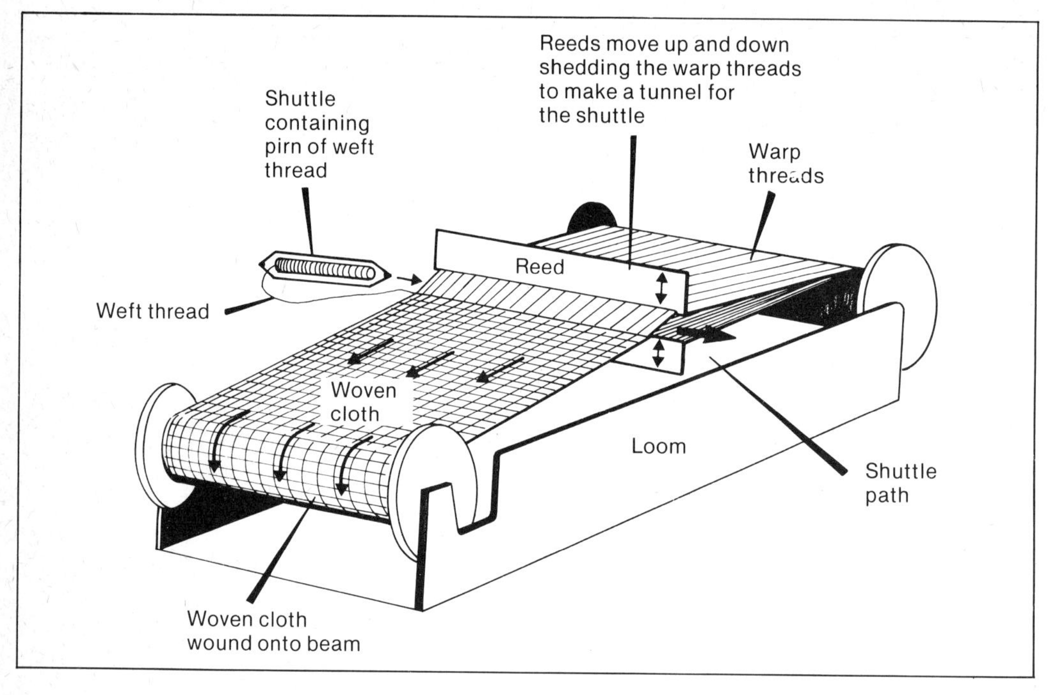

Figure 4.2 Main components of a weaving loom

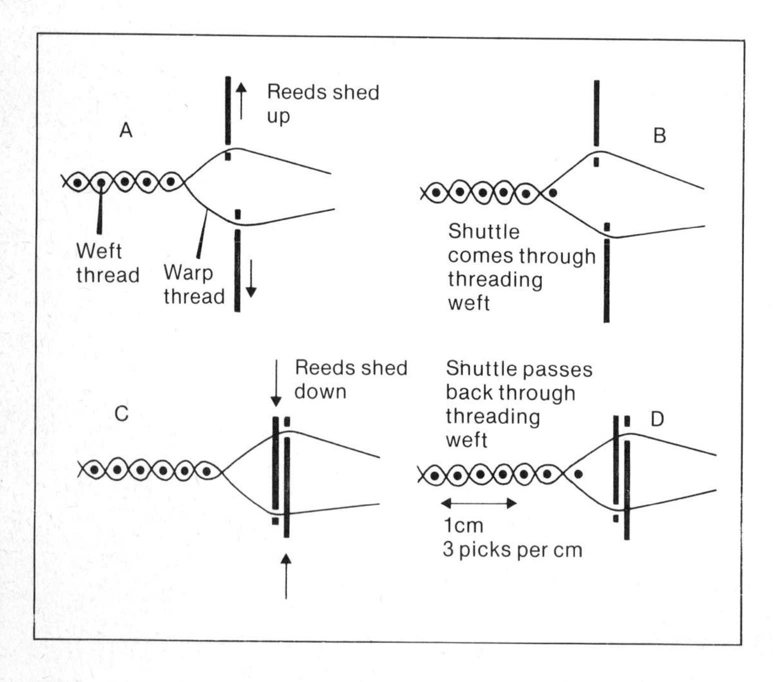

Figure 4.3 The action of the reeds sheds the warp threads so that the shuttle can insert the weft

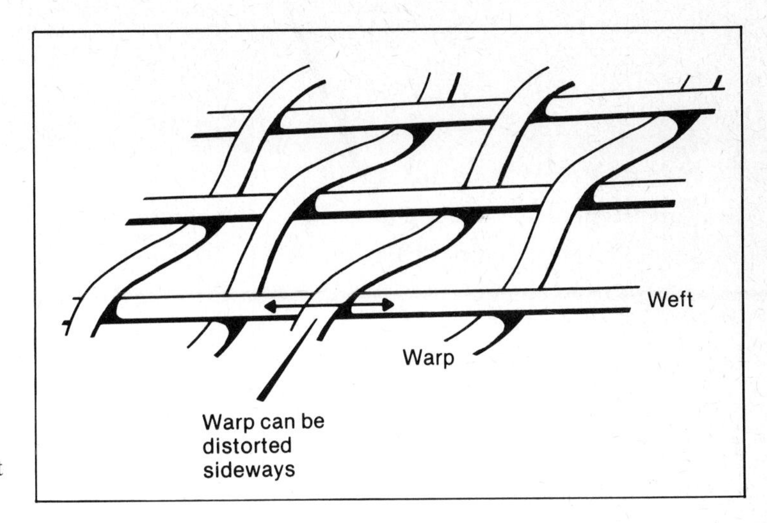

Figure 4.4 Loose warp/weft structure

(b) the influence of the structure on the pore spaces in the fabric, i.e. permeability;

(c) the highly anisotropic properties that are possible in woven fabrics.

In the first case, woven fabrics display generally the lowest extensibility and highest strengths of all fabrics. This should not automatically be assumed to be an advantage in all cases, since it is becoming common for modern documents[17] to recommend more extensible fabrics (in preference to stronger, less extensible ones) for use on soft ground. Furthermore, it should be recognised that, according to sources such as J.H. van Leeuwen,[18] although woven fabrics exhibit a slightly higher strength at 45° to the warp and weft directions than along them, they can also exhibit up to double the extensibility at 45° to warp/weft directions than parallel to the warp/weft (see Figure 4.5). Careful note should be taken of this, in view of the fact that invariably, published woven fabric data is for warp/weft direction properties. In calculating design stresses, only the minimum strengths should be taken. Furthermore, strength figures for wovens are taken up to the point of rupture, and do not highlight one aspect of woven membranes that has been experienced by practical engineers, that once a woven fabric rips, the rip can propogate very rapidly, and with only a low stress applied. The general guide here, therefore, is to be sure to obtain the *minimum* strength of the product to be used, and then to ensure that this stress is never exceeded. Owing to the relatively high strengths of woven fabrics, this safe situation is often arrived at by chance rather than by calculation. In those rare situations where a woven fabric has failed, in the author's

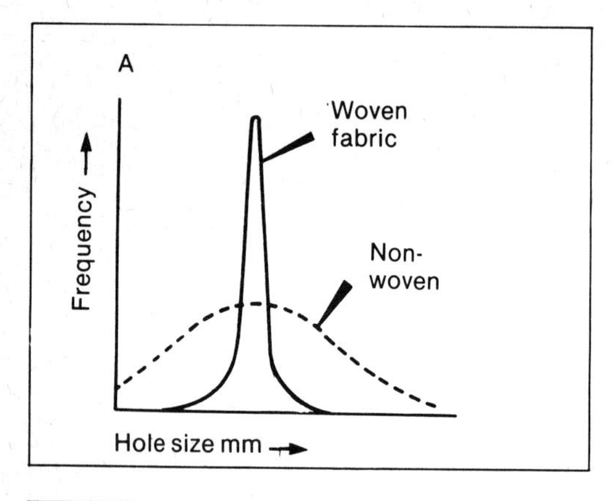

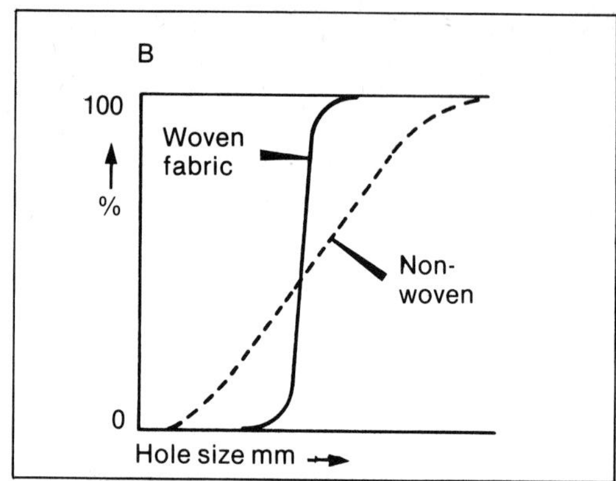

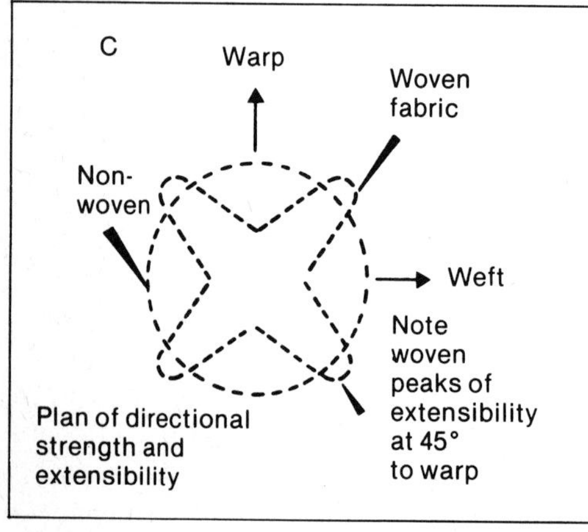

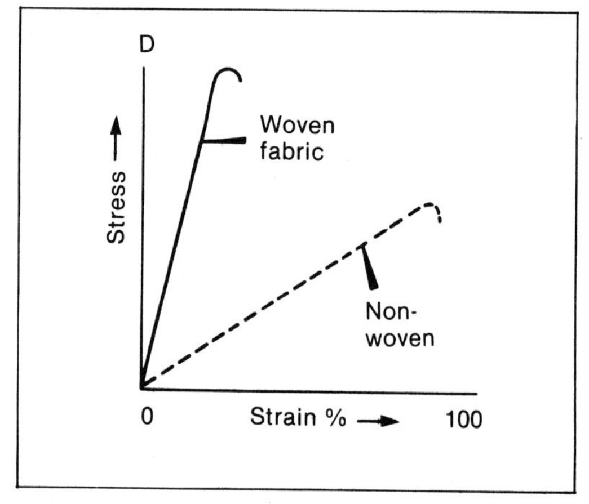

Figure 4.5 Description of fundamental properties of a woven fabric compared with non-woven

experience serious consequences have sometimes followed owing to the large size of resultant tears.

In the second case, the regular shape of the woven structure means that the holes between fibres are of a regular size and shape. They are virtually single sized, and instead of being able to use a 'pore distribution' diagram as for non-wovens, the woven fabric can only be described by (a) percentage of open area (% OA) and (b) effective opening size (EOS).

The % OA is the visible hole area measured under microscopic examination of a fabric as a percentage of the total fabric area. The EOS is commonly considered to be the hole diameter which is only 5 % smaller than the largest hole in the fabric.

In general descriptive terms, the strength and pore size properties of a woven fabric compared with non-wovens are shown in Figure 4.5. In Figure 4.5A, it can be seen that woven fabrics in general have a uniform hole size, which is also reflected in the steep curve of the hole distribution diagram shown in Figure 4.5B. Figure 4.5C shows how the strength of a woven fabric is higher at 45° to the warp and weft directions, but is lower parallel to the warp/weft, whereas non-woven fabrics tend to have a lower but more uniform strength in all directions. It seems logical that the woven warp/weft strength is the one which should be used for design purposes together with the 45° extensibility, but this is rarely published. It seems that there is a need for manufacturers of woven fabrics to publish their minimum and maximum directional strength and extensibility together with their angles to the warp. This could then be used for design purposes, and for direct strength/extension comparisons with non-woven fabrics. Indeed, in view of the possible anisotropies which can be introduced in non-woven fabrics, by processes such as stitching, it would be realistic for manufacturers of these fabrics to do the same, and publish a 'minimum directional strength'.

There are various methods used by manufacturers to adjust the effective permeability of woven membranes. The first is by varying the pick and weft insertion patterns (Figures 4.50C; 4.64B,C,D; 4.66A, B, C, D). Secondly, it is possible to use a different strength and cross-sectional material in the warp than in the weft (Figures 4.42C; 4.45; 4.54A; 4.62). This alters the three-dimensional geometry of the final fabric produced. Thirdly, there is a simple expedient of reducing the number of warps and wefts inserted in the fabric (Figures 4.59B, D, F; 4.48C; 4.50A, B). If, as a result of the adjustment of warp/weft patterns, the individual fibres are no longer in tight contact with each other, then they become free to move under local stress, and can easily be pushed aside to make larger holes. This is an undesirable feature, and therefore in order to minimise this possibility, such loose fabrics can be heat crimped to give greater stability, or chemically coated/lacquered to provide dimensional stability (Figures 4.42A, D; 4.50A, B; 4.59E, F).

Some of the woven fabrics available on the market at present are illustrated in the following pages (Figures 4.42-4.69). Samples and property figures are as published wherever possible.

Knitted fabrics

The use of knitted fabrics is really new in engineering technology. To the Author's knowledge, there is only one fabric of this type produced for civil engineering purposes — this is ICI's knitted fabric with the name RF/12 (see Figure 4.82 for properties). This knitted fabric combines high strength of the textile with a structure wherein the warp threads lie straight down the warp without direction change. This means that — unlike woven or non-woven fabrics — the RF/12 will take warp stresses immediately with the minimum of initial strain. The stress-strain line therefore becomes more linear since fibre readjustment does not need to take place. The fabric has no strength in its width, but exhibits an elastic property in this direction. The design technology of layered-soil reinforcement is still in its infancy, and there may be a place for the design facility of one-dimensional restraint that is not fully realised at the present time. The RF12 fabric is shown in Figure 4.82 and its properties listed there.

Non-woven fabrics

As can be seen in Figure 4.1, there are a wide variety of non-woven fabrics presently being used in Civil Engineering end uses. These fabrics are constructed by more modern techniques than the weaving process, and have really only been developed into Civil Engineering fields during the last ten years.

The main sub-division between different non-woven fabrics is that between relatively thin non-woven fabric and thick ones. In relation to the size of particles in most soils, the thin non-woven fabrics are commonly referred to as 'two-dimensional' membranes, whereas the thicker fabrics are termed 'three-dimensional' membranes or 'felts'. More rigid thick fabrics are often referred to as 'mats'.

Thin membranes (two-dimensional). (Figures 4.9-4.15; 4.22A; 4.24A, B; 4.29; 4.38)

The thin membranes are generally made on machinery similar to that shown in Figure 4.6. Continuous filaments of the particular polymer chosen are sprayed onto a moving belt, and then passed through a series of heated rollers which melt the polymer and bond the fibres together. The important points related to non-woven membranes of this kind are that the fabrics have the same strength properties in all directions by virtue of the random directional nature of the fibres, and also they have a wide range of pore sizes which tends to make them very effective in building up a good soil filter. On the whole, they are generally not as strong as the thicker non-woven fabrics, but they have greater extensibility and their two-dimensional structure makes them less susceptible to internal clogging by particles being washed through the fabric during filtration.

Melded fabrics. (Figures 4.9-4.15)

There is only one fabric of this type on the market and that is ICI's product 'Terram'. In fact this comprises an extremely wide range of heat-bonded fabrics covering most Civil Engineering applications.

The melding process developed by ICI involves the technically advanced production of heterofilaments. As can be seen in Figure 4.7, a heterofilament is a continuous filament of chemical with a core made inside a coaxial sheath. The core comprises one

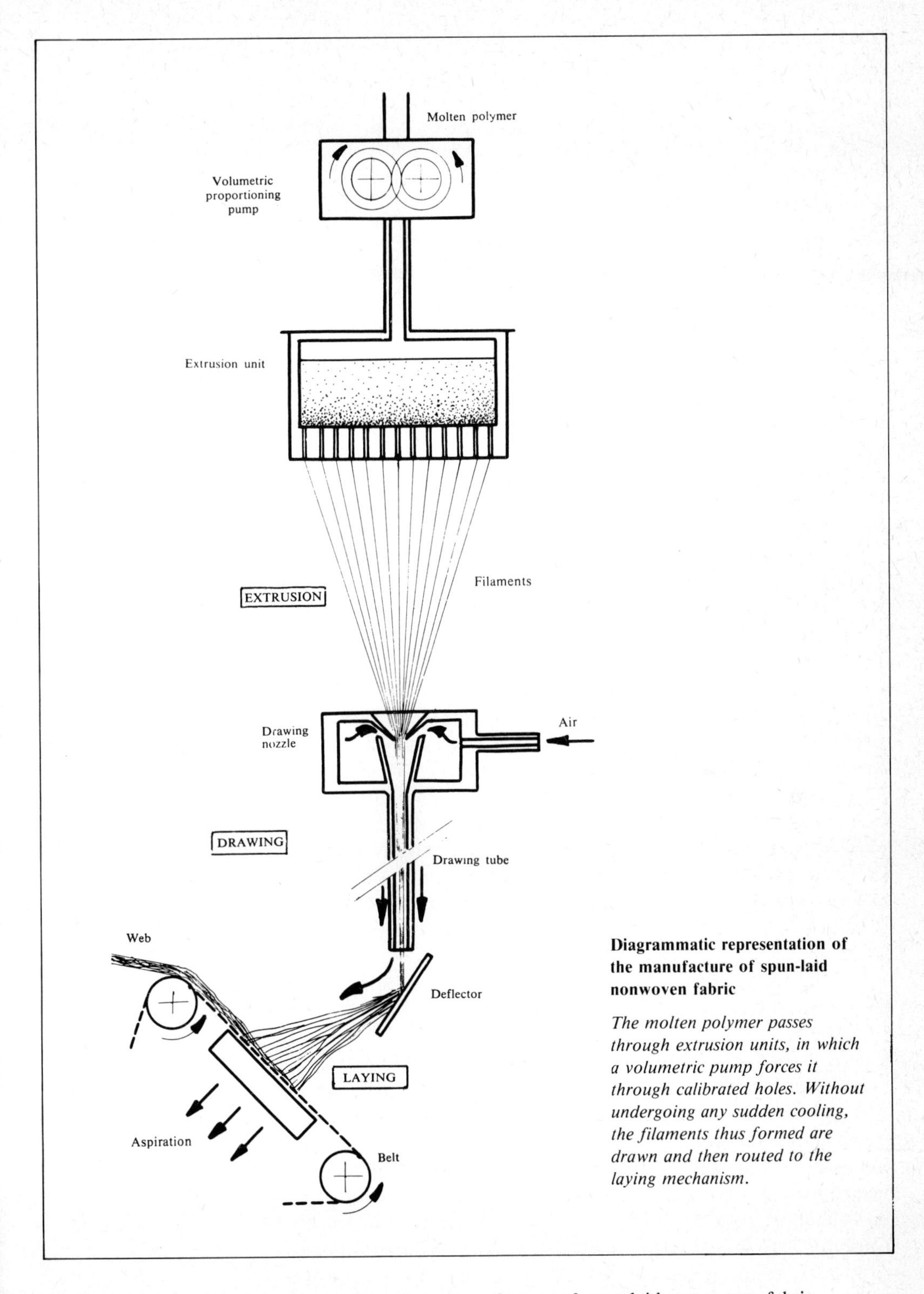

Figure 4.6 Diagrammatic representation of the manufacture of spun-laid non-woven fabric

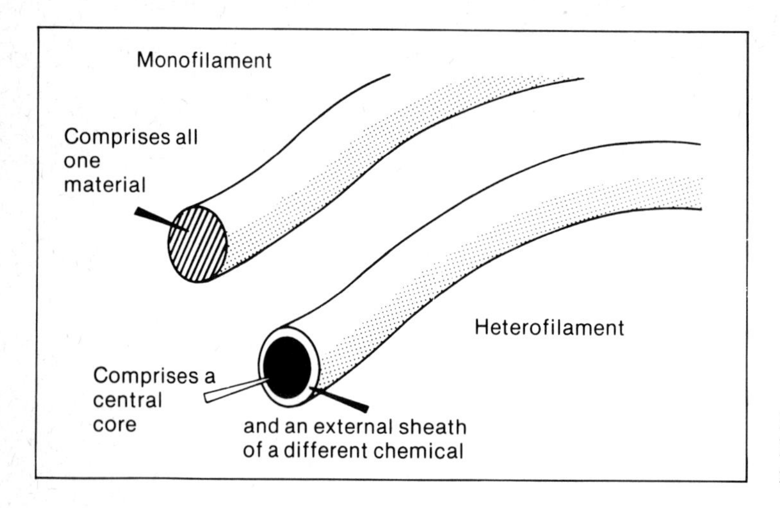

Figure 4.7 Construction of heterofilament and monofilament fibres

chemical while the sheath comprises another. In the case of ICI's latest 'Terram', the central core is polypropylene while the sheath is polyethylene.

The melding process involves the interlaying of heterofilaments and monofilaments: when these are passed through the heating chamber, only the outer sheaths of the heterofilaments melt and bond to other heterofilaments and monofilaments. Consequently the aligned molecules within the central core are not disturbed and retain their original strength. This overcomes the problem of molecule reorientation which can occur during a normal heat-bonding process. It is apparent that this technique is effective in producing a well bonded fabric in so much as the Author believes 'Terram 70' to be the lightest and most permeable filter membrane available on the market (Figure 4.9A). It has extremely high permeability and can be used in low-duress, high-flow applications, such as in connection with permeable sands, etc. If produced in the reinforced 'M' series (Figure 4.14), then it can be used for high-stress, high-flow situations.

The full range of the 'Terram' products is shown in Figures 4.9-4.15, and their properties tabulated there.

Spun bonded fabric. (Figures 4.22A; 4.24A, B; 4.29; 4.38)

Spun-bonded fabrics are manufactured in much the same way as the melded products, but use monofilaments of the same chemical material as one another. They exhibit a wide range of pore sizes within each membrane, and look very similar to melded membranes, having a lower extensibility in general.

Thick fabrics (felts, mats, three-dimensional). (Figures 4.16; 4.18; 4.20; 4.22B; 4.26; 4.28; 4.32; 4.34; 4.36; 4.40)

These thicker fabrics are produced by a process called 'needle-

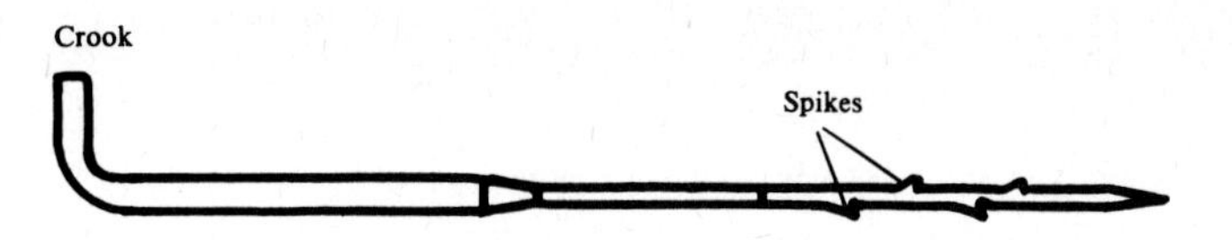

Needle used for needling

This instrument is in two parts:
— body for fixing on the table;
— active part, with hooks or spikes of different sizes depending on the degree of needling required.

The table is actuated by a reciprocating movement synchronized with the advance of the web

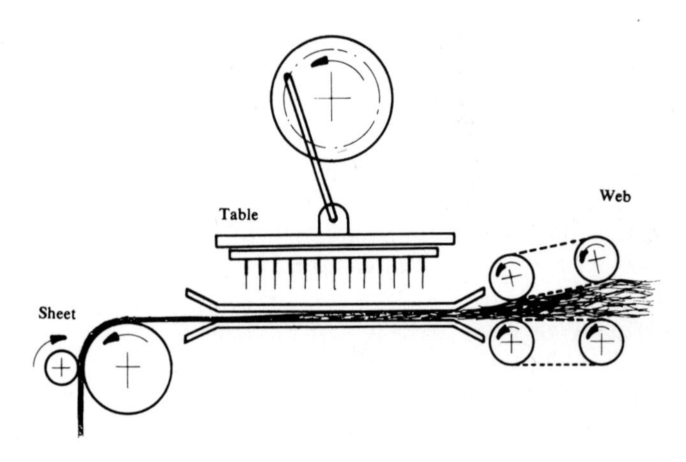

Diagrammatic representation of needling

The web is introduced in the active part of the needling apparatus by means of an endless belt.
The needles perforate the web at a certain density, which confers a certain strength on it, and it becomes a "sheet".

Figure 4.8 Diagrammatic representation of needling

punching'. A thick layer of randomly-orientated continuous fibres or staple fibres is laid down in the felting machine. Barbed needles are then driven through the fibres repeatedly, causing a thorough intermixing as shown in Figure 4.8. Different manufacturers apply variations on this basic theme in order to enhance their product. For example, Fibertex appear to insert continuous warp-strengthening threads in their felts, and also pass them through heated rollers to 'weld' the outside fibres to impart extra stability to the fibre mat. Terrafix have their felts bonded onto a strong warp/weft mesh to provide extra strength and stability. The Colbond felt is resin dip stabilised. Some manufacturers, such as Bidim, rely on the low-creep/high abrasion resistance characteristics of their polyester fibres to sell their product against competitive polypropylenes. The relevant published characteristics of each product are given in the subsequent pages, and the requirements for any particular project must be carefully assessed before choosing the correct membrane type.

Open Meshes

(Figures 4.70-4.80)

Open meshes are available in a wide variety of materials and hole sizes. Materials available are the usual range of synthetics, and cover a variety of end uses ranging through grass growing, soil stabilisation, sand precipitation, snow fencing, and many others. As can be seen from the given samples and specifications, one is even available with a degradable paper sheet filling which is used to protect the ground against soil erosion (Figure 4.81).

The difficulty in making an open mesh is that the relative position of each strand of the mesh has to be maintained in relation to the other strands in order to keep an even spacing of openings. In some of the examples given (Figures 4.70B, C), the warp threads are twisted around the wefts which only stabilises the weft positions, and allows the warps to move sideways. At the large end of the range, Linear Composites Ltd have stabilised their open mesh by heat welding the mesh intersections, and Netlon's product is joined into a mesh whilst still in a semi-liquid state immediately subsequent to polymer extrusion.

Interwoven webbings

Figures 4.85-4.88)

To the Author's knowledge, the only permeable plastic webbing system available is manufactured by Linear Composites Ltd, of Harrogate in the UK. This system is shown in Figure

4.85A. The strength of the webbing is derived from the internal bundles of polyester filaments within the flat polyethylene sheath. The thousands of filaments within each bundle have been stretched and their molecules aligned along the length of the filament in the direction of potential stress. This gives the webbing strip a low extensibility and low creep characteristics. The polyethylene sheath is treated with carbon black which gives it a claimed resistance to ultra-violet light of 30 years with no measurable deterioration. In its natural colourless state, or in other colours such as orange or green, the exposed ultra-violet life of polyethylene would not exceed three or four years at failure, and measurable deterioration could be apparent within six to twelve months. The properties of the overall web depend upon the strength characteristics of the individual strips from which it is made. Basically there are commercially available two strip strengths — 165 kg breaking strength and 400 kg breaking strength. This gives an overall web strength of about 3 tonnes and 8 tonnes per linear metre.

As shown in Figure 4.85B the Paraweb Interwoven Webbing is usually used with a filter membrane welded to one side. The side of the webbing with the filter attached becomes the back of the webbing and is usually presented to a soil surface, whereas the front of the webbing can be exposed to the sun or overlain with stones in a variety of designs, as discussed later in the book. The properties of the webbing and strip are given in Figure 4.86.

Impermeable membranes

There are two basic types of impermeable membrane available on the market. In the first case there are sheets of extruded plastic material which are continuous and extruded as a single sheet. Secondly, there are interwoven membranes and webbings overlaid and bonded to a continuous impermeable plastic sheet. Of the first kind, ICI's 'Visqueen' sheet (Figure 4.89) is typical, and Figure 4.90 shows a product by Sarlon Industries which represents the second type.

Some uses of impermeable membranes are described later in Section 2 of this book.

Summary

It can be seen that membranes come in a wide variety of types whose various features and attributes must be balanced against costs and availability for any particular application. Although some extremely tough and durable membranes are available at the present time, they are obviously more expensive than their weaker counterparts. It is part of a sensible cost-effective engineering design to use the least expensive product compatible with the essential physical requirements of any particular job. This brings one to the most difficult part of the design process — the evaluation of those factors which *are* important to the operation of the finished design. For example, it might automatically be assumed that the strength and permeability of the membrane are the most important factors for a coastal defence design, but for a given particular contract, for reasons of raw materials and contractual procedure, it may well turn out that extensibility and ultra-violet resistance may be far more critical. It is these aspects that we hope to explore in more detail in the coming chapters, by looking at design objectives and construction techniques on a structure-by-structure basis.

The following pages contain cross-referenced lists of over 130 individual membranes from some 30 different manufacturers or suppliers. It must be recognised that this list and the accompanying photographs and descriptions are not exhaustive, and the Author takes this opportunity to apologise for any error or omissions. In a few cases — owing to technical data being unavailable — the Author has tried to give some comments based on visual examination.

Alphabetical list of some manufacturers/suppliers of Civil Engineering membranes

Amoco, UK [15]
Carthage Mills, USA [16]
Celanese Corporation, USA [9]
Chemie Linz, Austria [4]
Coisne & Lambert, France [19]
Du Pont, Switzerland/USA [6]
Enka Glanzstoff, The Netherlands [5]
Fibertex, Denmark [2]
Gouderak, The Netherlands [23]
Huesker Synthetic, W. Germany [30]
Hoechst, W. Germany [12]
Hold/Gro E.C.S., USA [26]
ICI Fibres, UK [2]
ICI Plastics, UK [29]
Kenross Naue, Canada [3]
Kuraray, Japan [20]
Linear Composites, UK [27]
Low Brothers, UK [17]
Lutravil Spinnvlies, W. Germany [13]
Menardi-Southern, USA [8]
Mitsui Petrochemicals, Japan [7]
Netlon, UK [24]

Nicolon, The Netherlands [21]
Robusta, The Netherlands [14]
Rhone-Poulenc, France [11]
Sarlon Industries, Australia [18]
Sodoca, France [10]
Soil Structures, UK [28]
Taiyo Kogyo, Japan [20]
Takiron, Japan [25]
Tonen Petrochemical, Japan [22]

List of manufacturers or sales organisations mentioned in this book

[1.] ICI Fibres 'Terram'
Pontypool,
Gwent
NP4 0YD
GREAT BRITAIN
Tel: Pontypool (04955) 57722
Telex; 668411 ICIMSC G

Terram is sold in Australia under the trade name of 'Terra Firma' by:

Fibremakers,
69 Maquarie Street,
Sydney 2000,
New South Wales,
Australia.

[2.] A/S Fibertex,
Svendborgvej 16,
DK-9220, Aalborg Øst,
DENMARK
Tel; 45 -(08) 158600
Telex: 69600 FIBER DK

[3.] Kenross-Naue Canada Limited
1329 Martingrove Road,
Rexdale,
Ontario M9W 4X5
CANADA
Tel; 416-745-7290

[4.] Cehmie Linz AG,
A - 4020 Linz,
St Peter Strasse 25,
AUSTRIA
Tel: 56471
Telex; 02-1324

[5.] Enka Glanzstoff BV.,
Industrial Yarns Product Group,
Velperweg 76,
Arnhem
THE NETHERLANDS
Tel: (085) 664600
Telex: 45204

[6.] E.I. Du Pont de Nemours International S.A.,
Spunbonded/Spunlaced Products,
50-52 Route des Acacias,
CH-1211 Geneva 24
SWITZERLAND
Tel: (022) 278111
Telex: 23713

[7.] Mitsui Petrochemical Spunbond Co. Ltd,
(Mitsui Sekiyukagaku Supabond KK),
Kasumigaseki,
Chiyoda-Ku,
Tokyo,
JAPAN

[8.] Menardi-Southern Division,
Soil & Erosion Control Department,
US Filter Corporation
3908 Colgate,
Houston,
Texas 77017
USA
Tel: (713) 643-6513

[9.] Celanese Fibers Marketing Company
1211 Avenue of Americas
New York,
NY 10036,
USA

[10.] Sodoca S.a.r.l.,
Zone industrielle Est,
Biesheim,
N.P. No 28,
68600 Neuf-Brisach
FRANCE

[11.] Rhône-Poulenc Textile,
Non-Woven Department,
Public Works Sector,
69 rue Casimir-Périer,
9 5870 Bezons
FRANCE
Tel: 982 3340
Telex: 697802 RHONE F.

[12.] Hoechst Aktiengesellschaft,
Trevira technisch Service,
6230 Frankfurt (m) 80
W. GERMANY
Tel: (0611) 3055833

[13.] Lutravil Spinnvlies,
Zweigwerk der Carl Freudenberg,
Postfach 1220,
D-6750 Kaiserslautern
W. GERMANY
Tel: (0631) 5311
Telex: 45813 LUTRA D.

[14.] Robusta B.V.,
Genemuiden
HOLLAND
Tel: 05208 1916

[15.] Amoco Fabrics,
Lynnfield House,
Church Street,
Altrincham,
Cheshire
GREAT BRITAIN
Tel: 061-928 8616
Telex: 668301

[16.] Carthage Mills,
Erosion Control Division,
124 W. 66th Street,
Cincinnati,
Ohio 45216
USA
Tel: (513) 242-2740

[17.] Low Brothers & Co. (Dundee) Ld,
P O Box 54,
Dundee DD1 9JQ
GREAT BRITAIN
Tel: 0382 27311
Telex: 76270

[18.] Sarlon Industries Pty. Ltd,
47 McEvoy Street,
Waterloo,
New South Wales 2017
AUSTRALIA
Tel: 699 2099
Telex: 24887

[19.] Coisne & Lambert,
11-68 Rue de Lille,
F 59280 Armentieres
FRANCE
Tel: (20) 770707
Telex: 820515

[20.] Taiyo Kogyo Co Ltd., — supplier,
(Public Works Division),
8-4 Kigawahigashi 4 Chome,
Yodogawa-ku,
Osaka,
JAPAN
Tel: (06) 302-5131
Telex: 523 3818

Kuraray Co Ltd., — manufacturer,
(Industrial Textile Department),
8 Umeda Kita-ku,
Osaka,
JAPAN
Tel: (06) 346-1351
Telex: 523 3725

[21.] Nicolon B.V.,
Department of Marine Works & Road Building,
Postbus 440,
Richtersweg 140,
Enschede
NETHERLANDS
Tel: 053-355 455
Telex: 44440

[22.] Tonen Petrochemical Co Ltd.,
(Tonen Sekiyukagaku kk),
Togeki Building,
4-1-1 Tsukiji,
Chuo-ku,
Tokyo
JAPAN
Tel: 03-542-7131

[23.] Gouderak B.V.,
Middelblok 154,
Gouderak
HOLLAND
Tel: 01827 3344
Telex: 24662

[24.] Netlon Limited,
Mill Hill,
Blackburn,
Lancashire BB2 4PJ
GREAT BRITAIN
Tel: 0254 62431
Telex: 63313

[25.] Takiron Co Ltd.,
(Takiron Kagaku kk),
Saiwai Building,
2-45 Kitakyutaro-machi,
Higashi-ku,
Osaka
JAPAN
Tel: (06) 262 1161

[26.] Hold/Gro Erosion Control Systems,
Gulf States Paper Corporation,
P O Box 3199,
Tuscaloosa,
Alabama 35401
USA
Tel: 205-553-6200

[27.] Linear Composites Ltd,
Hookstone Road,
Harrogate,
Yorkshire HG2 8QN
GREAT BRITAIN
Tel: (0423) 68021
Telex: 57947 ICIFIBRES

[28.] Soil Structures Ltd,
Anchor House,
Guildford Road,
Lightwater,
Surrey GU18 5SA
GREAT BRITAIN
Tel: 0276 73701
Telex: 858880

[29.] Imperial Chemical Industries Ltd,
Plastics Division,
'Visqueen' Marketing Department,
Welwyn Garden City,
Hertfordshire
GREAT BRITAIN
Tel: 07073 23400

[30.] Huesker Synthetic KG.,
Postfach 47,
D-4423 Gescher
W. GERMANY
Tel: (02542) 876
Telex: 0892328

Alphabetical list of membranes mentioned in this book

Amoco
Bildim
Colbac
Colbond
Fibertex
Filter-X
Filtram
Gouderak
HaTe
Hold/Gro
Krafter
Kuralon
Lotrak
Lutradur
Mirafi
Monofelt
Monofilter
Mypex
Netlon
Nicolon
N.Y.
Paraweb
Polybac
Polyfelt
Poly-Filter
Polymesh Leno
Polyweave
Robusta
Sarglas
Sodoca
Stabilenka
T.E.R.
Terrafix
Terram
Tissu XXX
Tone Sheet
Toughnell
Trevira Spellbound
Trical-Net
Typar
Visqueen
Websol

List of membranes illustrated in this book

1. *Non-woven membranes*

Terram 70 (ICI Fibres, UK) [1]
Terram 140
Terram 210
Terram 280
Terram 350
Terram 700
Terram 1000
Terram 1500
Terram 2000
Terram 3000
Terram 4000
Terram 30 m15 (Reinforced)

Terram has been sold in the UK in the past under the trade names PRF 140 and Cambrelle. Neither of these are used any longer in connection with Civil Engineering applications.

Fibertex S-170 (Elephant Felt) (A/S Fibertex, Denmark) [2]
Fibertex S-300
Fibertex S-400
Fibertex PPR 433
Fibertex PORA 73
Fibertex F-70/30

Terrafix 200NA (Kenross-Naue Canada Ltd, Canada) [3]
Terrafix 300NA
Terrafix 500NA
Terrafix 701N (Reinforced)

Polyfelt TS-200 (Chemie Linz AG, Austria) [4]
Polyfelt TS-300
Polyfelt TS-400

Colbac (Enka Glanzstoff bv, The Netherlands) [5]
Colbond

Typar 136 g/m^2 (E.I. Du Pont de Nemours International S.A., Switzerland) [6]
Typar 200 g/m^2
Toughnell TS-40(Mitsui Petrochemical Spunbond Co. Ltd., Japan) [7]
Toughnell TS-80

Monofelt (Menardi-Southern Division, US Filter Corporation, USA) [8]

Mirafi 140S (Celanese Fibers Co., USA) [9]

Sodoca NS 115 (Sodoca S.A., France) [10]
Sodoca AS 200
Sodoca AS 320
Sodoca AS 420

Bidim U14 (Rhone-Poulenc — Textile, France) [11]
Bidim U24
Bidim U34
Bidim U44
Bidim U64

Trevira Spunbon (Hoechst Aktiengesellschaft, W. Germany) [12]

Lutradur (Lutravil Spinnvlies, W. Germany) [13]

HaTe K-5/T (Huesker Synthetic K.G., W. Germany) [30]
HaTe H-5/T

2. *Woven Membranes*

Robusta 175 (Robusta B.V., Holland) [13]
Robusta 265
Robusta 500
Robusta 600
Robusta 750
Robusta 830
Robusta 1200

Stabilenka N99 (Enka Glaszstoff bv, The Netherlands) [5]

Monofilter (Menardi-Southern Division, US Filter Corp., USA) [8]

Mypex 2400 (Amoco Fabrics, Great Britain) [15]
Mypex FLW
Polybac C133

Filter-X (Carthage Mills, USA) [16]
Poly-Filter X
Poly-Filter GB

Lotrak 16/15	(Low Brothers & Co (Dundee) Ltd, Great Britain) [17]
Lotrak 56/46	
Lotrak Needleweave	
Polyweave F (40007)	(Sarlon Industries Pty Ltd, Australia) [18]
Polyweave R (41613)	
Terram W/20-20	(ICI Fibres, UK) [1]
Terram W/30-30	
N.Y. 149 (Polyamide)	(Coisne & Lambert, France) [19]
N.Y. 169 (Polyamide)	
N.Y. 169 (Polyester)	
N.Y. 171 (Polyamide)	
N.Y. 171 (Polyester)	
N.Y. 180 (Polyamide)	
N.Y. 188 (Polyamide)	
N.Y. 378 (Polyester)	
N.Y. 409 (Polyamide)	
N.Y. 721 (Polyamide)	
N.Y. 4298 (Polyamide)	
N.Y. 4298 (Polyester)	
TER. 344 (Polyamide)	
TER. 515 (Polyamide)	
TER. 610 (Polyamide)	
TER. 903 Renforce (Polyamide)	
TER. 903 (Polyester)	
TER. 903 Renforce (Polyester)	
Krafter 7540	(Taiyo Kogyo Co. Ltd, Japan) [20]
Krafter KF	
Krafter 7510	
Kuralon NF	
Kuralon H	
Nicolon	(Nicolon B.V., The Netherlands) [21]
Tone Sheet S	(Tonen Petrochemical Co. Ltd, Japan) [22]
Tone Sheet F	
Amoco 1211	(Gouderak B.V. Holland) [23]
Amoco 6060	
Amoco 6062	
Amoco 6061 (or called 6261 with handling stitching)	

Amoco 6063 (or called 6263 with handling stitching)
Amoco 6064 (or called 6264 with handling stitching)
Amoco 6065 (or called 6265 with handling stitching)
Amoco 6066 (or called 6266 with handling stitching)
Amoco 4121 TH

HaTe C50.002 (Huesker Synthetic, West Germany) [30]
HaTe F00.520
HaTe C10.340

3. *Open Mesh Fabrics*

Sarglas (2130) (Sarlon Industries Ptg. Ltd, Australia) [18]
Polymesh Lens (40802)
Polymesh Lens (40803) — no sample
Polymesh Lens (40804)

Netlon CE 111 (Netlon Ltd, Great Britain) [24]
Netlon CE 121
Netlon CE 131
Netlon CE 151/152
Netlon 161

Gouderak 1266 (Gouderak BV, Holland) [23]

Stabilenka N33 (Enka Glanzstoff BV, The Netherlands) [5]

Trical-Net 10-527/94-80 (Takiron Co. Ltd, Japan) [25]
Trical-Net 20-200/16-20
Trical-Net 30-902/09-03
Trical-Net 30-902/08-04
Trical-Net 20-600/02-04
Trical-Net 20-900/04-08
Trical-Net 20-906/94-81
Trical-Net 10-320/10-15
Trical-Net 40-708/94-82
Trical-Net 20-920/10-25
Trical-Net 20-900/25-37
Trical-Net 20-400/34-48

HaTe 50.145 (Huesker Synthetic KG, W. Germany) [30]

4. *Webbings and Miscellaneous Membranes*

Hold/Gro (Hold/Gro Erosion Control Systems, USA) [26]

Terram RF/1[2] (ICI Fibres, Great Britain) [1]

Filtram (ICI Fibres, Great Britain) [1]

Tissu XXX (Rhone-Poulenc Textiles, France) [10]

Paraweb Interwoven Mat (Linear Composites Ltd, Great Britain) [27]

Paraweb Wind Fence
Paraweb Seaweed Mat

Websol Soil Reinforcing Strips (Soil Structures Ltd, Great Britain) [28]

5. *Impermeable Membranes*

Visqueen 1200DPM (ICI Plastics, Great Britain) [29]

Visqueen 2000T.DPC

Polyweave LX (50506) (Sarlon Industries Ptg.Ltd, Australia) [18]

Non-woven membranes illustrated

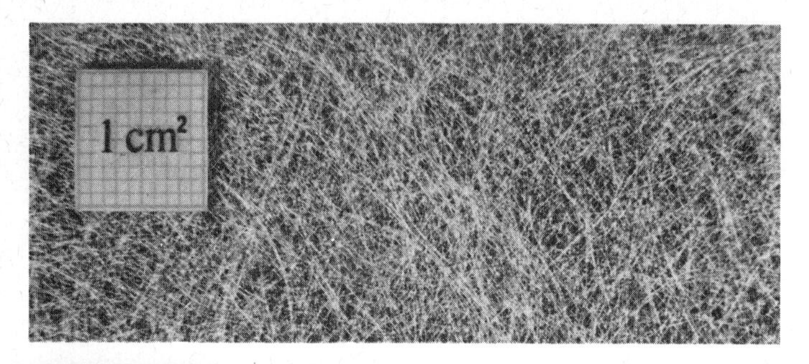

Terram 70

70 g.s.m.
25% Polyamide
75% Polypropylene
Melded White

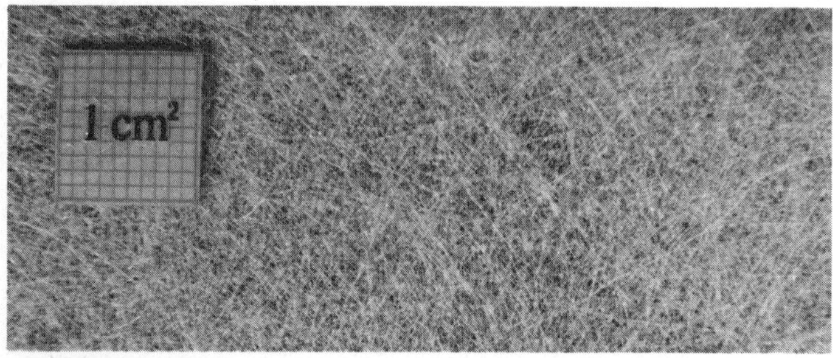

Terram 140

140 g.s.m.
25% Polyamide
75% Polypropylene
Melded White

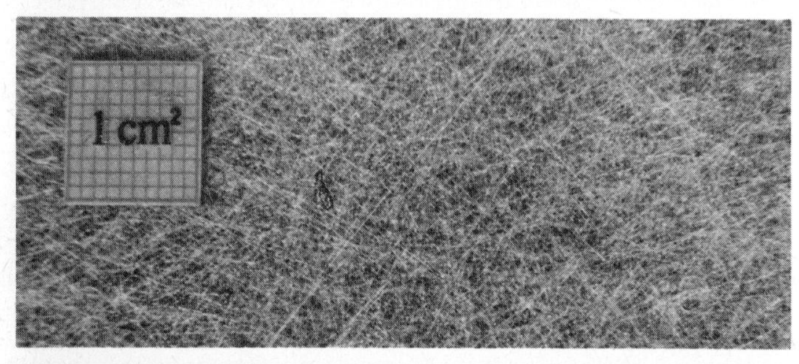

Terram 210

210 g.s.m.
25% Polyamide
75% Polypropylene
Melded White

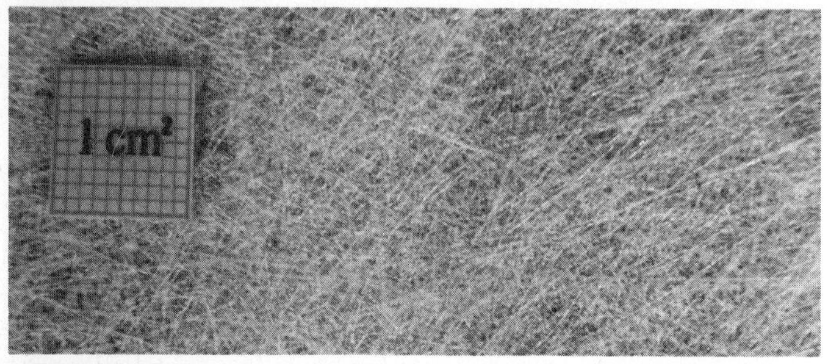

Terram 280

280 g.s.m.
25% Polyamide
75% Polypropylene
Melded White

Terram 350

350 g.s.m.
25% Polyamide
75% Polypropylene
Melded White

Published Data

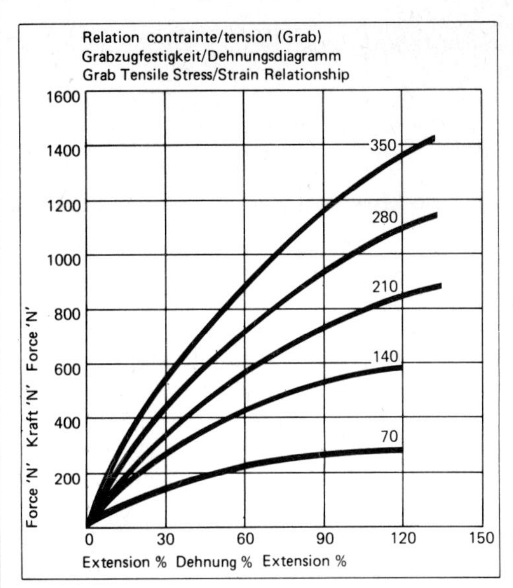

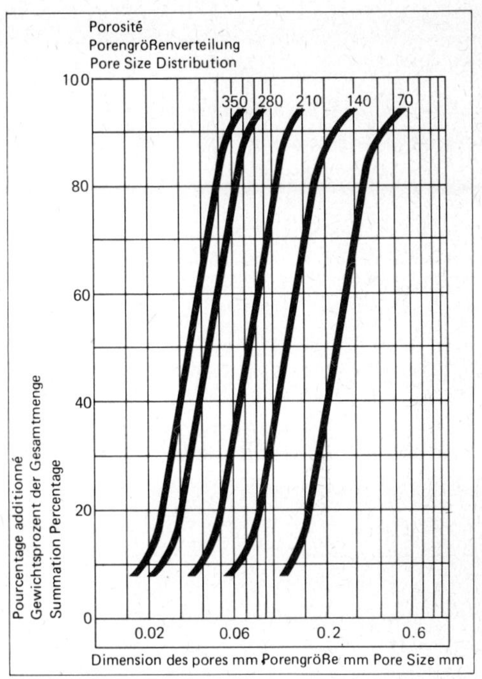

Fabric range: 70gm/m², 140gm/m², 210gm/m², 280gm/m², 350gm/m².

Fabric thickness range: from 0.5 to 1.5 mm.
Standard width of roll: 4.5 m.

Maximum width of roll: 5.5 m.

Special widths available down to 10 cm.

Standard length of roll: 100 m.
(70gm/m² fabric is supplied in standard rolls of 200 m).

Fabric can be supplied to order in rolls from 50 m to 500 m.

Caractéristiques de perméabilité
Durchlässigkeitseigenschaften
Permeability characteristics

'Terram'	Perméabilité "k" Durchlässigkeitsbeiwert "k" Permeability "k"	Ecoulement en litres avec charge de 0.10m d'eau Wasserdurchlässigkeit (Liter) bei 10cm WS Flow rate in litres at 10cm head of water
70	3.2×10^{-3} m/sec	150 L/m²/sec
140	1.7×10^{-3} m/sec	50 L/m²/sec
210	1.1×10^{-3} m/sec	44 L/m²/sec
280	0.7×10^{-3} m/sec	38 L/m²/sec
350	0.5×10^{-3} m/sec	30 L/m²/sec

Figure 4.9 (Left) ICI 'Terram' products

Figure 4.10 (Above) Terram characteristics

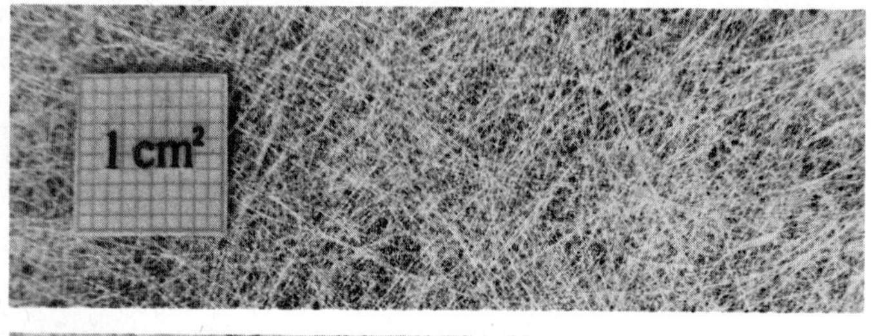

Terram 700

71-100 g.s.m.
33% Polyethylene
67% Polypropylene
Melded White

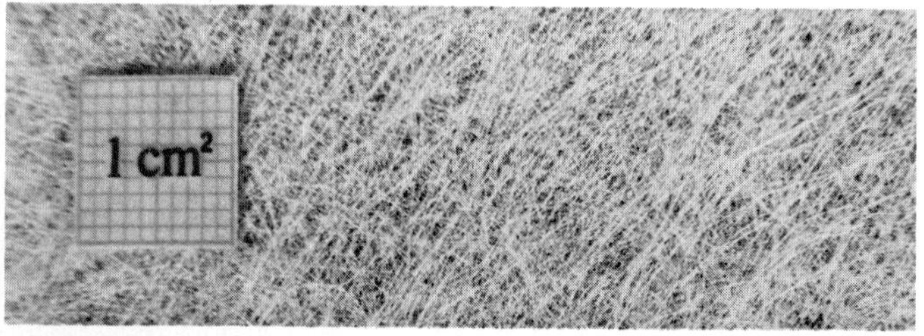

Terram 1000

101-170 g.s.m.
33% Polyethylene
67% Polypropylene
Melded White

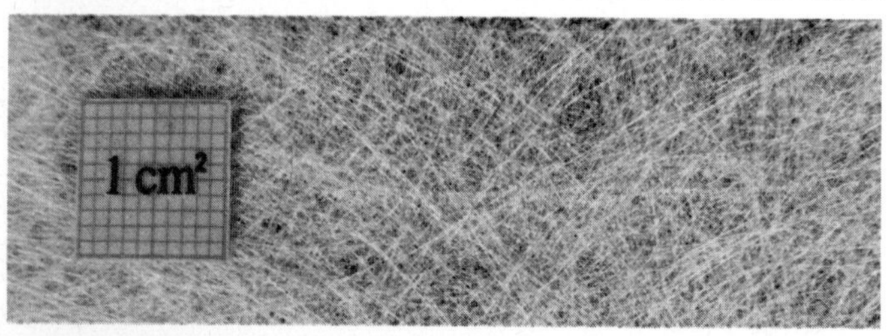

Terram 1500

171-222 g.s.m.
33% Polyethylene
67% Polypropylene
Melded White

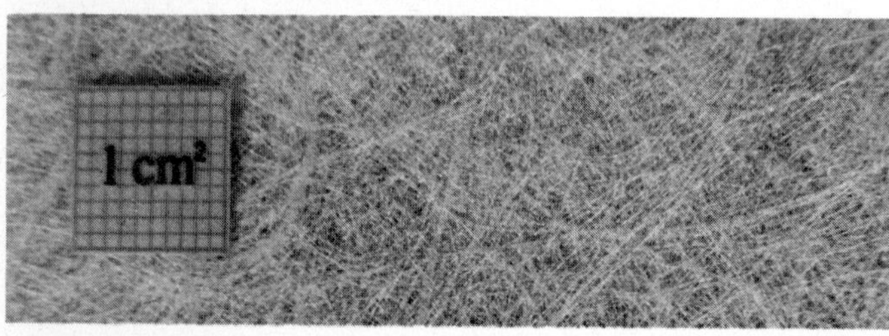

Terram 2000

223-253 g.s.m.
33% Polyethylene
67% Polypropylene
Melded White

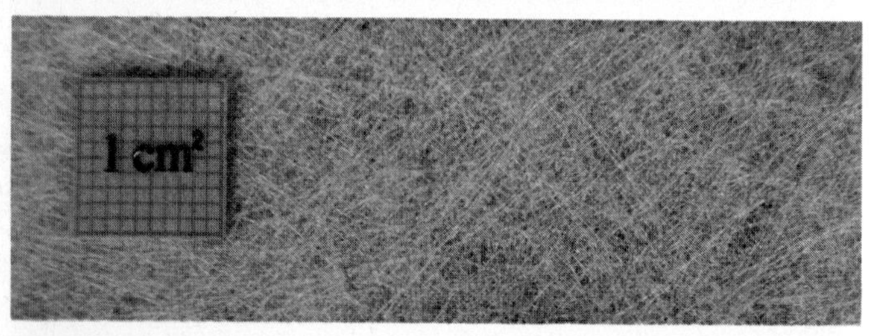

Terram 3000

254-300 g.s.m.
33% Polyethylene
67% Polypropylene
Melded White

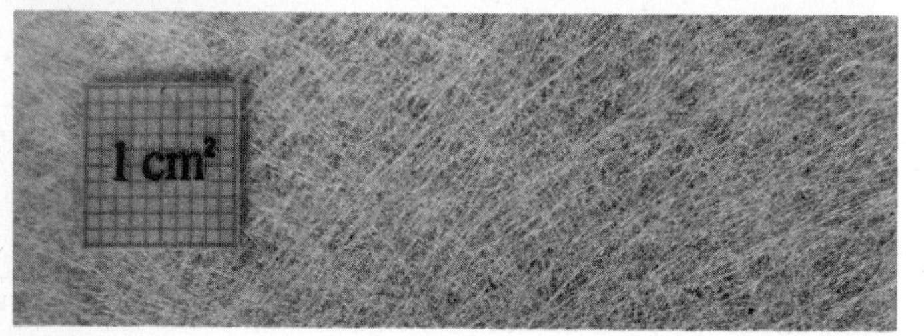

Terram 4000

301 g.s.m.
33% Polyethylene
67% Polypropylene
Melded White

Structural characteristics

'Terram' Product	500	700	1000	1500	3000
Thickness mm	0.4	0.5	0.7	0.8	1.0
Fibre Diameter micron	35	35	35	35	35
Thickness in terms of fibre diameters	11	16	20	23	28
Porosity % at 250 kg/m^2	81	80	79	73	70
Porosity % at 2 x 10^4 kg/m^2	75	74	74	68	65
Weight variability %	<10	<10	<10	<10	<10

Mechanical properties

	'Terram' Product	500	700	1000	1500	3000
200mm Plane strain test	Tensile Strength					
	Max load Newtons/200mm	750	1200	1700	2200	2800
	Extension at Max. load –%	35	40	45	50	60
	Load at 5% ext'n –Newtons	200	400	500	600	750
	Rupture Energy – Joules	40	70	100	160	230
25mm Grab test	Max load Newtons	400	600	850	1200	1600
	Ext'n to Max load –%	70	75	80	80	80
	Load at 5% ext'n–Newtons	70	120	160	210	270
	Tear Strength Wing – Newtons	110	190	250	310	400
Burst test	Bursting Load Newtons	50	80	110	150	210
	Distension at Burst – mm	15	15	15	15	15

The above figures are comparable with the equivalent figures obtained by testing to DIN 53857 and DIN 53858.
'Terram' is isotropic and hence properties are essentially the same irrespective of direction of test.

Tensile tests

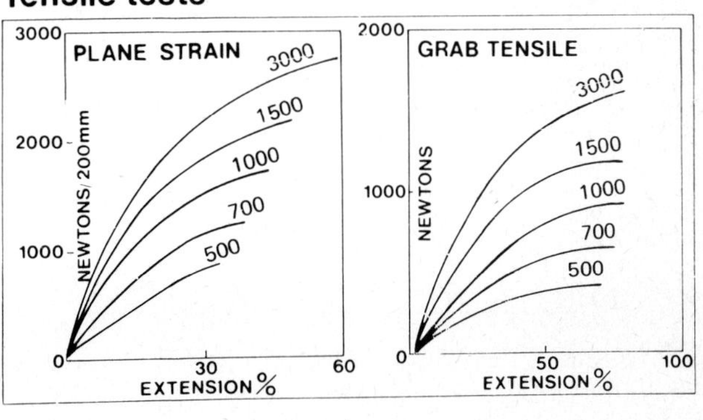

Figure 4.11 (Left) ICI 'Terram' products

Figure 4.12 (Right) Terram characteristics

Production data

'Terram' is a white fibre membrane developed by ICI for use in the Civil Engineering and Construction Industries. It is a thermally-bonded, non-woven material produced entirely from circular cross-section continuous synthetic fibres by ICI's unique 'melding' process.

'Terram' Product	500	700	1000	1500	3000
Standard Roll Width m.	4.5/5.3	4.5/5.3	4.5/5.3	4.5/5.3	4.5/5.3
Standard Roll Length m.	200	100	100	100	100
Standard Roll Weight kg	69/84	38/63	52/100	83/128	120/169
Standard Roll Diameter mm	310	270	300	330	370
Weight category gsm.	up to 70	71/100	101/170	171/222	254/300
Composition	67% polypropylene/33% polyethylene SG=0.92				

Hydraulic properties

Due to its 'melded' non-woven construction, 'Terram' retains essentially the same fabric geometry and hydraulic characteristics even under considerable planar extension.

'Terram' Product	500	700	1000	1500	3000
Water Permeability $L/m^2/sec$ 100mm head	150	80	40	35	30
Pore size O_{90} micron	350	180	100	60	40
Pore size O_{50} micron	200	120	70	40	20
The Darcy co-efficient (k) of the 'Terram' membranes shown in this table is of the order of 0.5×10^{-3} m/sec					

Pore-size distribution curves

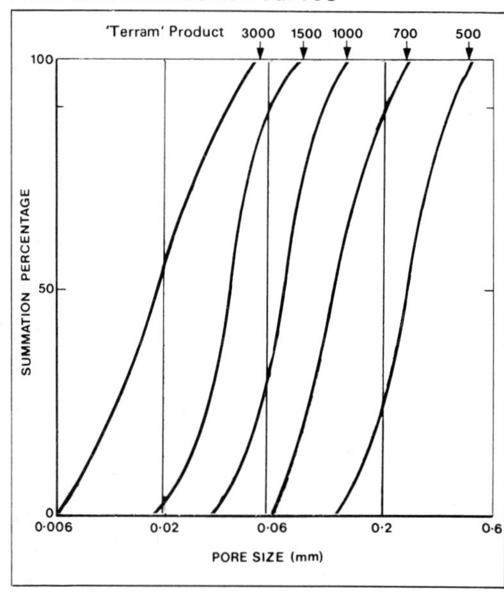

Figure 4.13 Terram characteristics

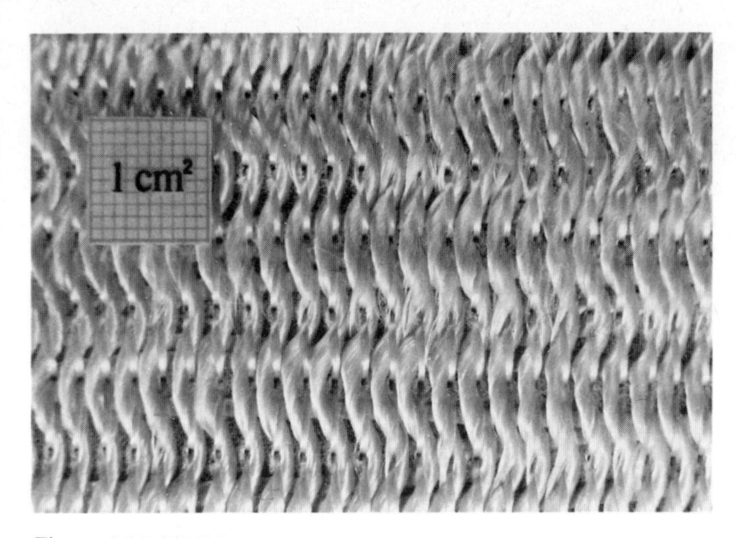

Figure 4.14 30M15

Extension at break
12% warp
13% weft

Water permeability
80 L/M^2/sec
at 10cm head

50% pore size 0.2mm

White

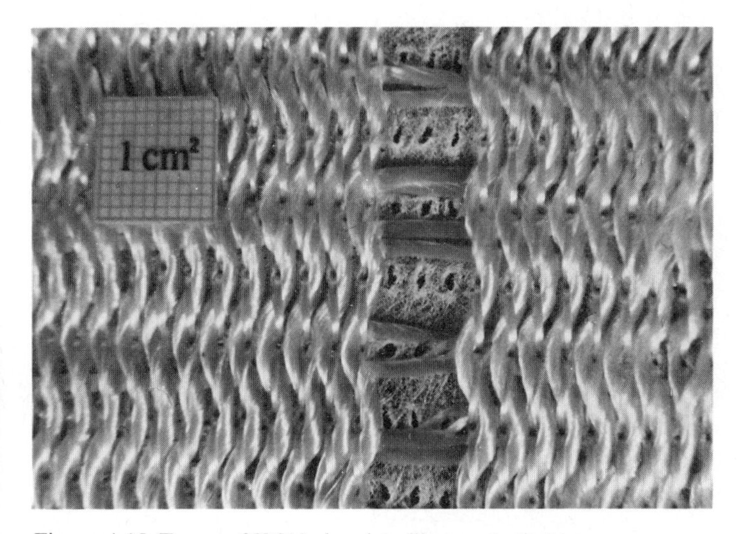

Figure 4.15 Terram 30M15 showing filter and stitching

Terram 30M15 is a combination meld-stitched product in which a Terram 70 filter membrane is stitched with heavy reinforcing polyester threads to provide high physical strength. It can be produced in a variety of strengths to order, and the numbers indicate properties as follows:—

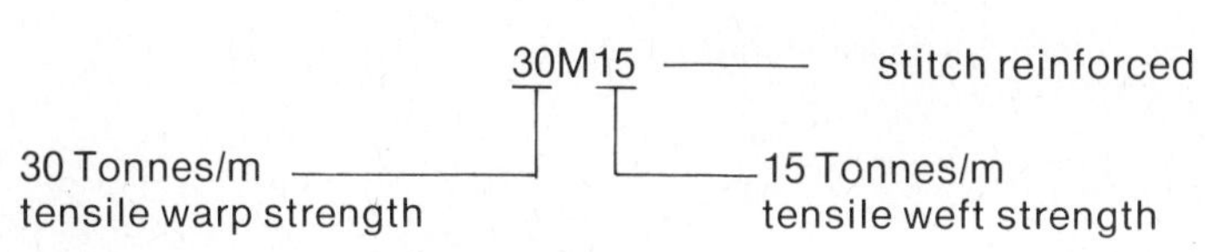

Civil Engineering Felt

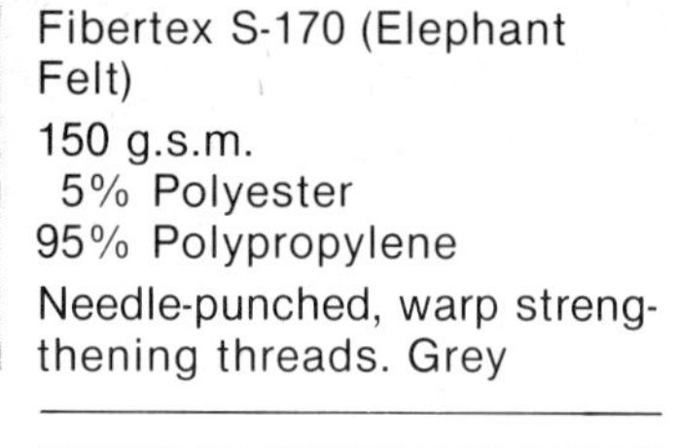

Fibertex S-170 (Elephant Felt)

150 g.s.m.
5% Polyester
95% Polypropylene

Needle-punched, warp strengthening threads. Grey

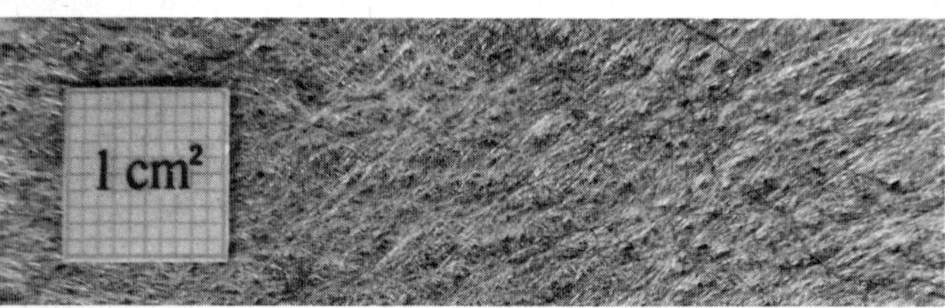

Civil Engineering Felt

Fibertex S-300

300 g.s.m.
2% Polyester
98% Polypropylene

Needle-punched, warp strengthening threads

Grey

Civil Engineering Felt

Fibertex S-400

400 g.s.m.
1% Polyester
99% Polypropylene

Needle-punched, warp strengthening threads

Grey

Irrigation Mat

Fibertex PPR 433

Properties not specified but construction similar to R-70/30

Light grey

Hot-house sun-absorber

Fibertex PORA 73

Properties appear to be the same as R-70/30, but coloured black

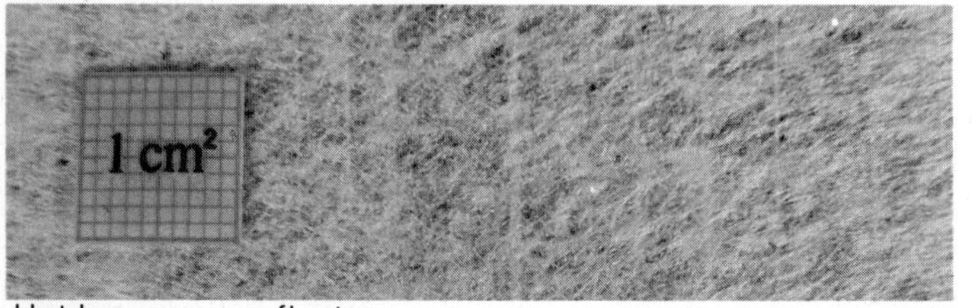

Hot-house sun-reflector

Fibertex R-70/30

155 g.s.m. White
30% Acrylic
70% Polypropylene

Needle-punched with warp thread strengthening

Fibertex S-170	Tensile strength as per DIN 53857: Longitudinally minimum 285N per 5 centimetres. Transversely minimum 390N per 5cm. Permeability at 10cm water head: 35 litres per sec. per sq. m.
Fibertex S-300	Tensile strength: Longitudinally minimum 590N per 5cm. Transversely minimum 785N per 5cm. Permeability 19 litres per sec. per sq. m.
Fibertex S-400	Tensile strength: Longitudinally minimum 590N per 5cm. Transversely minimum 1470N per 5cm.

	S-170	**S-300**	**S-400**
weight, approx.	150g/m^2	300g/m^2	400g/m^2
tensile strength kN/m — ASTM D-1682 (grab test)			
lengthwise	min. 11.6	min. 29.0	min. 36.7
crosswise	min. 15.4	min. 42.5	min. 57.9
elongation at burst	30-50%	40-70%	60-80%
permeability of water at 10cm wc	35 1/sec/m^2	19 1/sec/m^2	3,5 1/sec/m^2
pore size	20-60 microns	20-60 microns	20 microns
General properties			
specific gravity	0.9	0.9	0.9

properties practically the same in wet and dry condition
resistant to acids, bases and most solvents

contains no chemical binders

the material is pre-compressed. So, high mechanical load does not influence the pore size, and permeability is proportional to the water pressure

the material does not rot or support mildew

the material must be shielded from direct sunlight

PPR 433	Very efficient water dispersion — every single plant is secured an optimal supply of water and fertilisers Good remoisturing ability — also after repeated draining of the culture in the period of growth Does not shrink — it means a still better utilization of the production area Great mechanical strength — the irrigation mat can, e.g., easily be cleaned with a stiff broom/brush without destroying the cloth, and even towards obstinate roots the PPR 433 has excellent strength characteristics
PORA 73	Has high air permeability, which prevents condensed water from dripping on the plants Has a high insulation value — an important energy saving factor Is supplied in rolls of 100 metres, 460cm wide
R-70/30	Preventing overheating Shading plants from intense sunlight Helping to minimize heat loss at night Does not rot or mould U.V. stabilized Dry/wet strength uniform. Also does not shrink or otherwise change shape when wet Reflects about 85% of the sun's heat producing rays. This means easier climate control and helps assure constant, regulated plant growth Shade-value about 50% — another important factor in plant cultivation control Heat insulation value: 0.0404 kcal/hour/m/° Celsius (DIN 52612) Due to the non-woven construction, the fabric cannot ravel at all. Perfect to cut and sew Standard widths: 370, 400, 450 and 520cm. Other widths available on special order

Figure 4.16 (Opposite) Fibertex products

Figure 4.17 Fibertex characteristics

Terrafix 200 NA

350 g.s.m.

100% Polyamide

Needle punched non-woven with warp and weft reinforcing threads & resin bonded White

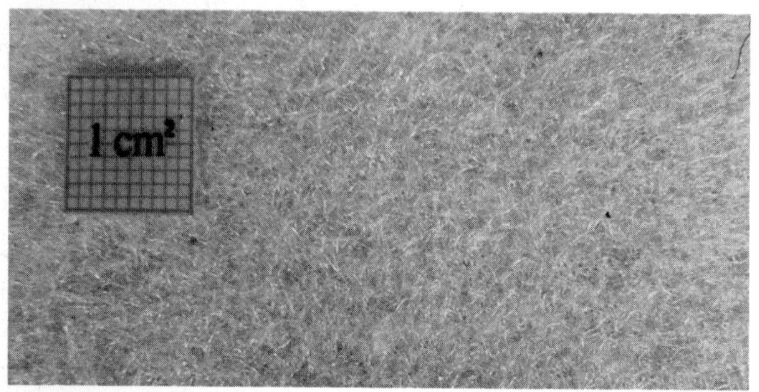

Terrafix 300NA

500 g.s.m.

100% Polyamide

Needle punched felt, resin bonded onto strong reinforcing warp and weft threads White

Terrafix 500 NA

700 g.s.m.

100% Polyamide

Needle-punched felt, resin bonded onto strong reinforcing warp and weft threads White

Terrafix 701 N

960 g.s.m.

100% Polyester

Needle-punched felt, resin bonded onto very strong reinforcing warp and weft threads Green

Figure 4.18 Terrafix products

200 NA

Specific weight		1.14 g/cm^3
Melting point		2.15°C
Stability		Rot proof, oil and seawater resistant
Colour		White
Roll dimensions	length:	100ft
	width:	6ft multiples
Material thickness		approx 2.8mm
Grab tensile, wet (ASTM 1682)		
maximum load		approx 50kp
elongation at max load		approx 15%
Resistance to sub-zero temperature as low as light and weather resistant		– 40°C
Abrasion test (rotary drum)		
40,000 revolutions		mark 2
80,000 revolutions		mark 4

Water permeability characteristics:

falling head	permeability ml/cm^{2}s	k-value cm/s
Δh1 (130-30) cm	100	0.3
Δh2 (50-30) cm	70	0.5
Δh3 (30-10) cm	45	0.6

500NA

Specific weight		1.14g/cm^3
Melting point		2.15°C
Stability		Rot proof, oil and seawater resistant
Roll dimensions	length:	100ft
	width:	6ft multiples
Material thickness		approx 4.5mm
Grab tensile, wet (ASTM 1682)		
maximum load		approx 80kp
elongation at max load		approx 15%
Resistance to sub-zero temperature as low as light and weather resistant		– 40°C
Abrasion test (rotary drum)		
480,000 revolutions		mark 1
560,000 revolutions		mark 2
680,000 revolutions		mark 3
720,000 revolutions		mark 4

Water permeability characteristics:

falling head	permeability ml/cm^{2}s	k-value cm/s
Δh1 (130-30) cm	65	0.4
Δh2 (50-30) cm	50	0.5
Δh3 (30-10) cm	30	0.7

300 NA

Specific weight		1.14 g/cm^3
Melting point		2.15°C
Stability		Rot proof, oil and seawater resistant
Colour		White
Roll dimensions	length:	100ft
	width:	6ft multiples
Material thickness		approx 3.5mm
Grab tensile, wet (ASTM 1682)		
maximum load		approx 60kp
elongation at max load		approx 15%
Resistance to sub-zero temperature as low as light and weather resistant		– 40°C
Abrasion test (rotary drum)		
160,000 revolutions		mark 1
200,000 revolutions		mark 2
280,000 revolutions		mark 3
320,000 revolutions		mark 4

Water permeability characteristics:

falling head	permeability ml/cm^{2}s	k-value cm/s
Δh1 (130-30) cm	75	0.3
Δh2 (50-30) cm	55	0.5
Δh3 (30-10) cm	35	0.6

701 N

Specific weight	1.38 g/cm^3
Melting point	260°C
Stability	Rot proof, oil and seawater resistant
Material thickness	approx 7mm
Tensile strength according to DIN 53 857	
maximum load (in wet condition)	approx 140kp/5cm
elongation at max load (in wet condition)	approx 12%
elongation at break (in wet condition)	approx 60%
Abrasion test according to BAW — method	
240,000 revolutions	mark 1
440,000 revolutions	mark 3
Water permeability (ml/s cm^2 = cm/s)	
Δh1 = (130-30) cm	approx 60
Δh2 = (50-30) cm	approx 30
Δh3 = (30-10) cm	approx 15

Figure 4.19 Terrafix characteristics

Polyfelt TS-200

Approx 150 g.s.m.
100% Polypropylene
Needle-punched continuous filament felt
Grey

Polyfelt TS-300

250 g.s.m.
100% Polypropylene
Needle-punched continuous filament felt
Grey

Polyfelt TS-400

350 g.s.m.
100% Polypropylene
Needle-punched continuous filament felt
Grey

Figure 4.20 Polyfelt products

	Measure-ment	Polyfelt TS 400	Polyfelt TS 300	Polyfelt TS 200	Standard	Test Conditions
Width	cm	250 500	250 500	240 480		2 standard widths
Thickness	mm	3.5 3.5	3.0 3.0	2.0 2.0	DIN 53.855	
Length	m	160 80	240 120	300 150		
Area	m^2	400 400	600 600	720 720		
Roll diameter	cm	80 80	80 80	80 80		
Resistance to tearing	N/5 cm	800	650	550	DIN 53.815	measured dry,
Elongation at break	%	80 40[1])	80 40[1])	90 45[1])	DIN 53.815	wet. wet at −20° C
Grab strength	N	1200	900	800	DIN 53.838	
Tear propagation strength	N	190	150	130	DIN 53.859	
					DIN 54.301	
Stitch point tearing-out resistance	N	120	100	90		
max. bursting pressure	N	4150	3100	2050		ø calotte 100 mm
max. vaulting burst	mm	70	65	55		ø fixing frame 110 mm
Re[illegible]tance to tearing w[illegible] impeded tranvers contraction	N/cm	110	85	68		
Puncture resistance		no perforation	no perforation	no perforation	BAW III 75/4	tamper 30 kg fall height 2 m
k-value	cm/sec	$8.2 \cdot 10^{-2}$	$5.9 \cdot 10^{-2}$	$5.5 \cdot 10^{-2}$		0[2])
	cm/sec	$3.9 \cdot 10^{-2}$	$3.2 \cdot 10^{-2}$	$2.1 \cdot 10^{-2}$		10
	cm/sec	$2.7 \cdot 10^{-2}$	$2.2 \cdot 10^{-2}$	$1.4 \cdot 10^{-2}$		30
	cm/sec	$1.7 \cdot 10^{-2}$	$1.5 \cdot 10^{-2}$	$7.7 \cdot 10^{-3}$		80
Rate of flow Q	l/dm² min	140	142	147		0[2])
	l/dm² min	104	114	114		10
	l/dm² min	86	91	97		30
	l/dm² min	61	70	78		80
k-value	cm/sec	$1.1 \cdot 10^{-1}$	$1.2 \cdot 10^{-1}$	$1.3 \cdot 10^{-1}$		0[2])
	cm/sec	$5.5 \cdot 10^{-2}$	$5.5 \cdot 10^{-2}$	$5.8 \cdot 10^{-2}$		10
	cm/sec	$2.7 \cdot 10^{-2}$	$1.8 \cdot 10^{-2}$	$1.2 \cdot 10^{-2}$		30
	cm/sec	$1.8 \cdot 10^{-2}$	$1.1 \cdot 10^{-2}$	$9.2 \cdot 10^{-3}$		80
Rate of flow Q	l/dm² min	162	169	180		0[2])
	l/dm² min	120	136	139		10
	l/dm² min	96	108	118		30
	l/dm² min	70	83	96		80

Test conditions (lower section): Loading vertical to fabric plane in N/cm²; first two groups: Flow direction vertical to the plane; last two groups: Flow direction in the plane; measured with 1 m water column.

Note: Polyfelt is supplied in folded rolls which can be more difficult to handle and unroll on site than wider unfolded fabrics.
Note: Polyfelt can be heat welded to join overlaps.

Figure 4.21 Polyfelt characteristics

Colback

130 g.s.m.
50% Polyamide
50% Polyester
Spun-bonded, continuous monofilament membrane
Grey

Colbond

250-450 g.s.m.
100% Polyester
Needle-punched staple fibre felt, resin bonded
White

Figure 4.22 Enka Glanzstoff products

Some relevant properties of Colbond®

Property	Unit		Test	P250	P350	P450
Weight	g/m^2		DIN 53854	250	350	450
Thickness	mm		DIN 53855	1,8	2,1	2.4
Tensile strength	N/5cm	m/c	DIN 53857	700/400	1000/600	1350/800
Resistance at 5% elongation	N/5cm	m/c	DIN 53857	300/110	480/160	600/230
Elongation at break	%	m/c	DIN 53857	25/55	30/55	30/50
Tear strength	N	m/c	DIN 53859	60/90	85/120	113/150
Mullenbursttest	N/cm^2	m/c		200	300	380
Puncture resistance	(Lastometer)					
Pyramid : Pressure	N			450	620	750
Impression	mm			9,5	11	11,5
Ball : Pressure	N			230	390	500
Impression	mm			6,5	7,5	8,5
Water permeability	$m^3/m^2/h$ 10cm column of water			190	120	90

m = machine direction
c = cross direction

Note that COLBOND BV do not recommend the use of COLBACK material for civil engineering use. The Author has included it for the sake of completeness since it is believed that it has been used in the past, and since it is mentioned in their civil engineering literature.

Figure 4.23 Enka Glanzstoff characteristics

Typar 136

136 g.s.m.

100% Polypropylene

Spun-bonded, continuous monofilament membrane

Grey

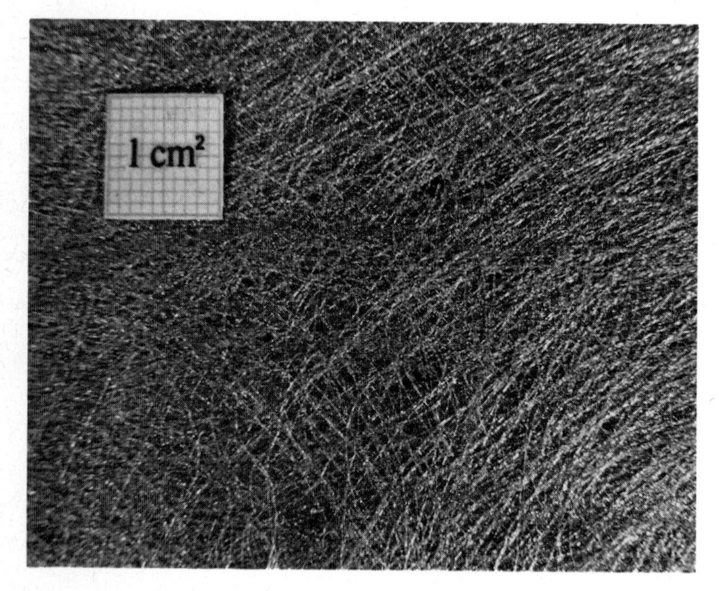

Typar 200

200 g.s.m.

100% Polypropylene

Spun-bonded continuous monofilament membrane

Grey

Figure 4.24 Typar products

Typar

technical data and grade specification

	'Typar' 136	'Typar' 200
Fabric Composition	100% polypropylene	
Fabric Weight	136 g/m²	200 g/m²
Fabric Thickness	0.46 mm	0.60 mm
Roll Width	3.5 m 4.2 m 5.2 m	 4.2 m 5.2 m
Roll Length	150 m 300 m 1000 m	100 m 300 m 800 m
Tensile Strength under Uniform Applied Stress (Multidirectional Test)	10600 N/m (1100 kg/m)	17600 N/m (1800 kg/m)
Elongation at Maximum Load under Uniform Applied Stress	72%	69%
Strip Tensile Strength	340 N/50 mm (35 kg/50 mm)	630 N/50 mm (64 kg/50 mm)
Elongation	39%	43%
Water Permeability 785 Pa (80mm water column)	100 litres/min/m²	50 litres/min/m²
Specific Gravity	0.90	0.90
Effect of Naturally Occurring Acids	none	none
Effect of Naturally Occurring Alkali	none	none
Effect of Bacteria	none	none
Service Temperature Range	Approximately —40°C to 130°C (melting point = 165°C)	
Effect of Ultra-Violet Light	Unaffected up to one month in strong direct sunlight but prolonged exposure can cause strength losses.	

tensile stress/strain relationship (uniform applied stress)

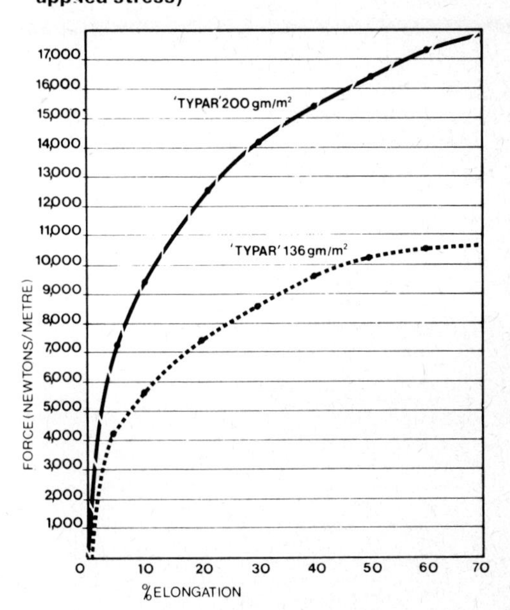

A good soil stabilisation mat must have, under uniform stress, high initial strength at low elongation, combined with high strength and elongation under maximum load.

The excellent tensile properties of 'Typar' are balanced in both the length and width direction and remain the same under wet and dry conditions.

Figure 4.25 Typar characteristics

Toughnell TS-40

Approx 200 g.s.m.
100% Polypropylene
Needle-punched felt
White

Toughnell TS-80

400 g.s.m.
100% Polypropylene
Needle-punched felt
White

Figure 4.26 Toughnell products

品質と性能

	品　名	TS-40	TS-80	(試験法)
Tensile strength	引張強さ(kg/5cm)	25	50	JIS-L1068
Extension	伸　度(%)	110	100	JIS-L1068
Tearing strength	引裂強さ(kg)	18	28	当社試験法
Weight	重　さ(g/m²)	200	400	当社試験法
Thickness	厚　さ(mm)	3	4	当社試験法
Porosity	(%)	91	91	
Temp stability	(°C)	−30 to +130	−30 to +130	
S.G.	of fibre (g/cm³)	0.91	0.91	
Roll width	(m)	2.7	2.7	

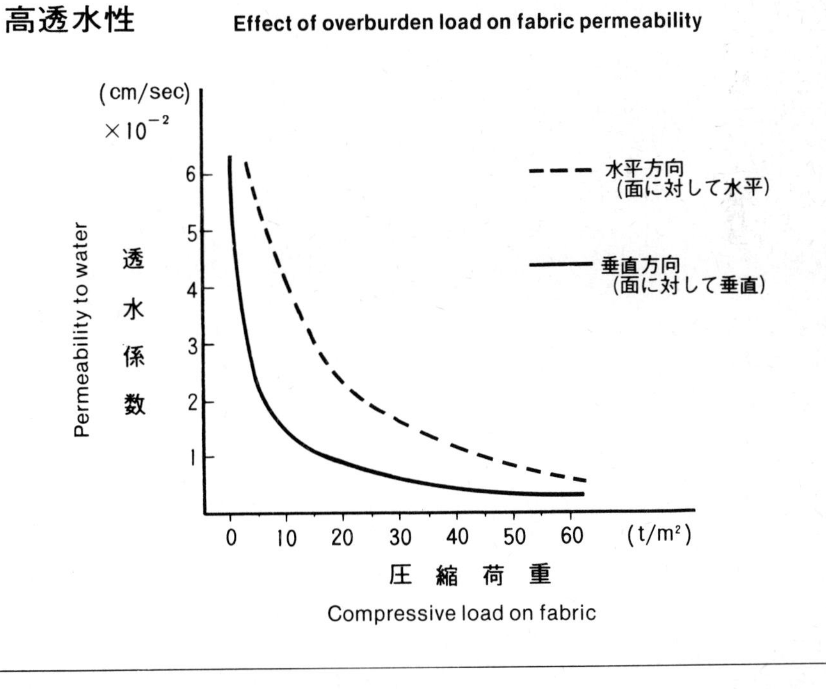

Figure 4.27 Toughnell characteristics

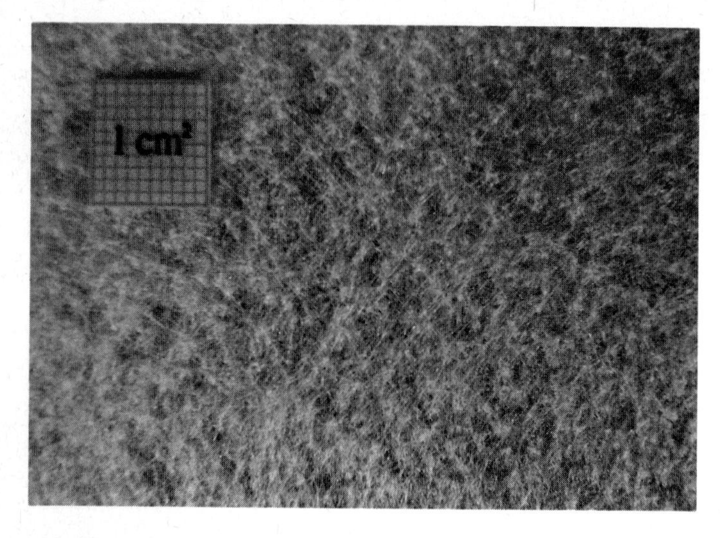

Monofelt

170 g.s.m.

100% Polypropylene

Appears to be needle punched with resin bonding

White

Figure 4.28 Menardi Southern products

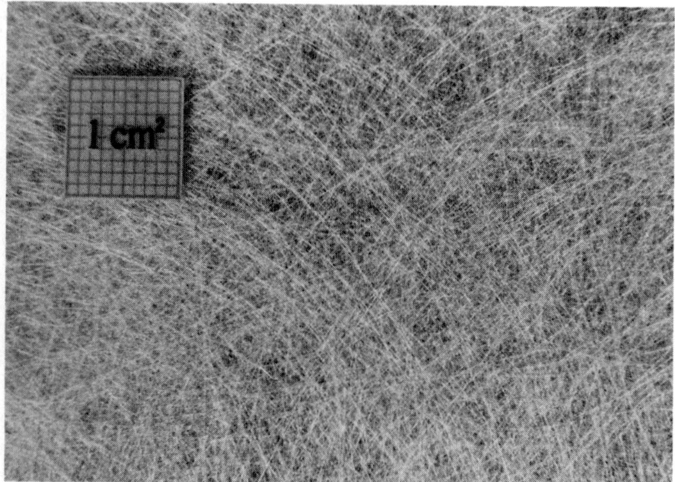

Mirafi 140S

140 g.s.m.

33% Polyethylene

67% Polypropylene

Spun-bonded continuous hetero- and monofilaments

White

Figure 4.29 Celanese product

Monofelt published data

Weight	5oz/sq. yd. (170 g.s.m.)
Thickness	44 mils (1.11mm)
Mullen burst strength	Gross 259 p.s.i. (1785kN/m^2)
Grab tensile (strip width unspecified)	143lbs (65kg) Machine direction W/59% elongation 187lbs (85kg) Cross direction W/45% elongation
Abrasion resistance (units not specified)	110 machine direction 124 cross direction
E.O.S. (meaning not specified)	80-100

Figure 4.30 Monofelt characteristics

Thickness mm	0.7
Fibre diameter micron	35
Thickness in terms of fibre diameters	20
Porosity % at 250kg/m^2	79
Porosity % at 2×10^4kg/m^2	74
Weight variability %	<10
Standard roll width m	4.5/5.3
Standard roll length m	100
Standard roll weight kg	52/100
Standard roll diameter mm	300
Weigh category gsm	101/170
200mm Plane strain test	
Tensile strength	
max load Newtons/200mm	1700
extension at max load %	45
load at 5% extension Newtons	500
Rupture energy Joules	100
25mm Grab test	
max load Newtons	850
Extension to max load %	80
load at 5% extension Newtons	160
Tear strength wing Newtons	250
Burst test	
bursting load Newtons/cm^2	110
distension at burst mm	15
Water permeability L/m^2/sec 100mm head	40
Pore size 0_{90} micron	100
Pore size 0_{50} micron	70

The Darcy co-efficient (k) is of the order of 0.5×10^{-3}m/sec

Figure 4.31 Mirafi 140S characteristics

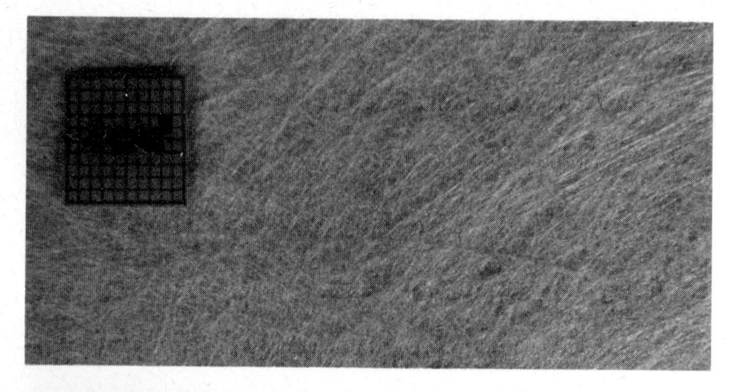

Sodoca NS 115
125 g.s.m.
100% Polypropylene
Needle punched continuous filament
White

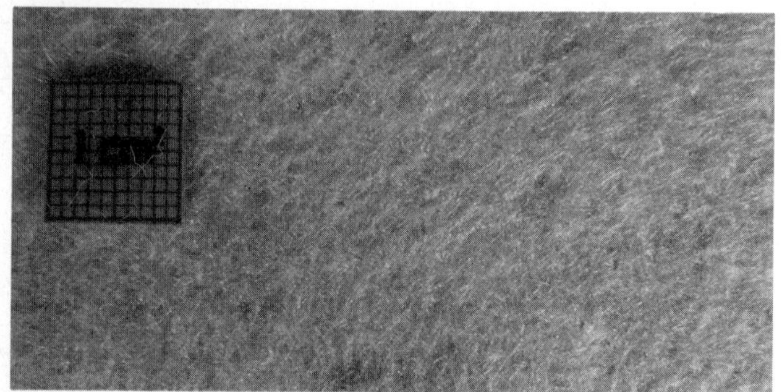

Sodoca AS 200
200 g.s.m.
100% Polypropylene
Needle punched continuous filament
White

Sodoca AS 320
290 g.s.m.
100% Polypropylene
Needle punched continuous filament
White

Sodoca AS 420
370 g.s.m.
100% Polypropylene
Needle punched continuous filament
White

Figure 4.32 Sodoca products

TESTS		Test Type	Units	NS	AS								FJ	
				115	150	200	250	300	320	400	420	600	120	250
5cm strip breaking strength		NF G.07.001	daN/5cm	35	35	50	60	70	75	85	100	165	13	55
Wet tearing strength		EDANA 70.0-75 NF G.37.104	daN	15	25	40	50	55	60	65	70	120	10	50
				9	10	17	20	25	30	33	35	55	2	25
Wet grab strength		NF G.07120	daN	50	55	90	100	110	130	150	170	250	20	90
Elongation at rupture		NF G.07.001	%		> 60 %									
Plain strain strength		L.R. St Brieuc	10^4 N/m	1,1	0,9	1,5	1,6	2,0⁺	2,4	2,7	3,0	4,9	/	/
				1,7	1,3	1,9	2,1	2,3	2,7	3,0	3,6	5,6	/	/
Permeability at right angles to membrane	< 0,05 bar	INSA Lyon	10^{-3} m/s	2,0	2,0	2,0	2,0	2,0	2,0	2,0	2,0	2,0	5,0	6,0
	< 2 bars		10^{-4} m/s	1,0	5,0	5,0	5,0	5,0	5,0	5,0	5,0	5,0	5,0	5,0
Permeability within the membrane	< 0,05 bar	INSA Lyon	10^{-3} m/s	1,0	4,0	4,0	4,0	4,0	3,0	3,0	3,0	2,0	4,0	5,0
	< 2 bars		10^{-4} m/s	1,0	5,0	5,0	5,0	5,0	5,0	5,0	5,0	5,0	5,0	5,0
Pore size distribution	< 0,05 bar	Calculated	Micron	/	117 à 73	108 à 67	104 à 63	105 à 64	102 à 62	96 à 58	92 à 55	80 à 47	100 à 63	166 à 104
	< 2 bars			/	46 à 23	47 à 23	32 à 13	40 à 18	44 à 22	44 à 21	44 à 21	50 à 25	45 à 23	67 à 34
Fires retained		Limon 2 à 60μ	% retained	95	20	30	50	50	60	70	80	95	10	5
Weight of fabric			g/m²	125	130	200	240	270	290	340	370	600	120	240
Thickness	< 0,005 bar	Measured	mm	1,4	1,8	2,5	2,8	3,0	3,1	3,5	3,7	4,8	1,9	3,5
	< 2 bars			0,2	0,5	0,6	0,8	0,9	1,0	1,2	1,4	2,6	0,6	1,0
Porosity	< 0,005 bar	Calculated	%	91	92	91	90	90	90	89	89	86	93	92
	< 2 bars			35	71	63	67	67	68	69	71	74	78	73
Apparent volumetric mass	< 0,005 bar	Calculated	kg/m³	89	72	80	86	90	93	97	100	125	63	69
	< 2 bars			625	260	330	300	300	290	280	260	230	200	240

Figure 4.33 Sodoca characteristics

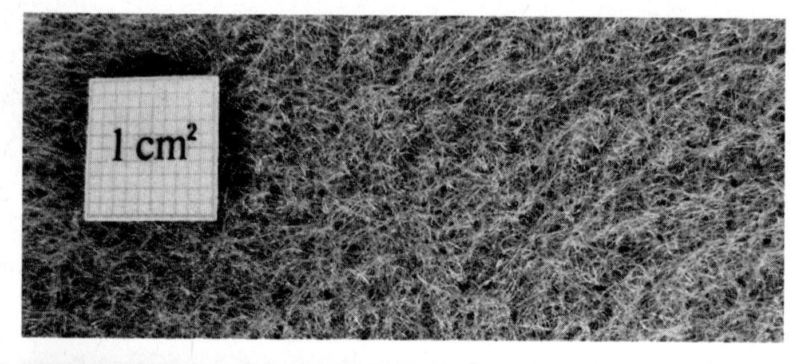

Bidim U14
150 g.s.m.
100% polyester
Needle punched,
continuous filament felt
Grey

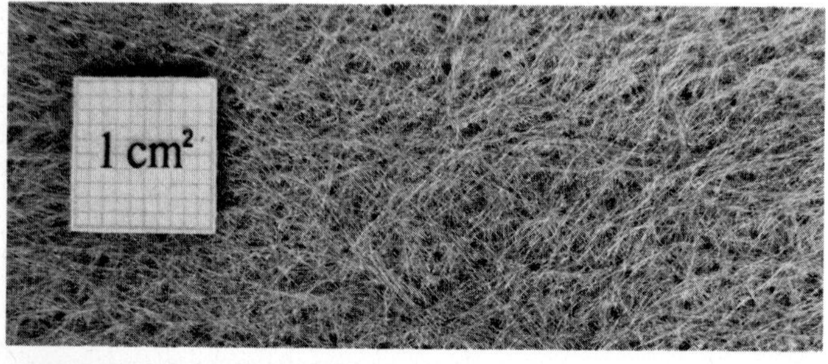

Bidim U24
210 g.s.m.
100% polyester
Needle punched,
continuous filament felt
Grey

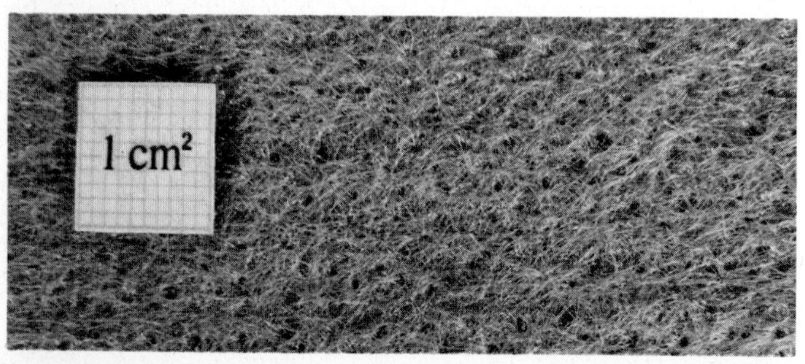

Bidim U34
270 g.s.m.
100% polyester
Grey

Bidim U44
340 g.s.m.
100% polyester
Needle punched,
continuous filament felt
Grey

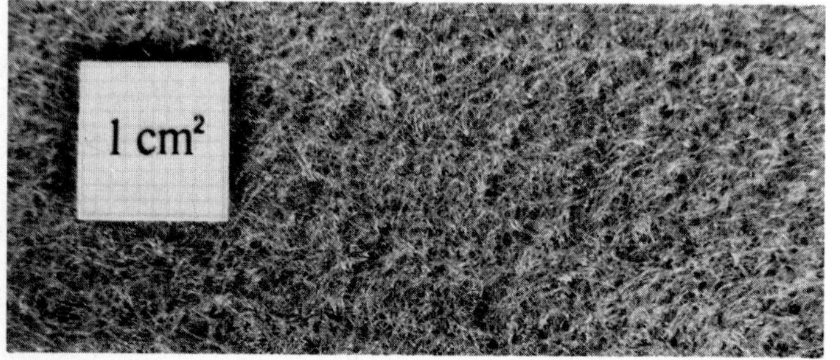

Bidim U64
550 g.s.m.
100% polyester
Needle punched,
continuous filament felt
Grey

				U 14	U 24	U 34	U 44	U 64	Remarks
Physical characteristics	Mass per unit surface		kg/m^2	0.15	0.21	0.27	0.34	0.55	Ref. AFNOR standard G 07-104
	Thickness under	0.005 bar	mm	1.5	1.9	2.3	2.8	4.4	Ref EDANA 30074
		2 bar	mm	0.6	0.8	1.05	1.3	2.1	
	Bulk density under	0.005 bar	kg/m^3	100	111	117	121	125	Calculated
		2 bar	kg/m^3	250	260	260	260	260	
	Porosity under	0.005 bar	%	93	92	91	91	91	Calculated
		2 bar	%	82	81	81	81	81	
	Specific surface	Superficial	m^2/m^2	16.1	22.5	28.9	36.4	58.9	Calculated
		In the mass	m^2/kg	107.2	107.2	107.2	107.2	107.2	
Permeability	Normal permeability under	0.020 bar	m/s	$3\ 10^{-3}$	$3\ 10^{-3}$	$3\ 10^{-3}$	$3\ 10^{-3}$	$3\ 10^{-3}$	Measured by LCPC and INSA Lyon
		2 bar	m/s	$7\ 10^{-4}$	$7\ 10^{-4}$	$7\ 10^{-4}$	$7\ 10^{-4}$	$7\ 10^{-4}$	
	Radial permeability under	0.020 bar	m/s	$6\ 10^{-4}$	$6\ 10^{-4}$	$6\ 10^{-4}$	$6\ 10^{-4}$	$6\ 10^{-4}$	Measured by LCPC and INSA Lyon
		2 bar	m/s	$4\ 10^{-4}$	$4\ 10^{-4}$	$4\ 10^{-4}$	$4\ 10^{-4}$	$4\ 10^{-4}$	
	Theoretical porometry (retention of fines)	0.005 bar	μm	97/61	94/59	90/56	89/56	86/53	Ref. LCPC
		2 bar	μm	54/30	50/28	52/29	52/29	52/29	
Mechanical characteristics	One-directional tension	Resistance	kgf/5cm	30	50	70	95	165	Ref. AFNOR standard G 07001
			N/m	$5.9\ 10^3$	$9.8\ 10^3$	$1.4\ 10^4$	$1.9\ 10^4$	$3.2\ 10^4$	
		Stretch	%	50/70	50/70	50/70	50/70	50/70	
	Multi-directional tension	Resistance	N/m	$1\ 10^4$	$1.6\ 10^4$	$2.1\ 10^4$	$2.6\ 10^4$	$4.4\ 10^4$	Ref. AFNOR
		Stretch	%	27/30	27/30	27/30	27/30	27/30	Measured by
		Dry and wet modulus	N/m	$3.1\ 10^4$	$4.1\ 10^4$	$5.5\ 10^4$	$7.0\ 10^4$	$1.7\ 10^5$	L.R. St Brieuc
	Bursting strength (indicative)		N/m^2	$1.7\ 10^6$	$2.2\ 10^6$	$2.5\ 10^6$	$3.6\ 10^6$	$5.5\ 10^6$	Ref. AFNOR standard G 07112
	Continued tearing strength		N	$5\ 10^1$	$9\ 10^1$	$1.7\ 10^2$	$2.3\ 10^2$	$4.0\ 10^2$	Ref. AFNOR standard G 07055
	Plastic flow under one-dimensional tension		Under 20% of breaking load, stability attained after 1 hour Under 40% of breaking load, stability attained after 6 hours						

Figure 4.35 Bidim characteristics

Figure 4.34 (Opposite)
Bidim products

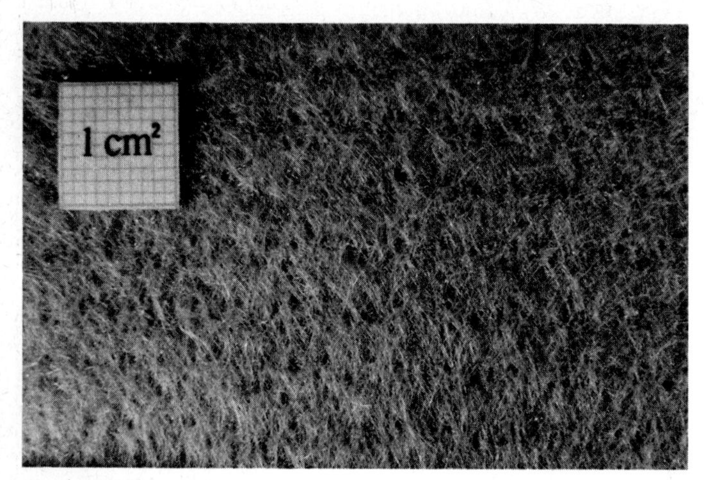

Trevira Spunbond

500 g.s.m.
100% Polyester
Needle punched felt
Grey colour

Also available in 200, 300, and 400 g.s.m. (see opposite)

Figure 4.36 Trevira products

Bodenmechan. Kennziffern

Spez. Gewicht γ_s Mp/m^3	1,37
Trockenraumgewicht γ_t	0,12
Porenvolumen n %	91
Durchlässigkeit bei	
0,0 m Wassersäule	10^{-3} m/s
0,5 m Wassersäule	$1{,}25 \cdot 10^{-4}$
1,0 m Wassersäule	$1{,}1 \cdot 10^{-4}$
Zusammendrückung %	
0,1 kp/cm^2	27
0,3 kp/cm^2	39
0,5 kp/cm^2	46
mittlere Fadendicke	$3{,}2 \cdot 10^{-3}$ cm
statistischer mittlerer Durchmesser des Porenraums	dj = 0,16 mm
wirksamer äquivalenter Korndurchmesser	$Dm_{äqu.}$ = 1 mm

Type	Fadentiter in dtex	Flächengewicht[1] in g/m^2	Dicke[2] in mm	Reißfestigkeit[3] in kp/5 cm	Reißdehnung in %	Schrumpf[4]
11	**8**	**200**	**2,5**	**45**	**65**	**<5 %**
		300	**3,0**	**75**	**65**	**<5 %**
		400	**3,5**	**100**	**65**	**<5 %**
		500	**4,0**	**135**	**65**	**<5 %**

1. Die angegebenen Flächengewichte werden im Rahmen des normalen Produktionsprogrammes hergestellt. Bei ausreichend großen Bestellungen können davon abweichende Flächengewichte produziert werden.

2. Die Dicke wird mit einem Auflagegewicht von 20 p/cm^2 bestimmt, nach DIN 53855.

3. Die Messung der Reißfestigkeit erfolgt an 100 mm breiten, auf 50 mm geschlauchten eingespannten Proben in Anlehnung an DIN 53857; der so gewonnene Meßwert wird halbiert.

4. Der Schrumpf wird in der Fläche bei 200°C in Heißluft gemessen.

Figure 4.37 Trevira characteristics

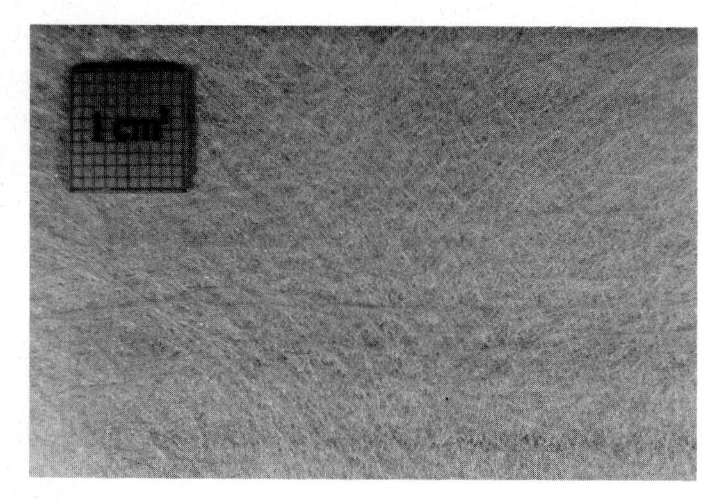

Lutradur

250 g.s.m.
100% Polyester
Continuous-filament spun-bonded membrane
White colour

Also available in 50, 70, 100, 140, and 200 g.s.m. (see opposite)

Figure 4.38 Lutravil products

Types	LUTRADUR®	7250	7270	H7210	H7214	H7220	H7225
Weight g/sqm DIN 53 854 Poids g/qm		50	70	100	140	200	250
Thickness mm DIN 53 855 Epaisseur mm		0,28	0,34	0,38	0,50	0,80	0,84
Breaking strength N/5 cm DIN 53 857	md chaine	80	130	200	330	485	730
Résistance à la rupture N/5 cm	cd trame	70	115	180	310	470	650
Breaking strength N/5cm (100 mm wide under stress in enclosure)	md chaine	230	250	370	550	950	1250
Résistance à la rupture N/5 cm (100 mm de large, enroulé et étiré)	cd trame	200	230	370	550	950	1100
Elongation % DIN 53 857	md chaine	35	40	40	45	50	50
Allongement %	cd trame	35	40	40	45	50	50
Tongue tear strength N DIN 53 859	md chaine	25	38	50	70	90	110
Résistance à la déchirure amorcée N	cd trame	25	38	50	70	90	120
Grabtensile strength N DIN 53 858 and in similar to ASTM 1682 — 64 GT	md chaine	140	200	290	480	750	1000
Résistance à l'enfouissement N DIN 53 858 et approchant la nome ASTM 1682 — 64 GT	cd trame	140	200	270	470	750	950
Heat shrinkage % (at 200 deg. C/2 min.)	md chaine	– 2,1	– 1,0	– 1,4	– 1,5	– 1,5	– 1,2
Rétrécissement à la chaleur % (200°C, 2 min.)	cd trame	– 1,1	– 0,4	– 0,5	– 0,2	– 0,2	– 0,2
Filtering capacity in microns Capacité de filtrage μ		600	450	300	180	85	70

Delivery units/Mode de livraison

Width in centimeters Largeur des rouleaux cm	420	420	420	420	420	420	420
Length in running meters Longeur des rouleaux ml		1000	200	500	100	100	100

Special widths and lengths available by request.
Des rouleaux de longueur inférieure à celles-indiquées ci-dessus sont livrables sur demande.

Figure 4.39 Lutravil characteristics

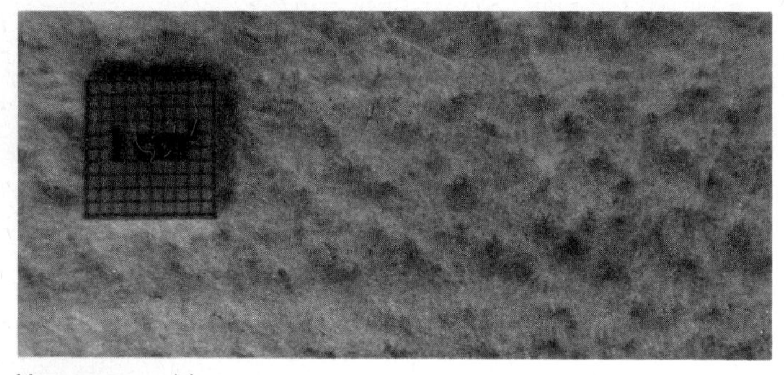

Non-woven side

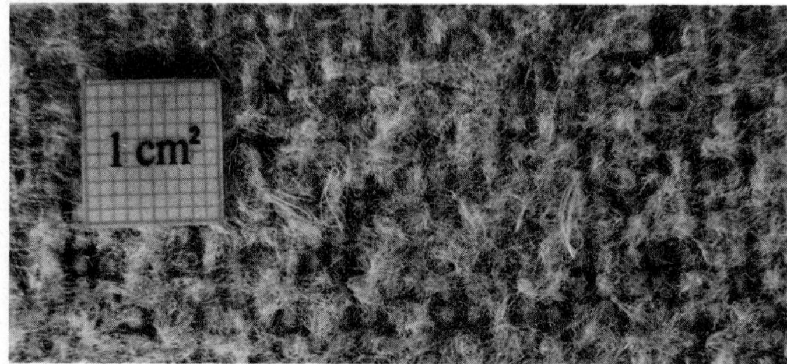

Woven side

Hate K-S/T

800 g.s.m.

100% polyester

Needle-punched non-woven bonded onto an open-mesh scrim, with chemical bonding

White non-woven
Black scrim

Non-woven side

Woven side

Hate H-5/T

500 g.s.m.

100% polyester

Needle-punched non-woven bonded onto an open-mesh scrim

White non-woven
Black scrim

Figure 4.40 Huesker synthetic KG products

Hate K-5/T
Thickness 6mm
Permeability at 10cm head: 30-40 $l/m^2/sec$
Standard width 200cm
Reisfestigkeit Warp: 125 daN/5cm
(Strip strength) Weft: 125 daN/5cm

Hate H-5/T
Thickness 5mm
Permeability at 10cm head: 60 $l/m^2/sec$
Standard width 200cm
Reisfestigkeit Warp: 140 da N/5cm
(Strip strength) Weft: 140 da N/5cm

Figure 4.41 Huesker synthetic KG characteristics

Woven membranes illustrated

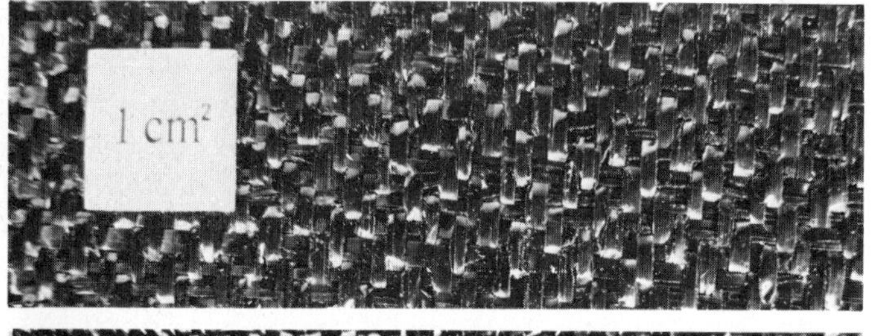

Robusta 175

175 g.s.m.
100% Polypropylene

Black Woven split tape
Appears to be resin stabilised

Robusta 265

265 g.s.m.
100% Polypropylene

Black Woven flat split tape woven in the warp, with strong twisted split tape bundles in the weft. Heavily coated with resin bonder

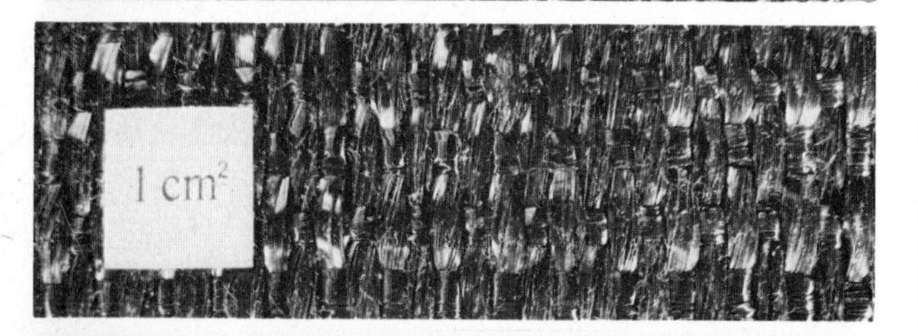

Robusta 500

500 g.s.m.
100% Polypropylene

Black Woven thinner split tape yarn in the weft with thick twisted split tape yarn in the warp.
No resin apparent

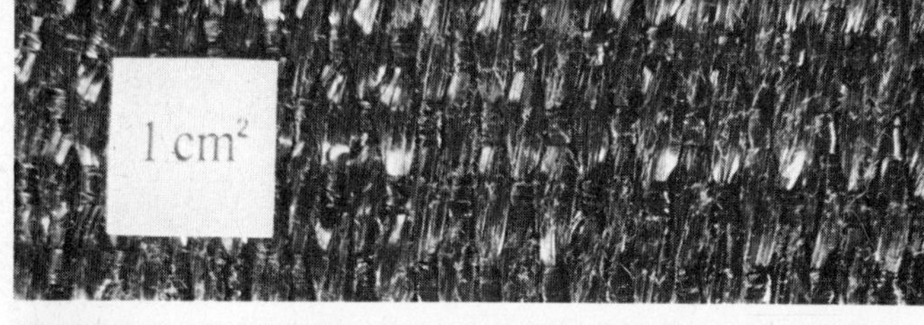

Robusta 750

750 g.s.m.
100% Polypropylene

Black Woven
As for 500, but resin dipped

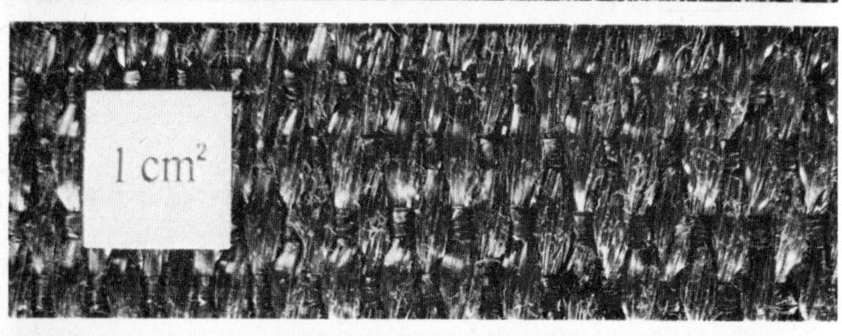

Robusta 830

830 g.s.m.
100% Polypropylene

Black Woven
As for 500

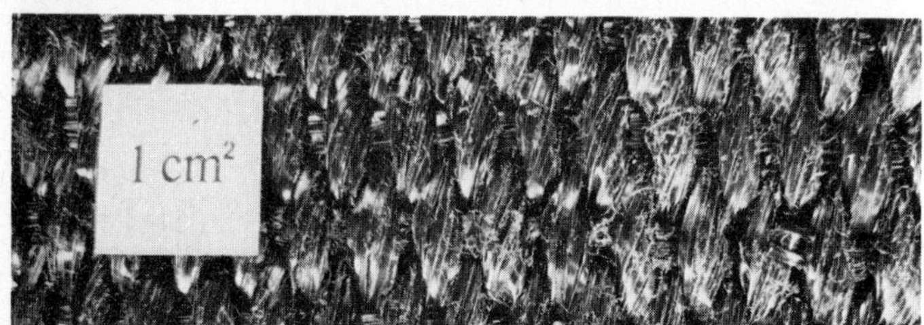

Robusta 1200

1200 g.s.m.
100% Polypropylene

Black Woven
As for 500

Published data

			175	265	500	600	750	830	1200
Treksterkte (tensile strength) for 10cm strip width	Ketting (warp)	kg	280	380	1100	1400	1800	2000	2800
	inslag (weft)	kg	170	480	300	300	300	300	600
Standard width	(m)		5	5	5	4-5	4-5	4-5	5
U.V. stabilised			√	√	√	√	√	√	√
Bestemming Zandkorreldiameter	(mm)		0.15	0.15	0.15	0.2	0.12	0.1	0.12

Figure 4.42 (Opposite)
Robusta products

Figure 4.43 (Above)
Robusta characteristics

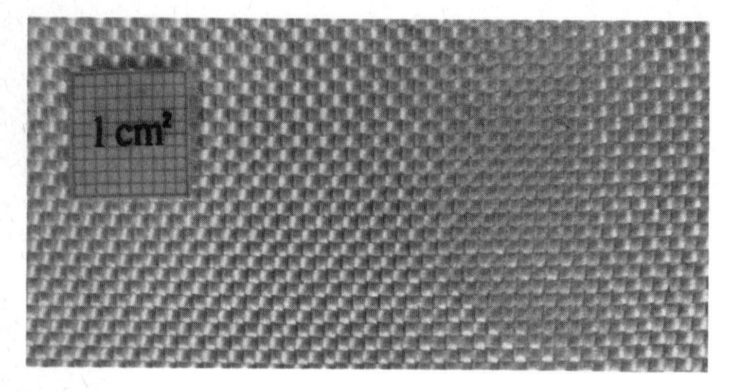

Stabilenka N99

Appears to be a woven colourless polyamide with multifilament warp and weft threads

Figure 4.44 Stabilenka product

Monofilter

240 g.s.m.
100% Polypropylene
Black
Woven monofilament 'wide' tape in the weft, with thin warp threads

Figure 4.45 Monofilter product

No details available
Visual inspection suggests 100-140 g.s.m.
with a thickness of 0.3mm

Figure 4.46 Stabilenka characteristics.

Monofilter

	Warp — 1,000 Denier F Fill — 930 Denier
Count	Warp — 30 ± 1 Fill — 20 ± 1
Weight	7oz/sq. yd. + (240 g.s.m.)
Thickness	.020 inches (0.5mm)
Equivalent opening size	70 sieve
Tensiles (width not specified!)	Warp 350lbs. minimum Fill 275lbs. minimum
Burst	545 PSI (gross)
Abrasion resistance (1000gms — 1000 revolutions)	Warp 110lbs. minimum Fill 151lbs. minimum
Elongation	Warp 27% Fill 29% (weft)
Weather-o-meter (1000 hrs. — continuous exposure)	Warp 350lbs. Fill 220lbs.
Permeability	125 CFM (conditions not specified)

Figure 4.47 Monofilter characteristics

Mypex 2400

113 g.s.m.
100% Polypropylene
Woven tape in both warp and weft
Grey

Non-woven side

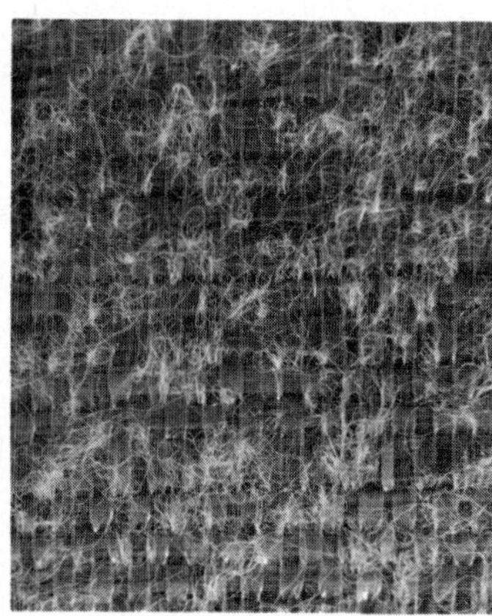

Woven side

Mypex FLW

195 g.s.m.
35% Polyamide
65% Polypropylene
A thin needle punched non-woven filter bonded onto a tape woven membrane (possibly onto 2400)
Light grey

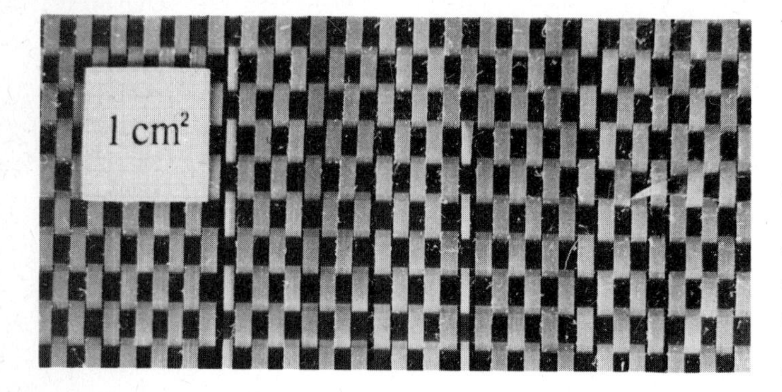

Polybac C 133

87 g.s.m.
100% Polypropylene
Woven tape in both warp and weft
Grey

Figure 4.48 Amoco products

	C133	2400	FLW
Fabric Weight	87g/m^2	113g/m^2	195g/m^2
Thickness	0.3mm	0.4mm	0.9mm
Specific Gravity	0.90	0.90	0.98
Tensile Strength (Warp)	700N/50mm	900N/50mm	600N/50mm
Tensile Strength (Weft)	575N/50mm	600N/50mm	480N/50mm
Elongation at Break (Warp)	15%	20%	9%
Elongation at Break (Weft)	11%	15%	6%
Effect of Light		One Month in Direct Sunlight	
Effect of Acid	Nil	Nil	Nil
Effect of Alkali	Nil	Nil	Nil
Roll Width		Widths up to 5.2 metres	
Roll Length c 4.2 mtrs	250 mtrs	175 mtrs	100 mtrs
Roll Weight x 4.2 mtrs.	91.5 Kg	83 Kg	82 Kg
Effect of Bacteria	Nil	Nil	Nil
Effect of Temperature	-10°C to 140°C	-10°C to 140°C	-10°C to 140°C
Fabric Composition		100% Polypropylene	65% P.P. 35% Nylon
Vertical Permeability (ltrs/sec)	180	187	67
Wing Tear (Warp)	280N	269N	143N
Resistance (Weft)	178N	255N	171N

Can be provided black for U.V. stabilisation

Figure 4.49 Amoco characteristics

Filter-X

393 g.s.m.

85% polyvinylidene chloride

15% U.V. stabilisers

Woven tape filaments calendered after weaving.

Green colour

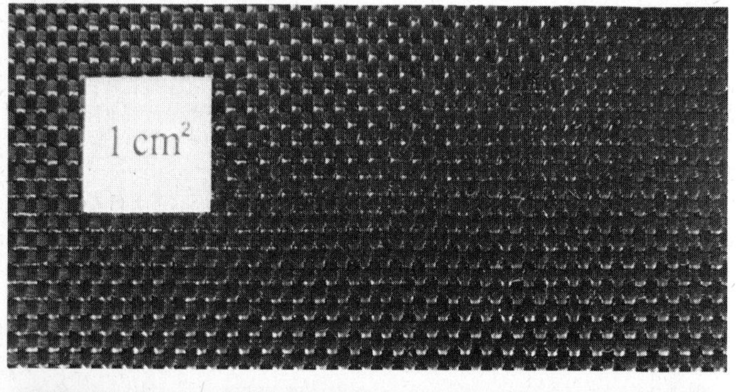

Poly-Filter X

243 g.s.m.

85% polypropylene

15% stabilisers for U.V.

Woven tape filaments calendered after weaving

Black colour

Poly-Filter GB

224 g.s.m. black

85% polypropylene

15% stabilisers & inhibitors for U.V.

Round cross-section threads in warp and weft, woven with warp stepping over two weft threads at a time, and therefore is very permeable

Figure 4.50 Carthage Mills products

		filter X	Poly-Filter X	Poly-Filter GB
Break Load. Grab. 1" jaws. ASTM D1682	Warp	200lbs (890 N)	380lbs (1690 N)	200lbs (890 N)
	Weft	110 lbs (489 N)	220 lbs (978 N)	200 lbs (890 N)
Burst strength ASTM D751	p.s.i.	260	540	500
	kN/m^2	1792	3723	3447
% open area	(%)	4-5	5-6	21-26
E.O.S. (equivalent opening size)	(mm)	0.149	0.210	0.420
Specific gravity	g/cm^3	1.70	0.95	0.95
Permeability	cm/sec $\times 10^{-2}$	4.8	3.8	Not published, but note the % open area

Figure 4.51 Carthage Mills characteristics

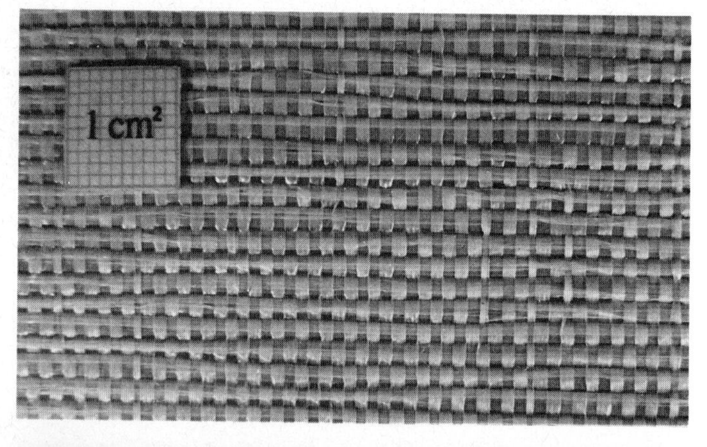

Lotrak 16/15

120 g.s.m.

100% extruded tape polypropylene

Woven tapes in warp and weft

Simple structure

Light brown

Lotrak 56/46

255 g.s.m.

100% polypropylene tape

Woven tapes in warp and weft. Simple structure.

Dark grey

Lotrak Needleweave

This is a light-weight needle punched felt bonded onto a woven base like Lotrak 16/15

The original Lotrak Needleweave fabric, with its finer pores, is still available for use on sites with a high silt content.

Figure 4.52 Lotrak product

Figure 4.53 (Opposite) Lotrak characteristics

Lotrak 16/15 Technical Data

Specific Gravity	0.91
Thickness	0.3mm
Tensile Strength — Warp	16 kN/m
— Weft	15 kN/m
Extension to Break — Warp	21%
— Weft	15%
Pore Size	90% finer than 300
Effect of naturally occurring acid	Nil
Effect of naturally occurring alkali	Nil
Effect of naturally occuring bacteria	Nil
Effect of ultra violet light	Unaffected by exosure to strong sunlight for periods up to one month. Continued exposure would reduce tensile strength.
Permability to water under head of 5cm	approx. 12 litres/sq.m./sec.
Standard roll width	4.50m (5.15m is possible)
Standard roll length	100m
Standard roll weight (including packing and centre)	4.50m roll — 65Kg. approx.

Lotrak 16/15 — Permeability Test

Test Condition — Water Temperature 15° ~ 1°C
Fabric fully submerged 65mm diameter test area
The curve is a mean of tests on at least three samples

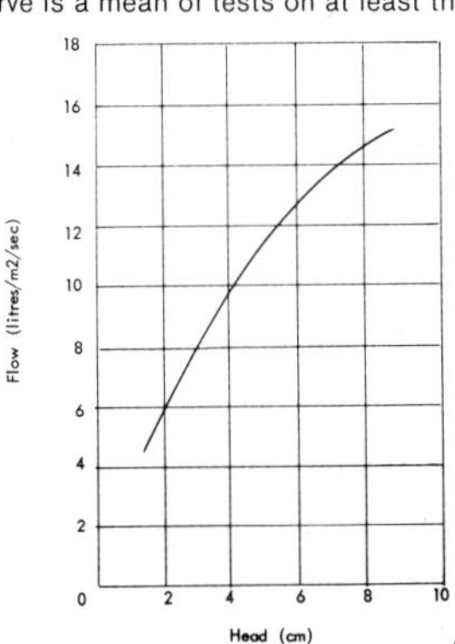

Lotrak 16/15 — Pore Size Tests

All points are average for 4 samples

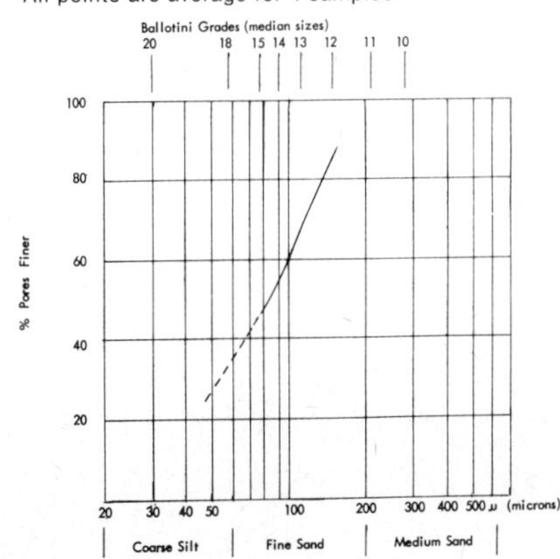

Lotrak 56/46 Technical Data

Specific Gravity	0.91
Thickness	0.6mm
Tensile Strength — Warp	56 kN/m
— Weft	46 kN/m
Extension to Break — Warp	10-15%
— Weft	10-15%
Pore Size	90% finer than 430 microns
Effect of naturally occurring acid	Nil
Effect of naturally occurring Alkali	Nil
Effect of naturally occurring bacteria	Nil
Effect of ultra violet light	Unaffected by exposure to strong sunlight for periods up to one month. Continued exposure would reduce tensile strength.
Permeability to water under head of 5cm	Approx. 12.0 litres/sq.m/sec.
Standard roll width	4.50m
Standard roll length	100m
Standard roll weight (including packing and centre)	125 kgs approx.

Lotrak 56/46 — Pore Size Tests

All points are average for 4 samples

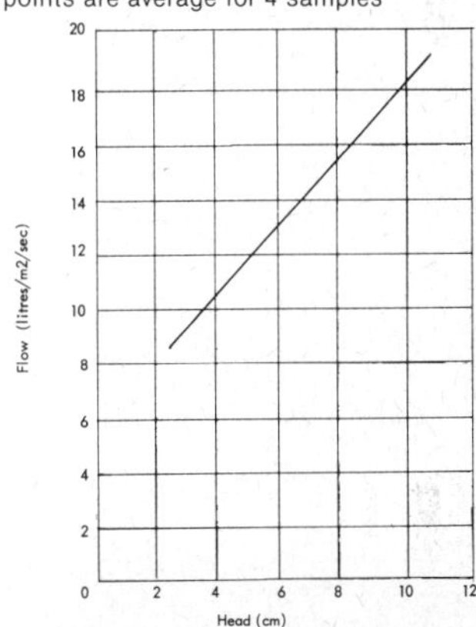

Lotrak 56/46 — Permeability Test

Test Conditions — Water Temperature 14°C
Fabric fully submerged 76mm diameter test area
The curve is a mean of tests on at least three samples

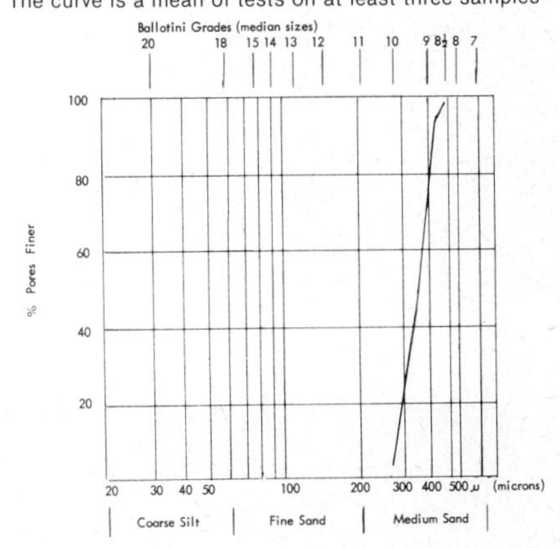

Polyweave F (40007)

190 g.s.m.

Material not specified but looks like U.V. stabilised polypropylene

Woven with tape in the warp and monofilament in the weft, resulting in high permeability and open area

Black colour

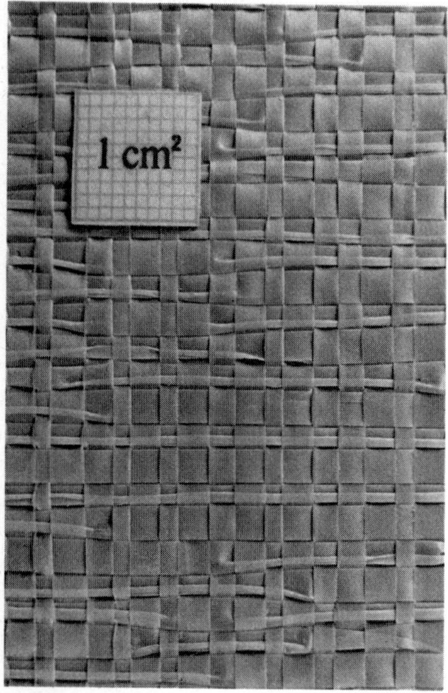

Polyweave R (41613)

103 g.s.m.

Material not specified, but looks like polypropylene

Woven with tapes both in the weft and warp

Light brown colour

Figure 4.54 Sarlon products

Polyweave F (40007)

Construction:	110 × 70 (per 10cm)
Roll Width:	1.83m
Roll Length:	Bulk rolls
Tensile strength kg/50mm:	Warp 238 Weft 144
Trapezoid tear strength (kgs):	Warp 45 Weft 35
Mullens Burst strength:	34kg/cm^2
Extension at break%:	Warp 20 + Weft 20 +
Loss of strength when wet%:	Nil
Effect of salt water:	None
Moisture regain %:	Nil
Moisture absorbency %:	Nil
Particle retention: Microns (actual hole size):	320
Permeability: m^2/sec/ 10cm column head:	170 litres
Effect of heat: Softening point	130°C
Thickness of woven sheet: Microns	670
Sieve size 90% retention:	50
Abrasion resistance: Taber CS17 wheel 1000gm load cycles to failure	5000 +

Polyweave R (41613)

Construction:	47 × 39 (per 10cm)
Roll Width:	3.86m
Roll Length:	250m
Shipping weight:	106kg
Specific gravity:	0.90
Tensile strength: per 50mm strip	Warp 90kg Weft 63kg
Mullens Burst strength:	13kg/cm^2
Extension at break:	20% +
Loss of strength when wet:	Nil
Moisture absorbency %:	Nil
Moisture regain%:	Nil
Permeability: m^2/sec/ 10cm column head	10-12 litres
Effect of heat:	Softens at 140°C
Thickness of woven sheet:	250 Microns
Effect of acids and alkalis:	Very resistant
Effect of organic solvents:	Resistant below 85°C
Elastic recovery:	Up to 98%

Figure 4.55 Sarlon characteristics

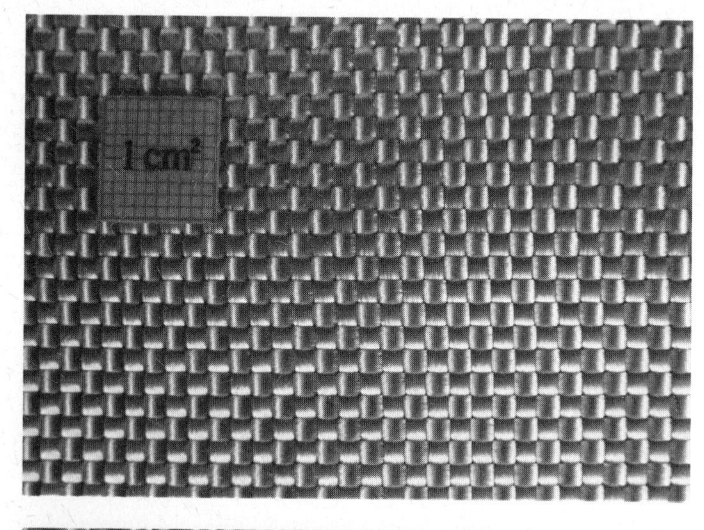

Terram W/20-20

Wt not specified
100% polyester
Woven multi-filament warp and weft threads
White colour

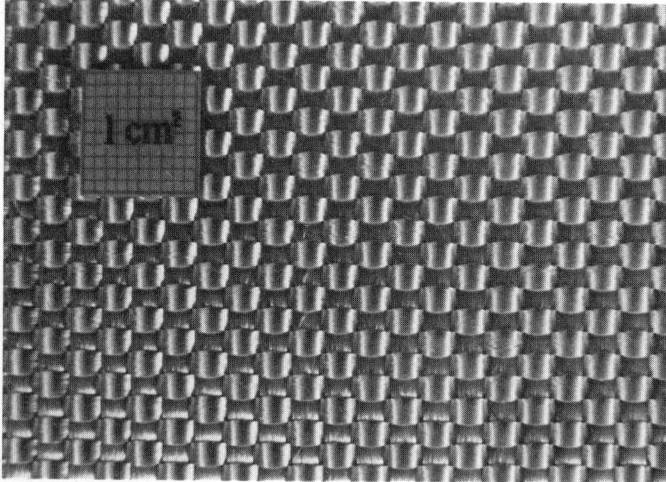

Terram W/30-30

Wt not specified
100% polyester
Woven multi-filament warp and weft threads
White colour

Figure 4.56 ICI products

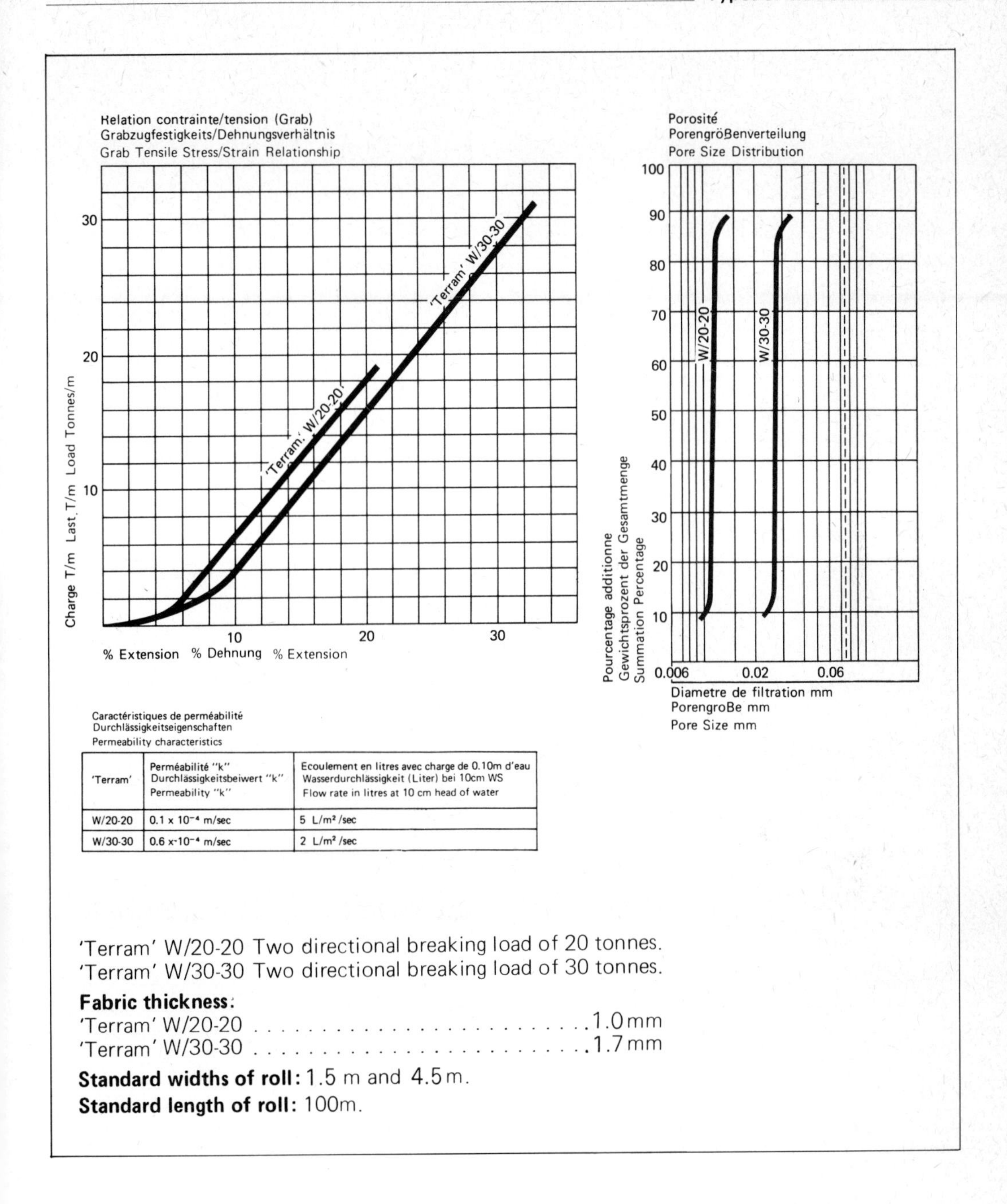

Caractéristiques de perméabilité
Durchlässigkeitseigenschaften
Permeability characteristics

'Terram'	Perméabilité "k" Durchlässigkeitsbeiwert "k" Permeability "k"	Ecoulement en litres avec charge de 0.10m d'eau Wasserdurchlässigkeit (Liter) bei 10cm WS Flow rate in litres at 10 cm head of water
W/20-20	0.1 x 10^{-4} m/sec	5 L/m² /sec
W/30-30	0.6 x 10^{-4} m/sec	2 L/m² /sec

'Terram' W/20-20 Two directional breaking load of 20 tonnes.
'Terram' W/30-30 Two directional breaking load of 30 tonnes.

Fabric thickness:
'Terram' W/20-20 . . . . .1.0 mm
'Terram' W/30-30 . . . . .1.7 mm

Standard widths of roll: 1.5 m and 4.5 m.
Standard length of roll: 100m.

Figure 4.57 ICI characteristics

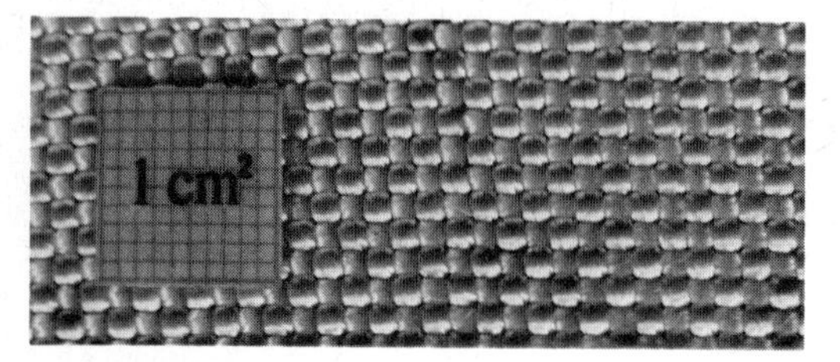

NY 149 Polyamide — White

NY 169 Polyamide — White

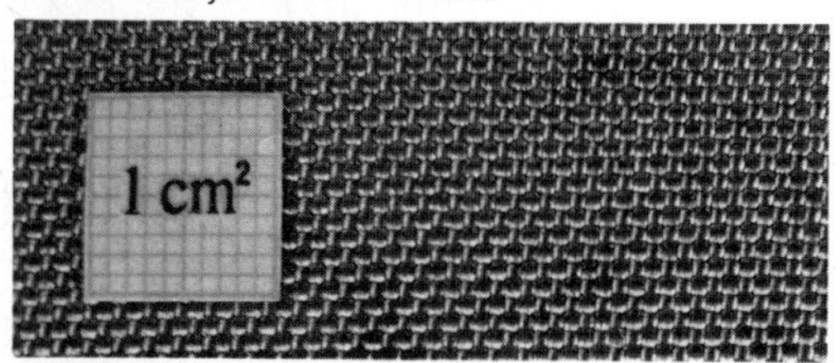

NY 169 Polyester — Brown

NY 171 Polyamide — White

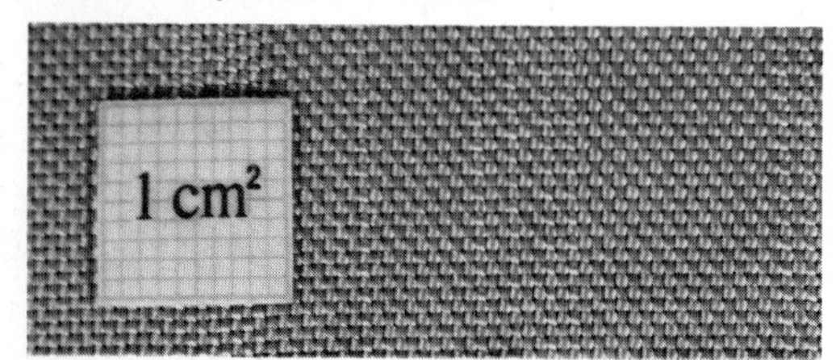

NY 171 Polyester — Brown

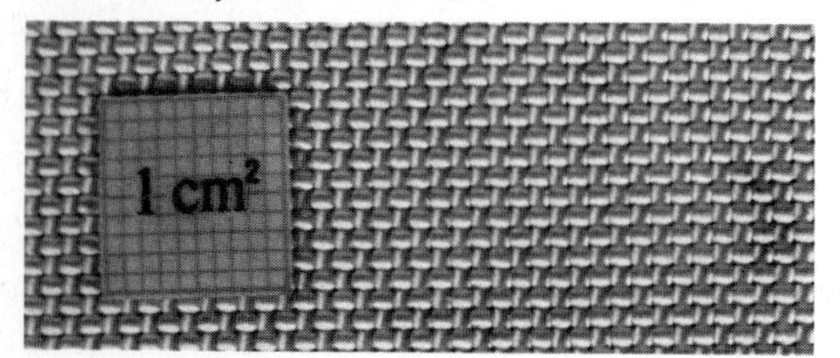

NY 180 Polyamide — White

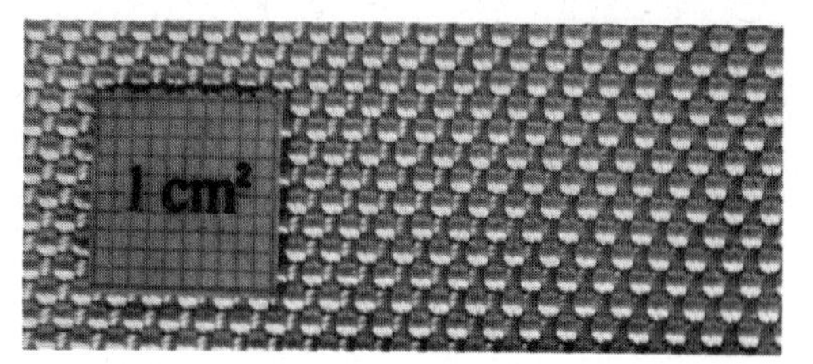

NY 188 Polyamide — White

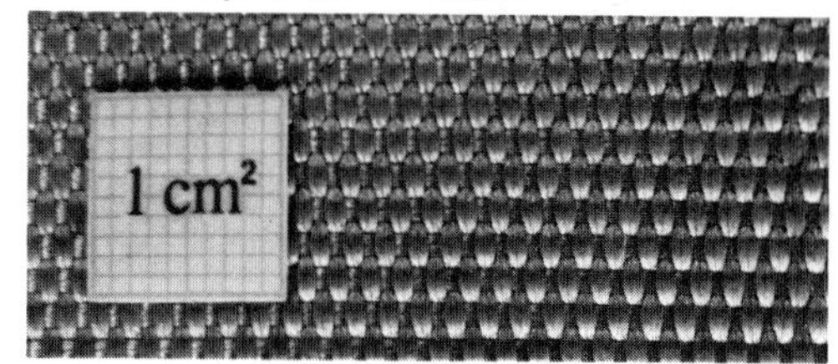

NY 378 Polyester — Brown

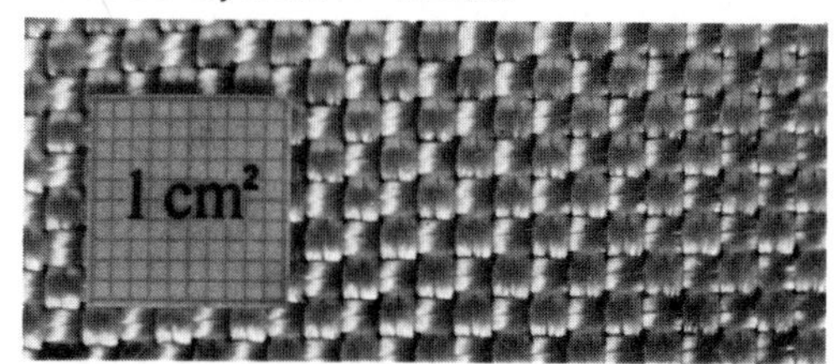

NY 409 Polyamide — White

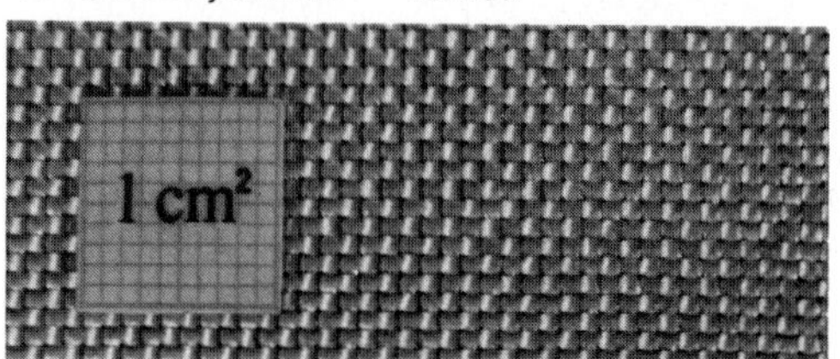

NY 721 Polyamide — White

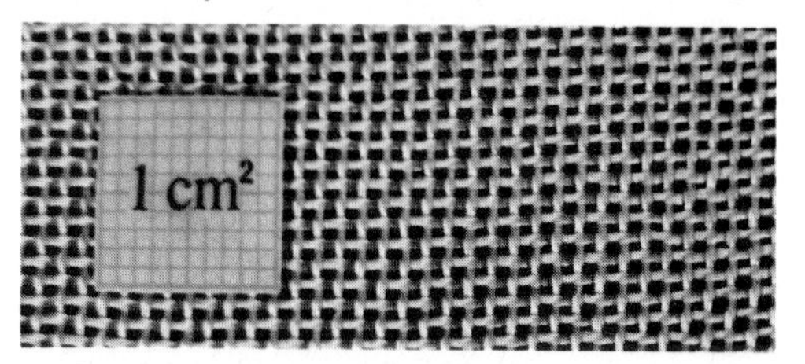

NY 4298 Polyamide — White

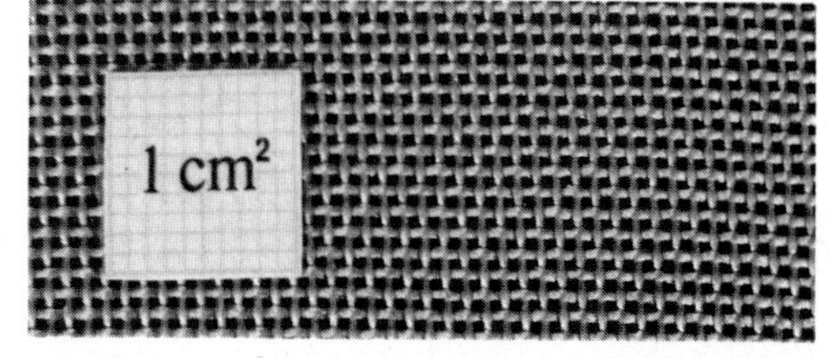

NY 4298 Polyester — Brown

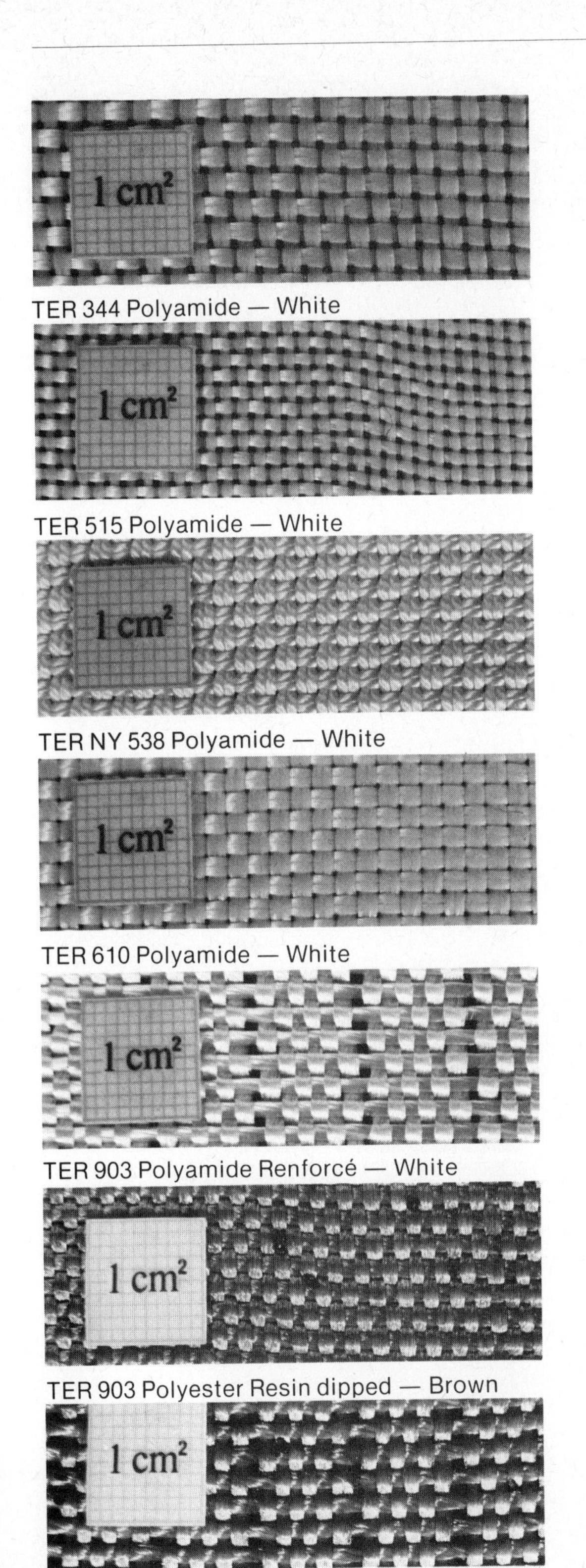

No technical details provided by Coisne & Lambert except that their full range of woven fabrics vary between 150-3000 g.s.m. and 1 ton - 150 ton/m strength, with the option for RFL resin dip

Figure 4.58 (Opposite page) Coisne and Lambert products

Figure 4.59 (This page) Coisne and Lambert products

Krafter 7540

100% Polyester?

Warp jumps two wefts, but still tight woven

White colour

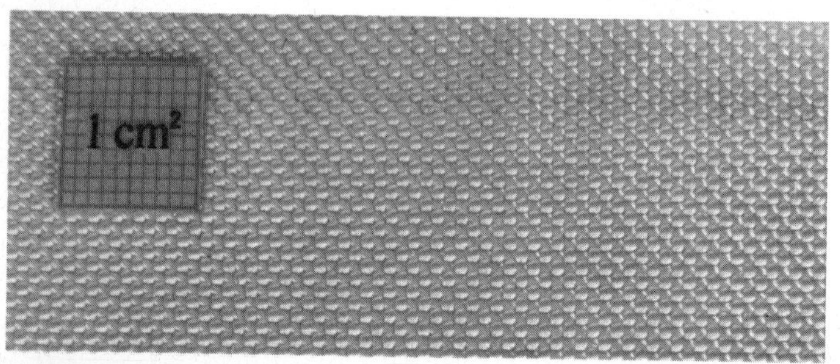

Krafter KF

100% Polyester?

Simple weave

White colour

Krafter 7510

100% Polyester?

Simple weave

White colour

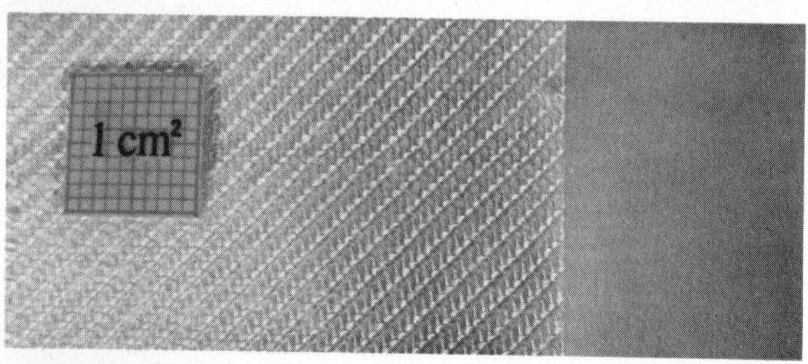

Kuralon NF

100% "Vinylon" (polyvinylidene?)

Weft jumps two warps

White colour

Kuralon H

100% "Vinylon"

White colour

(Taiyo Kogyo Co. also have literature mentioning a product called "New-NF" which is a *Polyester* woven at 180 g.s.m. No sample was available, but its properties are given below). *Note* that Taiyo use a 3cm wide strip for testing.

item	type	Kuralon H	Kuralon NF	Kuralon New NF	Krafter #7510	Krafter KF	Krafter #7540	Testing Method
Thickness	(m/m)	0.44	0.53	0.31	0.27	0.46	0.51	JIS L 1079—1966
Weight	(g/m^2)	198	230	180	167	280	350	JIS L 1079—1966
Tensile strength (kg/3cm)	dry	warp weft 140 × 115	warp weft 148 × 144	warp weft 160 × 167	warp weft 150 × 150	warp weft 230 × 230	warp weft 295 × 295	JIS L 1079—1966
	wet				150 × 150	230 × 230	295 × 295	
Elongation (%)	dry	20 × 17	19 × 15	16 × 16	14 × 14	21 × 13	16 × 15	JIS L 1079—1966
	wet				14 × 13	21 × 13	16 × 15	
Tongue tear strength (kg)	dry	11 × 9	27 × 24	34 × 31	37 × 32	49 × 42	74 × 68	JIS L 1079—1966
	wet				45 × 41	58 × 53	79 × 76	
Water permeability (cm/sec)		1.90×10^{-2}	2.30×10^{-2}	1.40×10^{-2}	1.40×10^{-2}	1.10×10^{-2}	1.46×10^{-2}	JIS A 1218

Figure 4.60 (Opposite) Taiyo Kogyo products

Figure 4.61 (Above) Taiyo Kogyo characteristics

Nicolon

Wt not specified
100% polypropylene?
Woven
Black filament warps
Green & black tape wefts
Green colour

Appears to be woven polypropylene tape in the weft and round filament in the warp, giving good permeability. It is not a tight weave, so we assume that stability has been imparted by calendering only. No other technical details available.

Figure 4.62 Nicolon product and characteristics

Tone Sheet F

Wt unspecified
100% polypropylene?
Woven split-tape
Tight weave
Appears to be resin dipped
Grey colour

There is also apparently a Tone Sheet S, which is a high-permeability product. Tone Sheet F is Tonen's high-strength product.

		Tone Sheet S	Tone Sheet F	
Tensile strength	kg/5cm	warp weft 98 × 90	warp weft 170 × 166	JISL 1068
Extension at break	%	8 × 9	16 × 15	JISL 1068
Stitch strength	Kg/5cm	90	155	JISL 1068
Permeability to water	cm/sec	1.2×10^{-1} JISA 1218	5.9×10^{-3} JISA 1218	

Figure 4.63 Tonen product and characteristics

Amoco 1211

90 g.s.m.

100% polypropylene (black)

Woven tape in both warp & weft
Simple woven structure

Amoco 6060

130 g.s.m.

100% polypropylene (black)

Woven tape in both warp & weft
Warp jumps two wefts.
Not simple construction

Amoco 6062

185 g.s.m.

100% polypropylene (black)

Woven tape in both warp & weft
Warp jumps two wefts

Amoco 6061

220 g.s.m.

100% polypropylene (black)

Woven tape warp & weft
Warp jumps two wefts

Figure 4.64 Gouderak BV products

	Fabric	Fabric			
	Fabric width (m)	Fabric length (m)	Tensile strength kN/5cm strip	Elongation at break %	O_{90} mesh size μ
			warp weft	warp weft	
1266 (not photographed)	2.10	1000	0.4 × 0.35	20 × 10	—
1211	4.20	1000	1.03 × 0.76	18 × 14	150
6060	5.20	400	1.21 × 1.18	17 × 15	220
6062	5.20	400-500	1.80 × 1.80	20 × 15	180
6061 (6261L)	5.20	250-400	2.90 × 0.90	24 × 12	330

Figure 4.65 Gouderak BV characteristics

Amoco 6063

240 g.s.m.

100% polypropylene (black)

Woven tape in warp & weft
Warp jumps two wefts

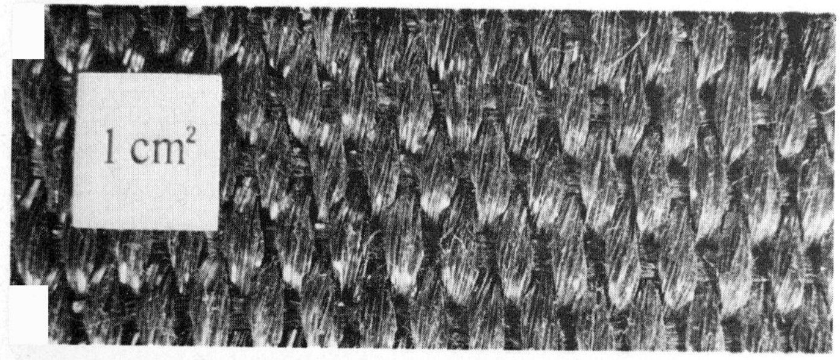

Amoco 6064

325 g.s.m.

100% polypropylene (black)

Woven tape in the warp and split tape yarn in the weft
Warp jumps two wefts

Amoco 6065

375 g.s.m.

100% polypropylene (black)

Woven tape yarn in the warp and tape in the weft
Therefore weft jumps two warps

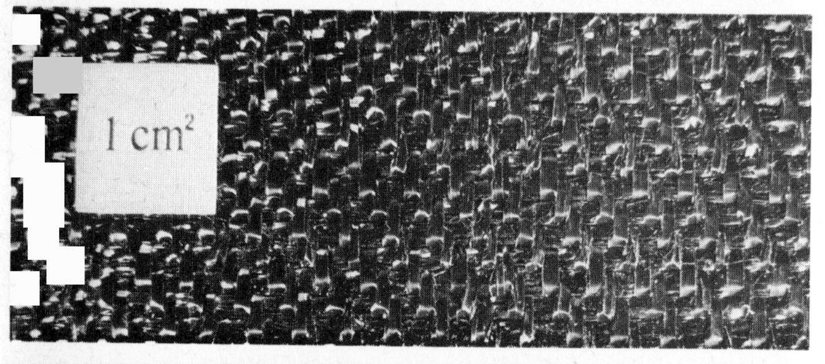

Amoco 6066

520 g.s.m.

100% polypropylene (black)

Woven tape yarn in both warp and weft
Weft jumps two warps

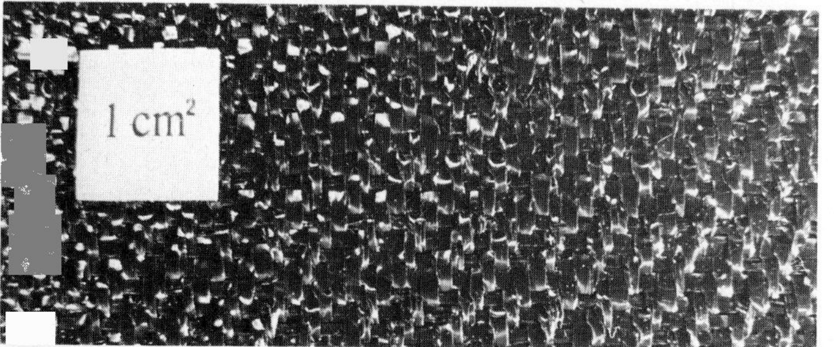

Amoco 4121 TH

750 g.s.m.

100% polypropylene (black)

Woven thinner split tape yarn in the weft with thick twisted split tape yarn in the warp
Simple woven construction

Published data

	Fabric width (m)	Fabric length (m)	Tensile strength kN/5cm strip	Elongation at break %	0_{90} mesh size μ
			warp weft	warp weft	
6063 (6263L)	5.20	250	2.49 × 1.23	28 × 13	185
6064 (6264L)	5.20	200-300	3.16 × 3.07	29 × 15	218
6065 (6265L)	5.05	200	4.15 × 1.84	18 × 22	670
6066 (6266L)	5.05	200-250	4.29 × 4.42	17 × 19	390
4121TH (4121thl)	5.00	150-1000	9.53 × 1.75	23 × 14	450

The Gouderak fabrics are also sold with stitched loops for various handling purposes. The number 2 is inserted into the reference to identify these. For example 6064 becomes 6264L with loops.

Figure 4.66 (Opposite)
Gouderak BV products

Figure 4.67 (Above)
Gouderak BV characteristics

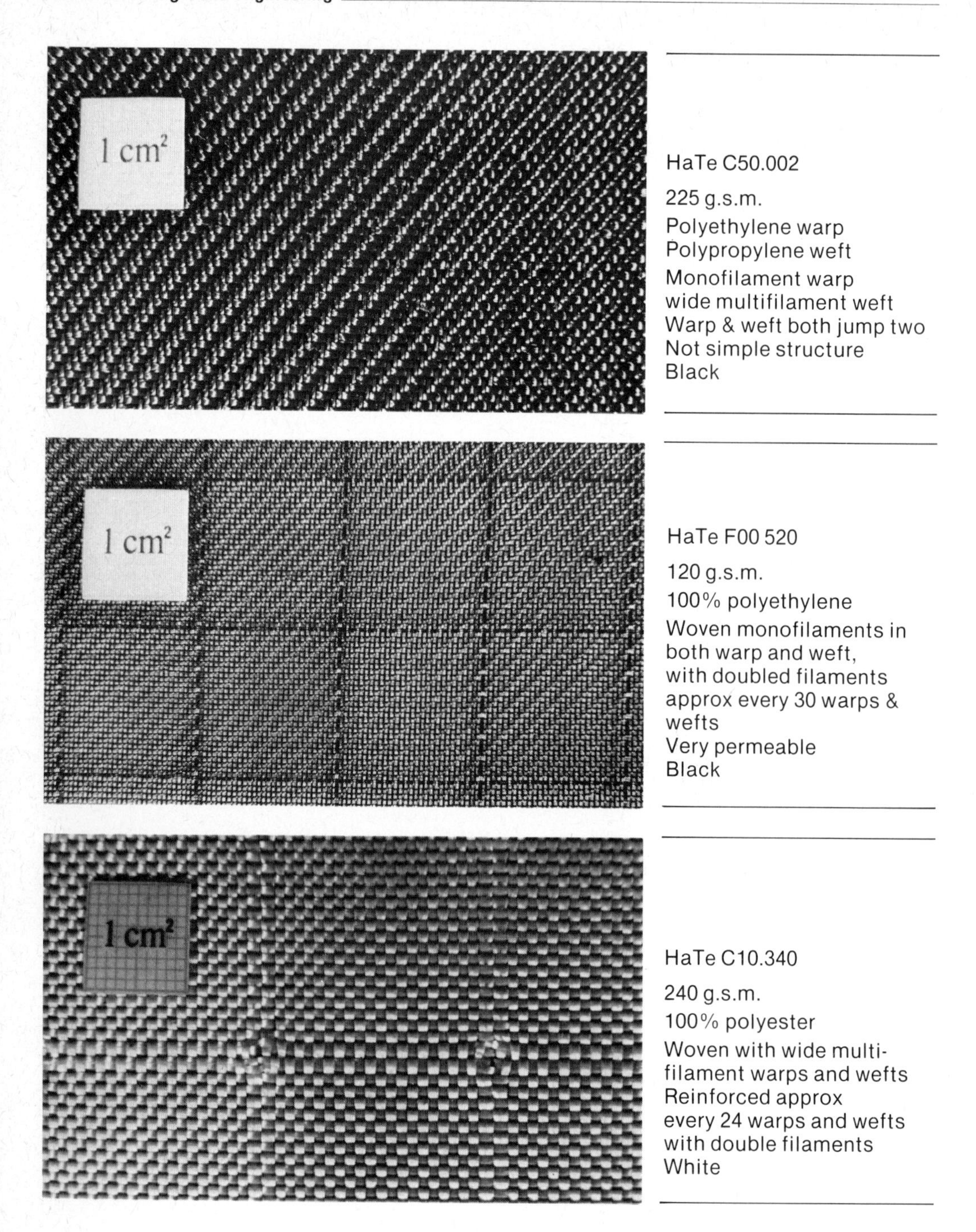

Figure 4.68 Huesker synthetic products

HaTe C50.002

Mesh size:		0.06-0.15mm
Permeability to water (10cm water column):		100-250l/m^2
Tensile strength	Warp:	±230daN
	Weft:	±260daN/5cm
Elongation to break	Warp:	±25%
	Weft:	±25%

HaTe F00.520

Mesh size:		±0.30mm
Permeability to water (10cm water column):		±440l/m^2s
Tensile strength	Warp:	±110daN/5cm
	Weft:	±110daN/5cm
Elongation to break	Warp:	±35%
	Weft:	±30%

HaTe C10.340

Mesh size:		±0.15mm
Permeability to water (10cm water column):		±50l/m^2s
Tensile strength	Warp:	±350daN/5cm
	Weft:	±320daN/5cm
Elongation to break	Warp:	±20%
	Weft:	±20%

Figure 4.69 Huesker synthetic characteristics

Open meshes illustrated

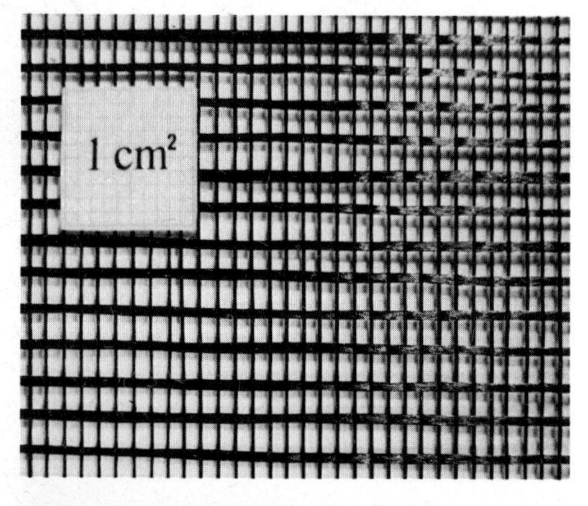

Sarglas (2130)

Open-weave glass filaments in warp and glass tape in weft
Stabilised by bitumen dip
Black colour

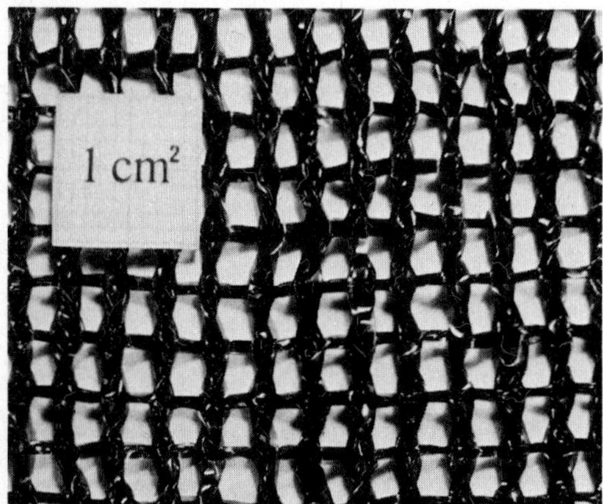

Polymesh Leno (40802)

45 g.s.m.?
Roll width?
Colour Black
Leno construction of twisted polypropylene(?) tapes

Polymesh Leno (40804)

45 g.s.m.?
Roll width 1.85m × 50m long
Colour light brown
Leno construction of twisted polypropylene(?) tapes

No product data to hand currently

Figure 4.70 Sarlon products

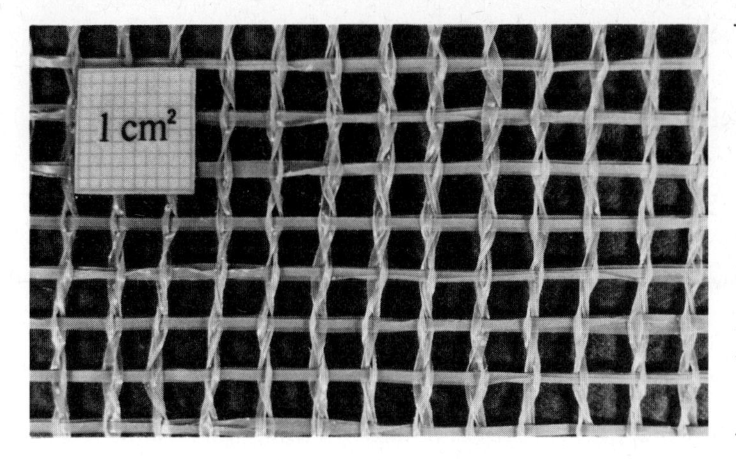

Gouderak 1266

Wt 40 g.s.m.?

Leno construction of twisted polypropylene(?) tapes
Light brown colour

Figure 4.71 Gouderak product

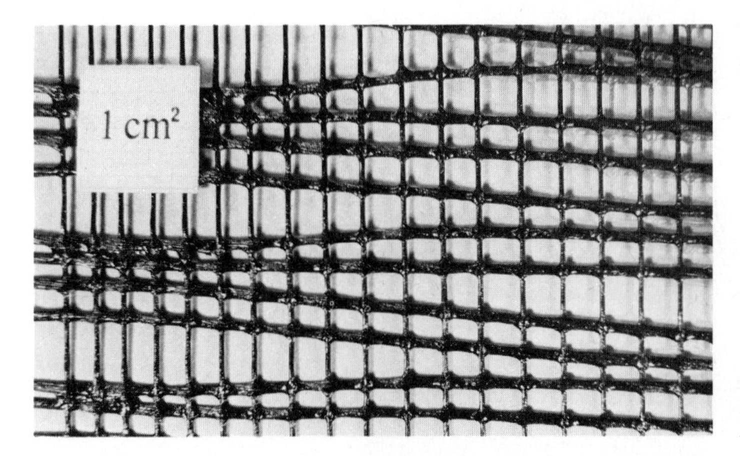

Stabilenka N33

Wt?

Material?

Open mesh weave
With filaments in the warp and multi-filaments in the weft
Dip-stabilised
Black colour

Figure 4.72 Enka Glanzstoff product

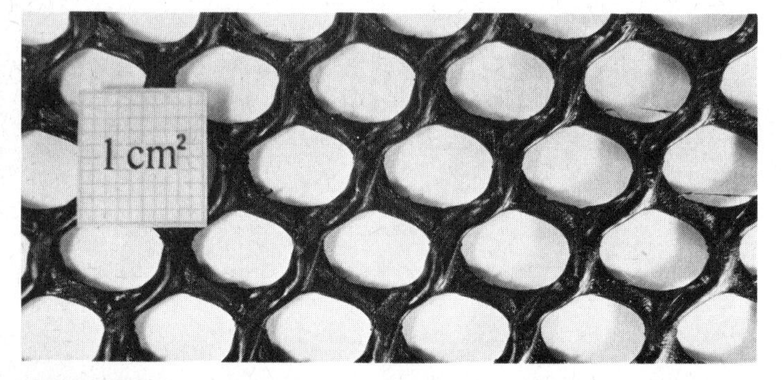

Netlon CE 111

450 g.s.m.

100% low density polyethylene

Black

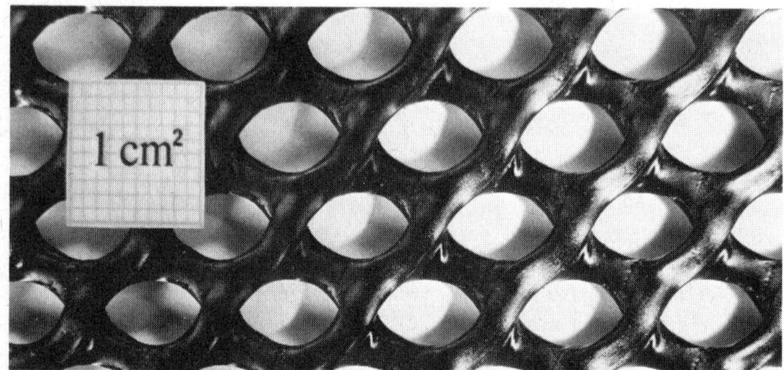

Netlon CE 121

730 g.s.m.

100% high density polyethylene

Black

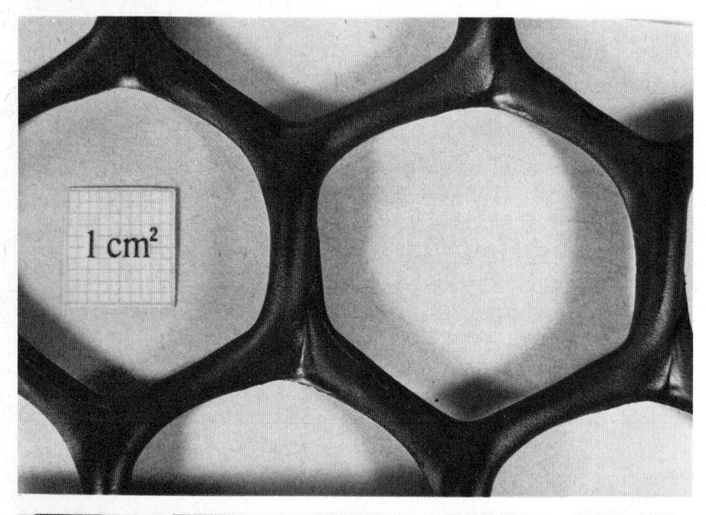

Netlon CE 131

660 g.s.m.

100% high density polyethylene

Black

Netlon 161

Black

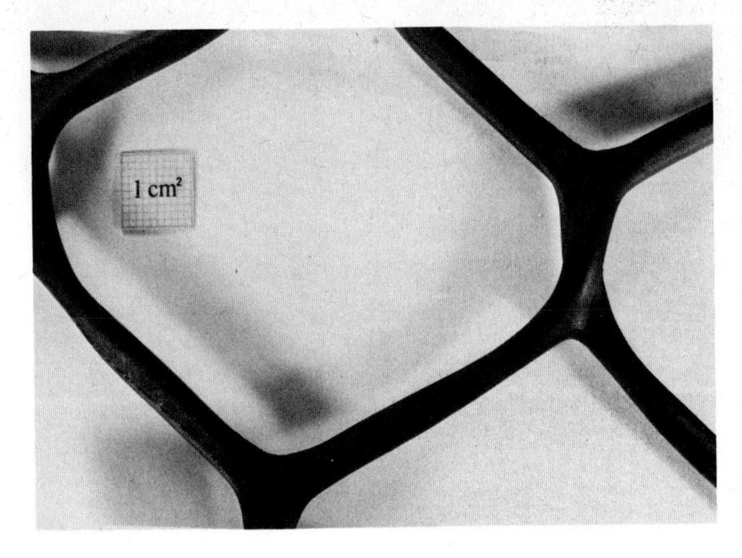

Netlon CE 152
(CE 151)

550 g.s.m.
100% high density polyethylene
Black

Figure 4.73 Netlon product

Specification data		CE111	CE121	CE131	CE151	CE152
Form		Rolled sheet	Rolled sheet	Rolled sheet	Tubular	Rolled sheet
Dimensions	Width	2m	2m	2m	1m lay-flat	2m
	Length	30m	15m	15m	5m	15m
Mesh aperture (width across)		6mm	6mm	27mm	74mm	74mm
Mesh thickness		2.9mm	3.3mm	5.2mm	5.9mm	5.9mm
Yield load 1m width	Machine direction	330kg/m	640kg/m	790kg/m	490kg/m	490kg/m
	Transverse direction	300kg/m	390kg/m	600kg/m	740kg/m	740kg/m
Break load at 1m width	Machine direction	500kg/m	764kg/m	790kg/m	530kg/m	530kg/m
	Transverse direction	360kg/m	445kg/m	600kg/m	760kg/m	760kg/m

Figure 4.74 Netlon characteristics

Trical-Net 10-527/94-80
Colour Blue

Trical-Net 20-200/16-20
Colour Orange

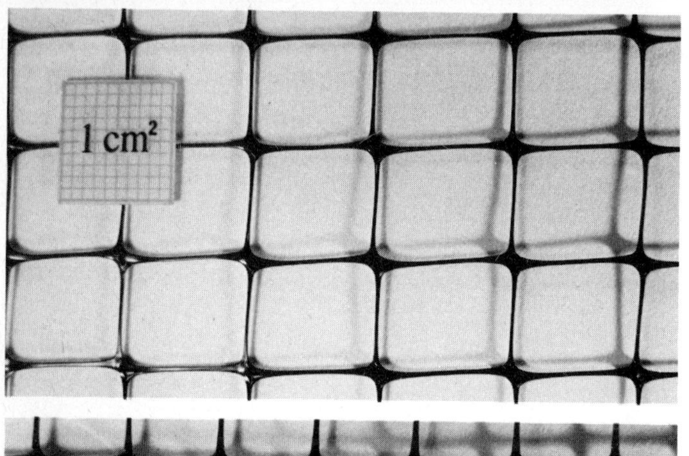

Trical-Net 30-902/09-03
Colour Black

Trical-Net 30-902/08-04
Colour Black

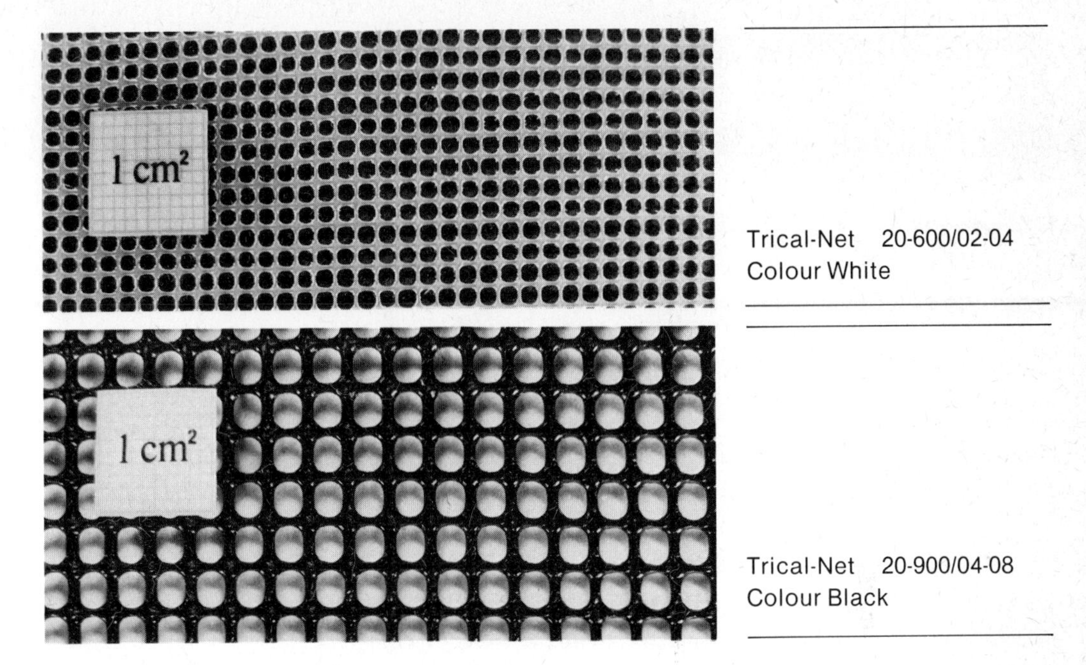

Figure 4.75 (Opposite and above) Takiron products

	Strip Strength kg/m (500mm/min pull)		Elongation at break %		% Open area
	warp	weft	warp	weft	
10-527/94-80	300 ×	200	350 ×	100	64
20-200/16-20	460 ×	560	100 ×	60	72
30-902/09-03	230 ×	240	10 ×	10	87
30-902/08-04	260 ×	380	10 ×	10	77
20-600/02-04	210 ×	220	500 ×	560	49
20-900/04-08	560 ×	580	120 ×	80	58

Figure 4.76 (Above) Takiron characteristics

Trical-Net 20-906/94-81
Colour Black

Trical-Net 10-320/10-15
Colour Yellow

Trical-Net 40-708/94-82
Colour White

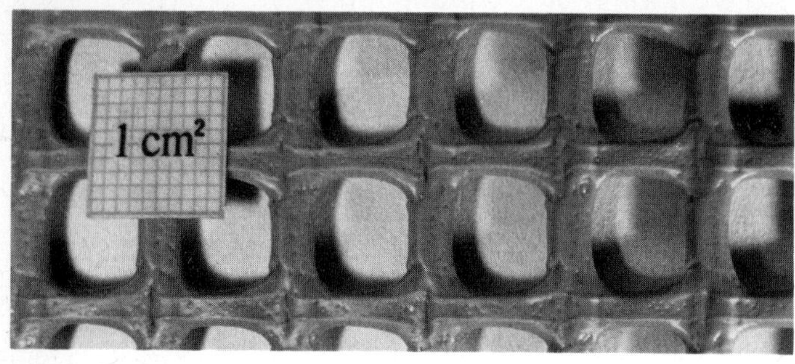

Trical-Net 20-920/10-25
Colour Grey

Figure 4.77 (Above and opposite top) Takiron products

Figure 4.78 (Opposite bottom) Takiron characteristics

Trical-Net 20-900/25-37
Colour Black

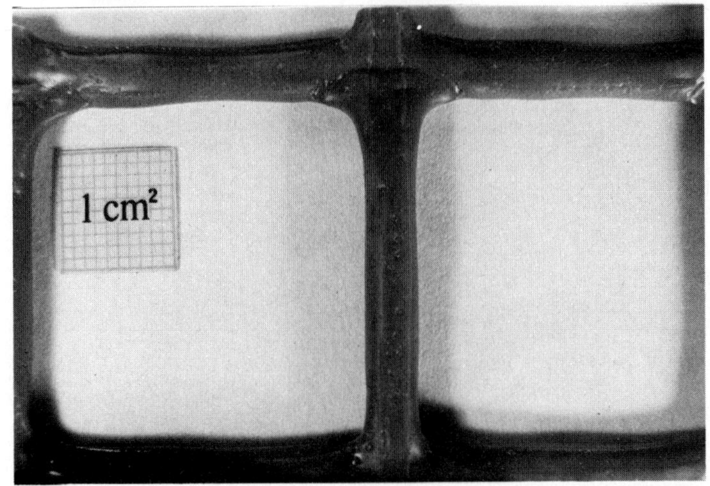

Trical-Net 20-400/34-48
Colour Green

	Strip Strength kg/m (500mm/min pull)		Elongation at break %		% Open area
	warp	weft	warp	weft	
20-906/94-81	810 ×	510	60 ×	50	60
10-320/10-15	330 ×	310	400 ×	300	64
40-708/94-82	280 ×	470	400 ×	400	39
20-920/10-25	1200 ×	1100	80 ×	50	51
20-900/25-37	1000 ×	900	80 ×	50	68
20-400/34-48	800 ×	780	40 ×	40	71

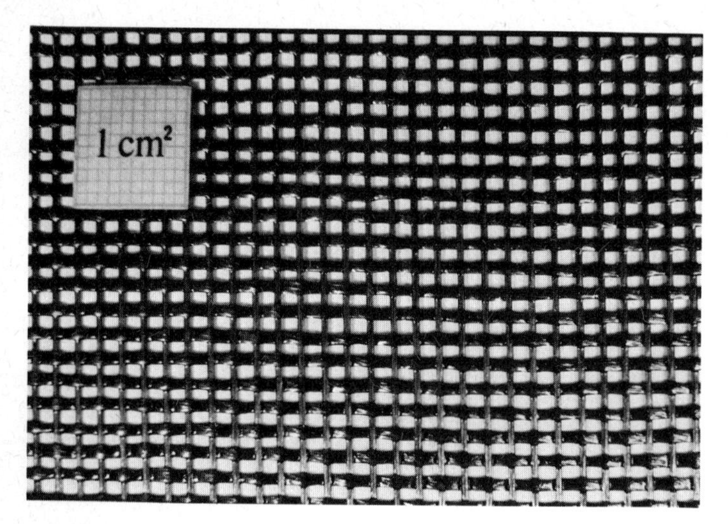

HaTe 50.145
225 g.s.m.
100% Polyester

Woven simple structure with narrow multifilament warp and wide multifilament weft

PVC dipped coating black

Figure 4.79 Huesker Synthetic product

HaTe Type 50.145

Mesh size:	1.00-1.20mm
Permeability to water (10cm water column):	±400l/m^{2}s
Tensile strength	
Warp:	±190daN/5cm
Weft:	±190daN/5cm
Elongation to break	
Warp:	±15%
Weft:	±15%

Figure 4.80 Huesker Synthetic characteristics

Miscellaneous products illustrated

Including soil reinforcing webs, soil filters, impermeable membranes, etc.

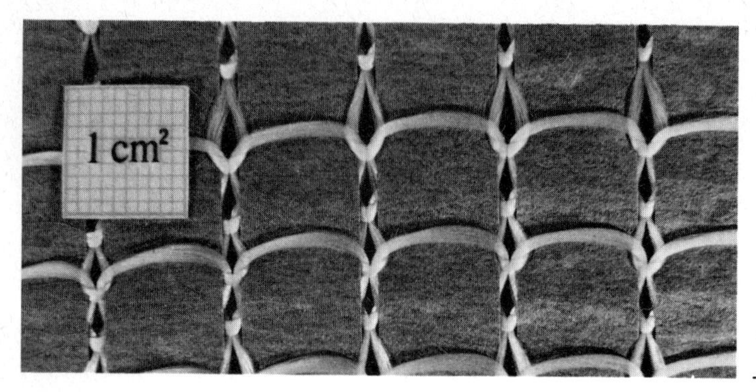

Hold/Gro

Paper and 100% polypropylene open-mesh. The polypropylene is used to hold the paper strips together. Neither is U.V. stabilised

Green/grey paper
White mesh

Hold/Gro membrane is used for covering newly-seeded soil slopes to protect against rain erosion. The product can be made to degrade over varying timescales, thus suiting it for different environments.

Hold/Gro Erosion Control Fabric gives control over erosion on graded slopes through enhancement of vegetation.

Hold/Gro is a combination of synthatic yarn netting interwoven with strips of paper. Its two-stage action gives maximum protection and mulching for freshly graded, planted or seeded surfaces when they are first needed. Then, as vegetation takes hold, the netting supports the new root structure.

The paper in newly installed Hold/Gro breaks the force of rainfall, while the slits between strips let the moisture seep into the soil. The same paper mulches the soil, holding soil and seed in place and guarding available soil moisture. This protection extends the growing period.

The yarn — a flexible knit of polypropylene — gives strength to the fabric and lets it be used on uneven surfaces. As the paper degrades the netting stays in place until vegetation is established. Then the netting, too, disintegrates and is absorbed into the soil.

Each erosion control job has its own challenges. Hold/Gro handles a wide range of applications because of its varying combinations of paper and yarn. Some degrade swiftly when vegetation can be quickly established. Others are long-lasting for protection over a longer period.

Width	Length	Coverage	Approximate gross shipping weight
5ft.	360ft.	200 sq. yds.	35lbs.
10ft.	360ft.	400 sq. yds.	70lbs.

Yarn Polypropylene. Denier and stabilizers vary according to permanence needed.

Filler Various papers according to permanence needed.

Packaging Opaque polyethylene bags provide protection from sun and moisture for up to three months in outside storage.

Figure 4.81 Hold/Gro product

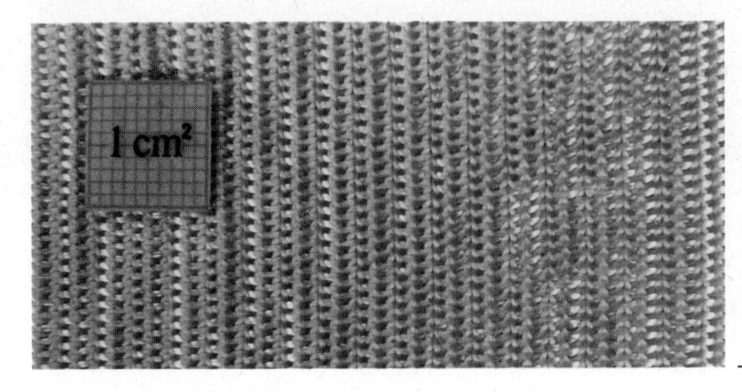

Terram RF/12

100% Polyester?
White

Knitted structure gives high strength in cross-machine direction

SOIL REINFORCEMENT- using 'Terram' RF/12

Fabric:
'Terram' RF/12 Uni-directional breaking load of 12 tonnes/m

Fabric thickness: 1mm.

Standard widths of roll: 1.5m and 3.0m.
Fabric can be supplied in widths of up to 6.0m.

Standard length: 200m.

Recommended working load: 4 tonnes per metre.

(each roll is accompanied by a certificate confirming stress/strain data and creep characteristics.)

When used as a soil-reinforcing membrane, strength is required usually only along the length of the fabric. Two-directional strength in such cases is unnecessary and expensive.

When two-directional soil-reinforcement is needed, the best solution is to cross-lay uni-directional fabrics.

Because of 'Terram' RF/12's special structure with aligned fibres, the initial high-strain, low-stress associated with conventional woven structures is greatly reduced. This enables the fabric to react to imposed load at low strain as illustrated below.

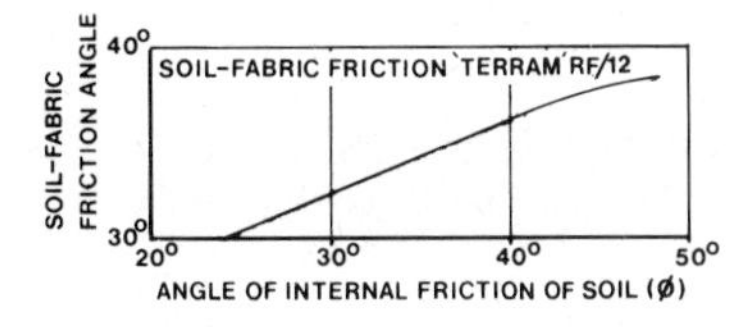

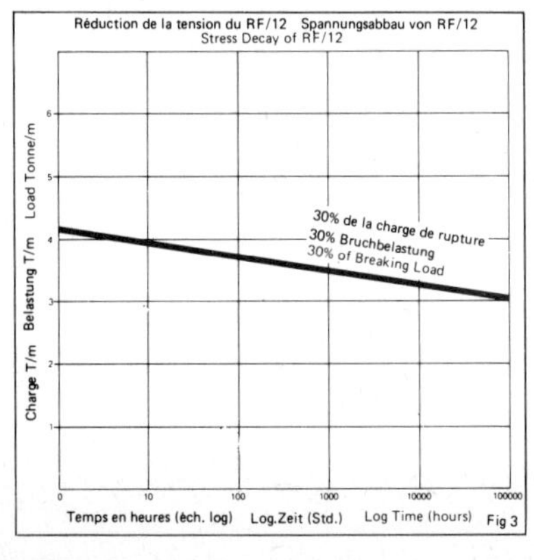

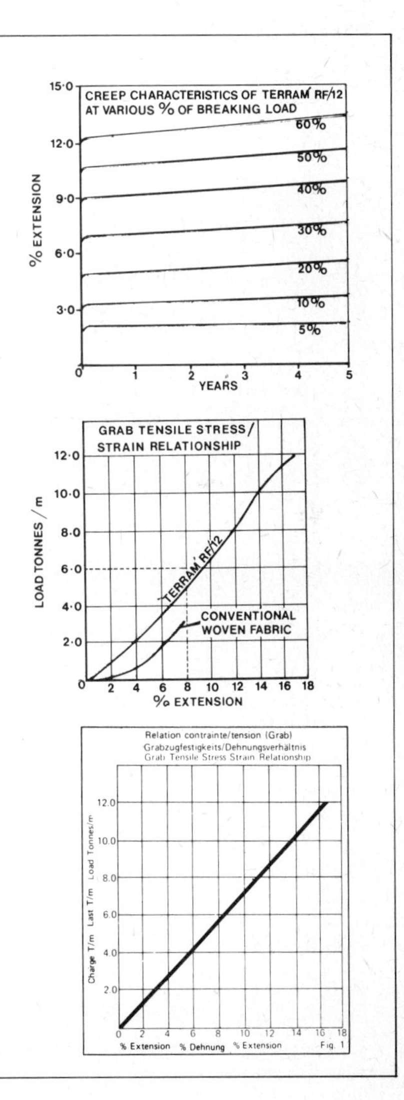

Figure 4.82 Terram RF/12 characteristics

Filtram is a plastic mesh with Terram 1000 filter membrane welded onto one or both sides. Filtram with Terram on one side is suffixed 'TN'. With Terram on both sides it is suffixed 'TNT'.

Filtram is used to filter soil and transport the water within its own structure, e.g.

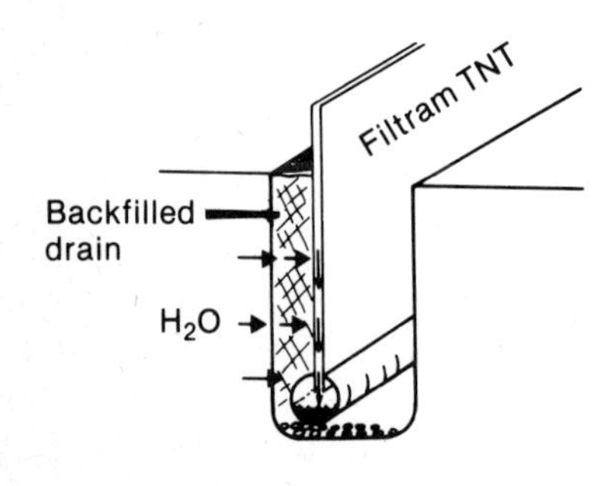

Figure 4.83 ICI product

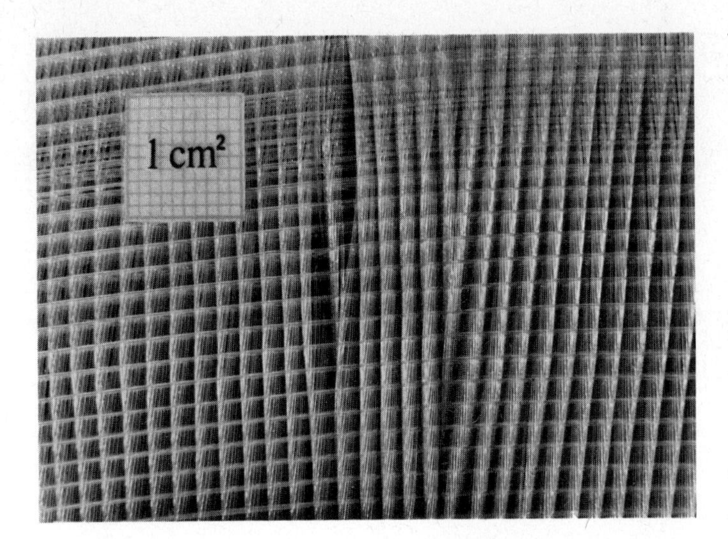

Tissu XXX

750-1000 g.s.m.

100% Polyester

Colourless (white)

A three-dimensional mat of interwoven monofilaments.
Internal voids of 75%
Tensile strengths of 10, 15, or 22 tonnes/metre width
Mechanical strength up to 20 tonnes/m^2 (compression?)

Figure 4.84 Rhône-Poulenc product

Paraweb Mat

Black

A composite woven web product. The outside sheath of each strip is polyethylene and is internally reinforced with continuous polyester filament fibres. The polyethylene is UV stabilised for over 30 years exposure.

The mat can be bonded with ICI's Terram 1000, to make a civil engineering filter material

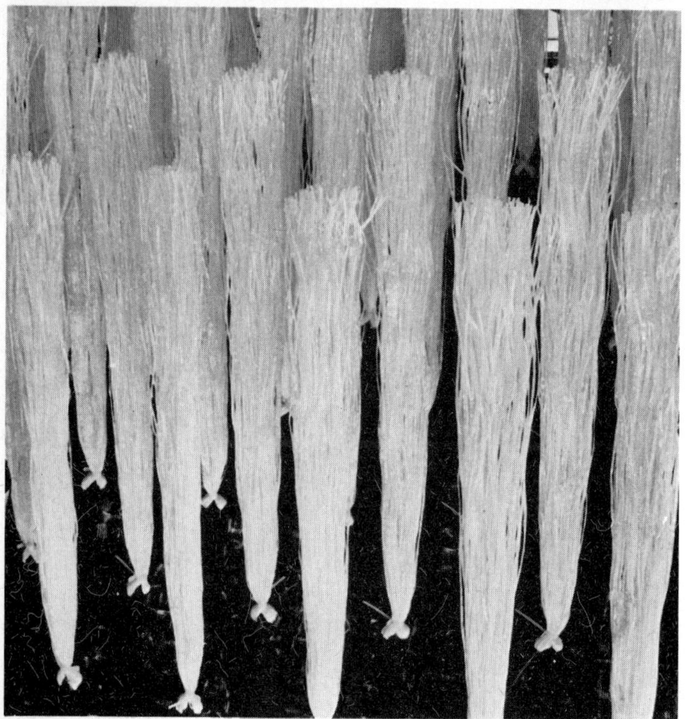

Paraweb synthetic seaweed mat

This is placed on the sea bed to prevent erosion by causing the precipitation of sand

Figure 4.85 Linear Composites products

Windbreak Design

Using the unique physical properties of PARAWEB to their best advantage a windbreak has been constructed, based on known scientific data and with the assistance of wind tunnel testing, to produce the maximum wind reduction over a large area. This has been achieved in a slatted structure of 58% density which slows down the wind whilst preventing the high degree of turbulence normally associated with solid walls or dense hedges.

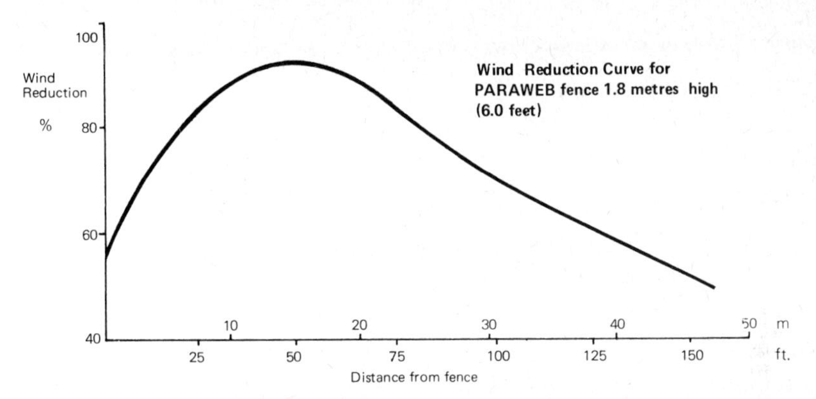

The above curve was obtained by wind tunnel tests carried out under controlled conditions in the Engineering Department of the University of Nottingham.

Nominal breaking strength of Paraweb strips		**165kg strip**	**400kg strip**
Width of PARAWEB strips	Warp Weft	47mm ± 1 75mm maximum	47mm ± 1 75mm maximum
Specific gravity		.85	.87
PARAWEB type		L	M
Tensile test — extension at break		15%	15%
Standard Black sheath		ICI ALKATHENE (Polythene) K1104 Black 904	
Core fibre		ICI Fibres TERYLENE (Polyester) Type 125, 1100dtx	
Non-standard clear or natural sheath		ICI ALKATHENE WNF 15/Q1388	
Non-standard coloured sheath		Full range of MASTERBATCH colours based on WNF 15/Q1388	
Resistance to ultra violet radiation (see text)			
Black	years	20 +	20 +
Brown	years	6 max	6 max
Green	years	5 max	5 max
Natural	years	2 max	2 max
Breaking strength of Paraweb Mat		3 tonnes/ metre width	8 tonnes/ metre width

Figure 4.86 Linear Composites characteristics

Websol
Soil-Reinforcing
Webbing

Black

Individual strips have breaking strengths in excess of 10 tonnes.

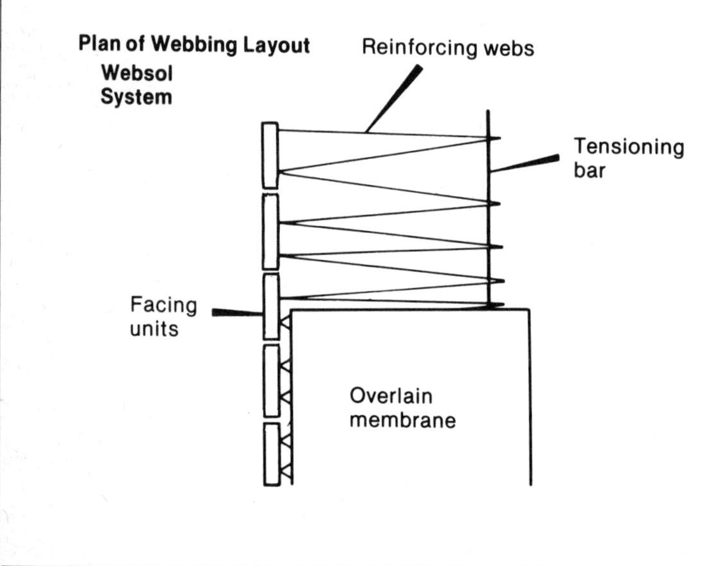

Figure 4.87 Soil Structures Ltd, Product

Figure 4.88 Vertical bridge abutment built using granular soil reinforced with a WEBSOL system

Visqueen 1200 DPM

Blue colour — transparent
Impermeable Plastic sheet
100% Polyethylene

Visqueen 2000T DPC

Black colour — opaque
Impermeable Plastic sheet
100% Polyethylene

Figure 4.89 ICI products

Polyweave LX (50506)

Blue colour

Appears to be a transparent film bonded onto a woven multi-filament polyamide membrane

Figure 4.90 Sarlon Industries product

Section 2
Design of permeable membranes

5

Drain filters in one-way waterflow situations

There are a number of different kinds of filter systems included in this category, but they all have the common feature that their intended design and functioning mode caters for a single direction water flow. However, although this waterflow is non-reversing, it can vary in quantity. For example, filters designed into a dam situation will be expected to cope with a consistent waterflow from a steady waterhead, whereas ground water drains designed to lower the water table in a particular area will be expected to cope with variable waterflow from the ground into the drain in relation to the amount of precipitation experienced at any given time.

The following is a brief list of one-way filter drains which do come within this general group:

(a) ground water-lowering drain trenches (both agricultural and industrial).
(b) drain filters on the downstream toe of dams.
(c) blanket membranes under roadways in areas of rising ground water.
(d) filters for soak-aways.
(e) filters for settling tanks and waste lagoons.

In these applications, it can generally be considered that flow rates are slow because filters are being placed on fine grained soils such as clays or silts. Variations in flow rates are also slow, being experienced more in terms of days than hours or minutes. It may be said in contrast that filters where alternating waterflows exist are often subject to much more rapid fluctuation in hydraulic gradient. For example repetitive wave action causing pressure fluctuations on a coastal defence, or the mechanical pressure oscillations induced by the passage of a train over railway sleepers.

The object of all single-direction flow drains is to allow water to pass out of the natural ground or low-permeability backfill material, whilst preventing finer particles of soil being washed out with the moving water. It can readily be understood that the removal of fine particles from the soil can cause both consolidating settlement of the soil mass and an increase in permeability within the soil. Settlement is damaging to structures, and permeability increase can lead to excess water loss in the case of dams or can lead to an increase of velocity of waterflow which in turn removes larger particles and thus sets up an accelerating cycle which can lead to the destruction of the soil mass by piping. Furthermore, soil particles washed out of the soil must enter the artificial drainage system, and will tend to block pipes and clog up filters.

Ground drains

Simple ground drains have been used over thousands of years for the purpose of removing excess water from agricultural land. Figure 5.1 shows the earliest simple form where a trench has been dug to a shallow depth, flat slates are placed down the sides of the trench and further series of flat slates placed over the top to form a cap. Sometimes the trench is backfilled, sometimes it is not. Water ingresses through the base of the trench and through the rough joints between the slates. Unfortunately, as may be expected, the water moving into the trench carries soil particles with it and the small trench soon 'silts up' and becomes inoperative. Although the drain can be easily cleared out and reinstated, it is extremely inefficient since it is small, shallow and requires constant maintenance. (This type of drain is often referred to as a 'French Drain'.)

As agricultural technology improved, and as crushed stone became available from quarrying activities, the simple trench drain became modified to the form shown in Figure 5.2. The object of the trench is to lower groundwater levels so that the root zones of agricultural crops are no longer saturated and therefore, no longer in an anaerobic condition.

Note that the concept is quite different from an open drain, since land drains permit the use of the agricultural land above them without causing a physical discontinuity. Open drains have the advantage that in areas of heavy rainfall — such as the tropics — they can cope with surface run-off, but they have the disadvantage that they break up the land into small parcels, and

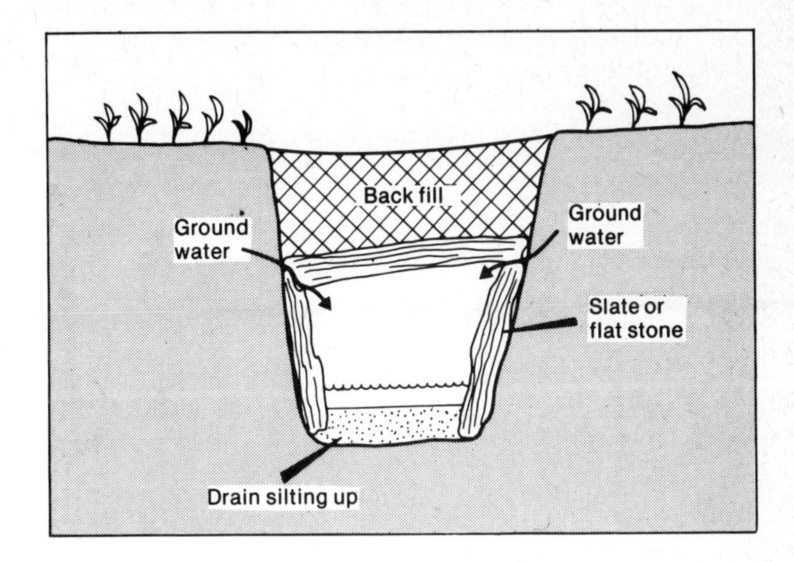

Figure 5.1 An early form of drain called a "French Drain"

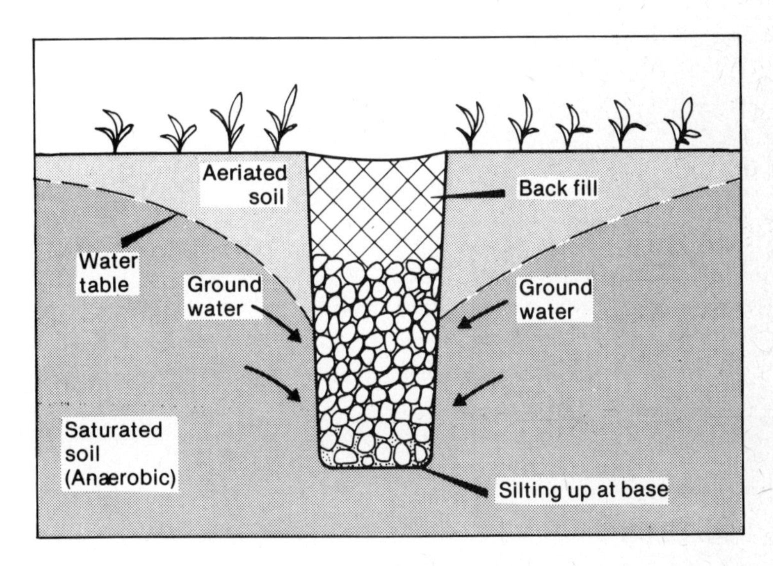

Figure 5.2 Simple stone-filled ground drain or land-drain

they get blocked up with weed growth and require constant clearing and maintenance.

The placing of crushed stone into a ground drain allows the drain to go deeper because the stone supports the sides of the trench thus preventing collapse, and by being covered with soil, there is no weed growth to clog the drain. However, it is still possible for fine particles to be washed into the drain from the surrounding soil, thus blocking it and making it useless. When this occurs, the drain is far harder to maintain than the old simple French Drain since the gravel would have to be removed, washed and replaced, which would be extremely time-consuming and expensive.

In principle therefore, the traditional design of a gravel-filled ground drain becomes self-defeating. In order to transport the water adequately down the drain the stone particles in the drain should be large. In order to prevent the ingress of fine materials from the soil, the stones should be small to act as a filter. Fortunately, the advent of membrane filters has allowed the rationalisation of design by removing the conflict between filter and transporter.

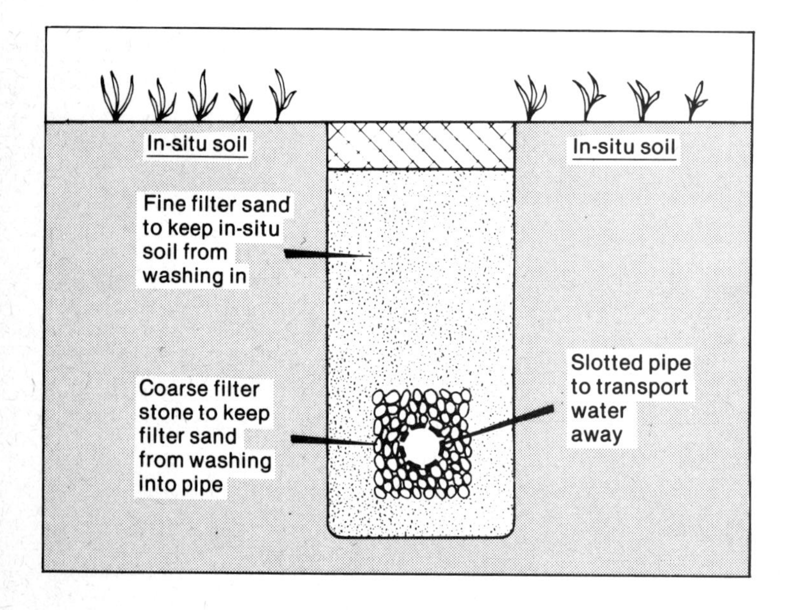

Figure 5.3 Design units of a granular filter drain

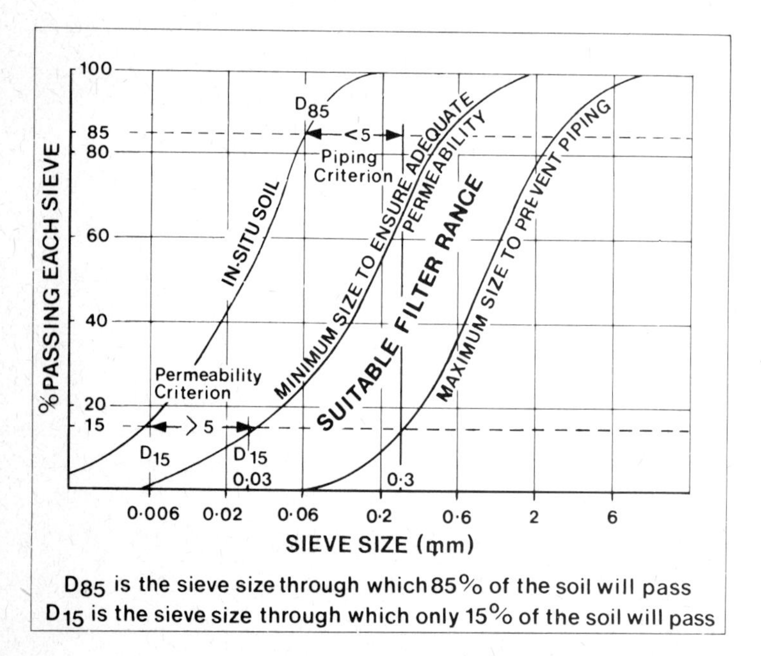

Figure 5.4 Terzaghi overlapping soil filter criteria as used in the choice of filter sands for granular drains as in Figure 5.3

The concept of using a 'filter' structure in a ground drain was developed by Terzaghi in 1925 and expanded upon by the United States Waterways Experimental Station and Cedergren. Its obvious culmination was to use a pipe in the centre to transport the water, instead of the stone. Figure 5.3 shows the basic engineering units of the design of a sophisticated ground drain of the type used before the advent of membrane technology. This type of drain was expensive because of the laying and levelling of the pipe. There were limitations in depth, and limitations in width because of having to shore up the sides of the trench to lay the pipe. As can be seen, such a drain is a relatively sophisticated engineering design, and the theoretical choice of filter materials is made upon the basis of the Terzaghi overlapping soil filter envelopes shown in Figure 5.4. The design requires a knowledge of the Particle Distribution Diagram of the *in situ* soil, and that of the filter materials available.

If the filter material is to allow water to pass through without allowing the soil to 'pipe', i.e. wash through, then the following criteria must be satisfied:

1. The D_{15} Filter $<$ (5 $\times$ D_{85} Soil) Piping Criterion
2. The D_{15} Filter $>$ (5 $\times$ D_{15} Soil) Permeability Criterion
3. The D_{50} Filter $<$ (25 $\times$ D_{50} Soil) Uniformity Criterion

The first criterion governs the fine particle sizes of the filter. It ensures that at least 15 % of the filter is of the same order of size as the coarser end of the soil distribution curve. Consequently, the soil particles can not wash in to the filter since there are a large number of fine filter particles available to hold them back.

The second criterion relates to permeability. It must operate effectively at the same time as the first, since although the filter is required to hold back soil particles as specified in the first criterion, it is also necessary that sufficient water be allowed out of the soil to prevent the build up of a back pressure within the soil itself. Since the overall permeability of the soil is governed by its own D_{15} size, and the overall permeability of the filter is governed by its D_{15} size, then providing the filter's D_{15} is more than five times that of the soil, then the filter will have ample excess permeability above the soil and will not sustain a backhead of water pressure. It will allow water to enter it and flow freely through it.

The third criterion — which is rarely applied — is intended to ensure that the overall principles of overlapping filters and relative particle sizes, are maintained relative to one another right across the particle distribution curve range.

Once the Particle Distribution Diagram of the natural soil i known, one can establish the criteria for a filter or even a series of filters to prevent 'piping' whilst still allowing adequate drainage. Multiple filters (or graded filters) simply have a series of curves conforming to the above rules.

As mentioned earlier, it is important to appreciate the granular design concepts, since the design of membrane wrapped drains is a technological extension of the soil principles developed by engineers during the last half century.

The introduction of a membrane into the design of a ground drain alters the engineering characteristics of the drain structure radically. Virtually all aspects of the design are changed, and, as can be seen in Figure 5.5, the functions of the various units are also altered. Figure 5.6 shows the advantageous comparison of a membrane-wrapped ground drain with that of a solely granular structure.

Naturally, there is a limit to the amount of water that can flow through the stone-fill of the membrane-wrapped drain, and if it is necessary to bring surface water run-off from storms in to the drain, then a pipe can be added as shown in Figure 5.7a.

Perhaps the ultimate extension of this logical development away from French Drains, is to eliminate the granular fill altogether. For this purpose, sheet filters have been developed with permeable cores, such as ICI's 'Filtram'. As can be seen in Figures 4.83 and 5.36, the Filtram is a central sheet of Netlon plastic mesh with Terram filter membranes welded to each side. It is placed vertically into an excavated trench, with a pipe at the base slotted on to the Filtram sheet (as shown in Figure 5.7b). The original excavated material is then replaced into the trench thus saving a considerable amount of transport and material cost. Ground Engineering Limited in the UK have a similar system called 'Trammel', which uses a woven fabric on the outside, but which is not purchased preassembled. Here, the Netlon and fabric are purchased and supplied separately, and are then assembled on the site. In this case, the fabric is fastened

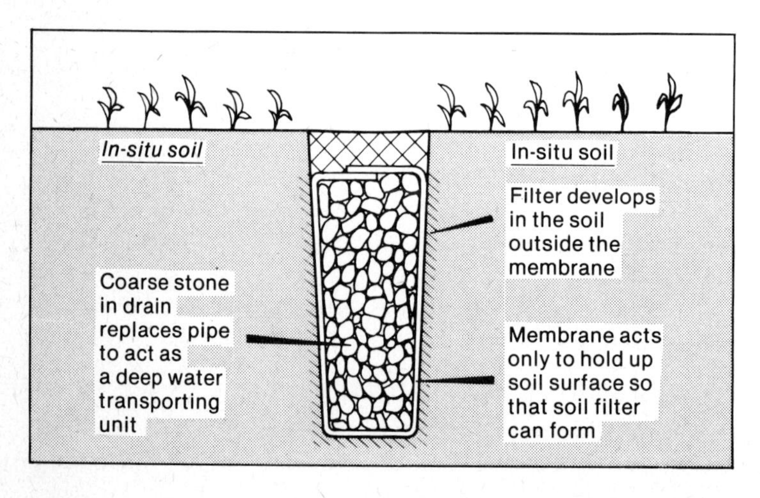

Figure 5.5 The construction and functional elements in a membrane-wrapped ground drain

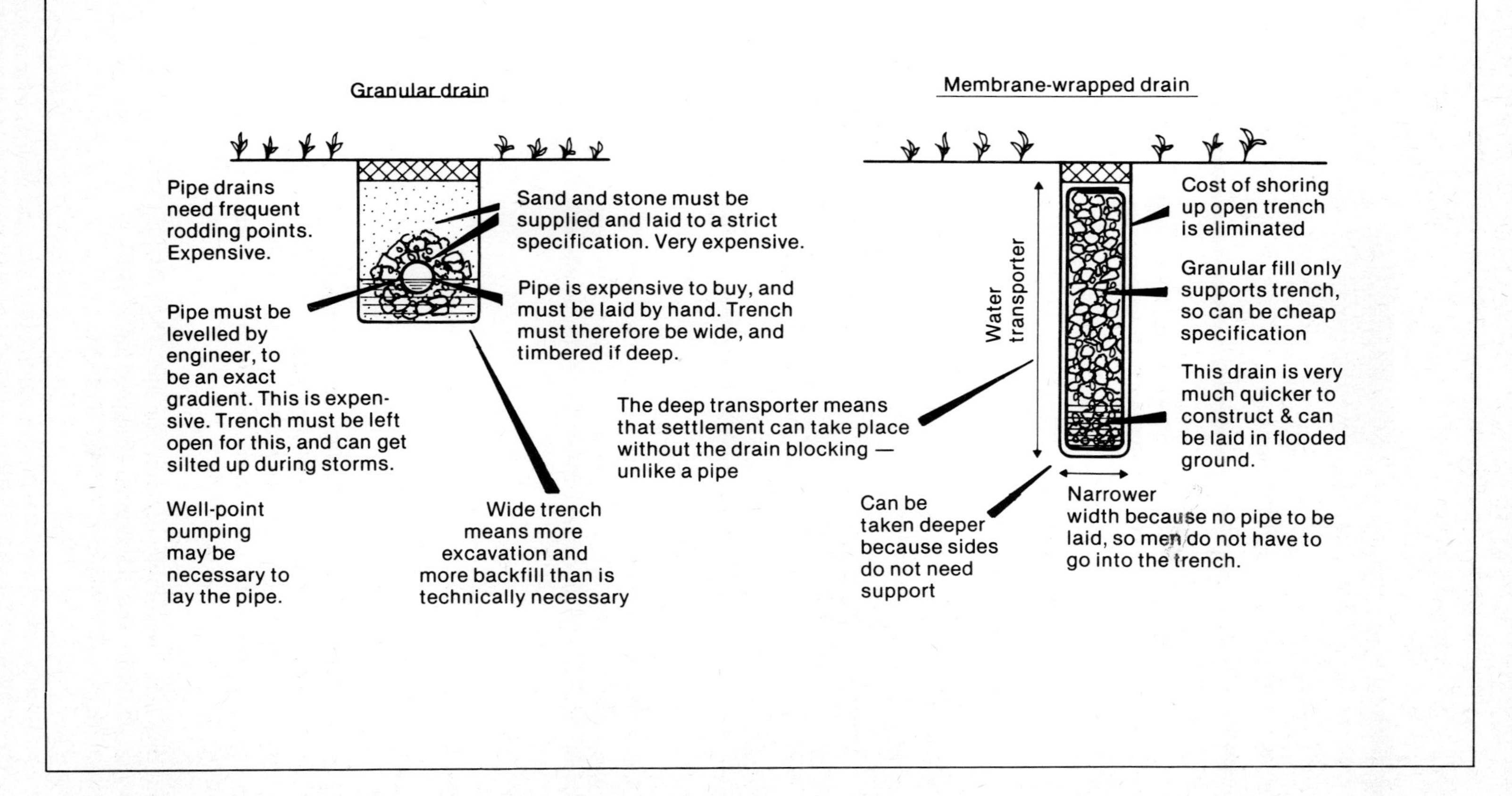

Figure 5.6 Comparison of some of the features that make membrane-wrapped filter drains so much more attractive to both the Engineer and Contractor than the granular type

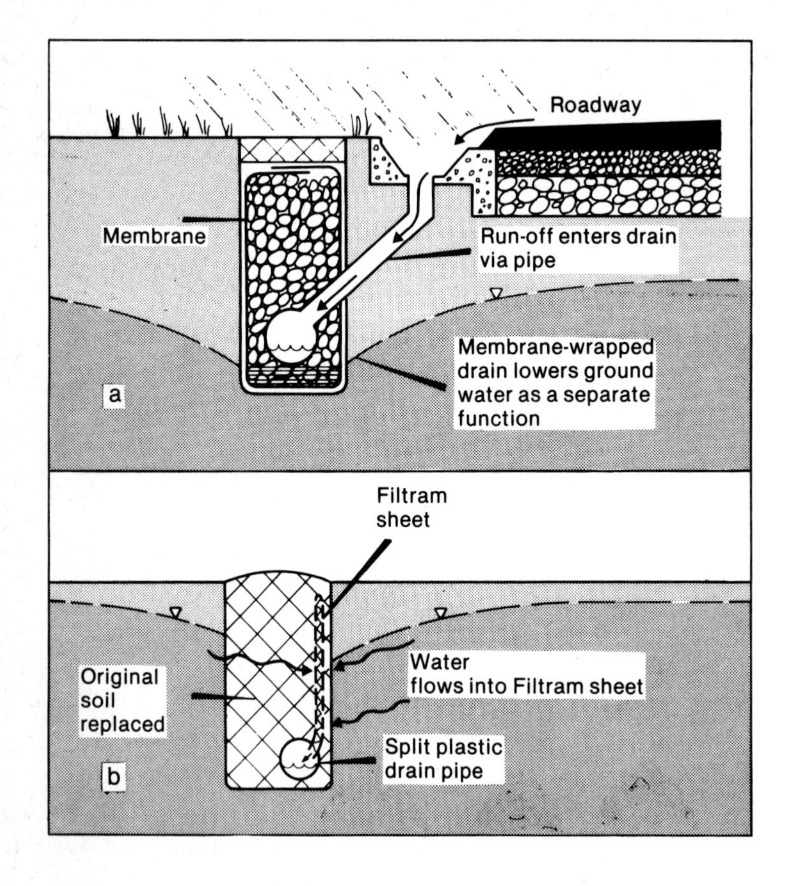

Figure 5.7 (a) Membrane-wrapped drain with a transporter pipe to carry high-volume flow from road run-off, whilst performing a separate ground water lowering function
(b) Filtram sheet drain provides rapid drain laying together with savings in raw materials

to the plastic mesh by plastic 'rivets' at spaced intervals. The Author also understands that similar products may be available in the USA, but has no immediate information or details on these.

The theoretical and practical design of membrane filters

Literature

It must be recognised that the amount of theoretical and practical design advice available in published form is small. In order to attempt to summarise the work available into a compact design guide, the following five papers are listed and their contributions subsequently summarised.

F.F. Zitscher. 'Recommendations for the Uses of Plastics in Soil and Hydraulic Engineering', Die Bautechnik, *52* (12) 1975, pp. 397-402 (Ref 19)

'Designing With Terram'. 1978. Published by ICI Fibres, Terram Section. Pontypool, Gwent, Great Britain NP4 8YD (Ref 20)

H.J.M. Ogink. 'Investigations on the Hydraulic Characteristics of Synthetic Fabrics'. Publication No.146, May 1975, Delft Hydraulics Laboratory, PO Box 177, Delft, Netherlands (Ref 7)

D.B. Sweetland, MSc Thesis. 'The Performance of Non-Woven Fabrics as Drainage Screens in Sub-Drains'. University of Strathclyde, 1977 (Ref 8)

C.C. Calhoun. 'Development of Design Criteria and Acceptance Specifications for Plastic Filter Cloths'. U.S.W.E.S. Technical Report, Vicksburg, Mississippi, USA June 1972 (Ref 5)

Notation Convention

In this book, the following conventions are used:

'O' represents the diameter of holes in a membrane, soil, or granular filter.

- 'O_m' represents the diameter of holes in a membrane.
- 'O_s' represents the diameter of holes in a soil.
- 'O_f' represents the diameter of holes in a granular filter.

'D' represents the diameter of particles.

- 'D_s' represents the diameter of particles in an *in situ* soil, or a soil being protected by a filter.
- 'D_f' represents the diameter of particles in a granular filter.

Therefore:

O_{50m} is the diameter of hole in a membrane having exactly half the holes in the membrane larger, and half smaller.

O_{85m} is the diameter of hole in a membrane having only 15 % of all other holes larger, and 85 % smaller.

O_{15s} is the diameter of hole in a soil having 85 % of all other holes larger, and only 15 % smaller.

O_{60f} is the diameter of hole in a granular filter having 40 % of all other holes larger, and 60 % smaller.

Also:

D_{85s} is the diameter of soil particle in a soil having 15 % of all other particles larger, and 85 % smaller.

D_{85f} is the diameter of particle in a granular filter material having 15 % of all other particles larger, and 85 % smaller.

The sizes of particles in a granular material are determined by sieving, and the sizes of holes in a membrane are determined by 'ballotini' tests as described later in this book.

Design Concepts

Zitscher[19] looked at the permeability of woven fabrics in relation to the filtering of sands. He considered the use of an 'effective width' or 'effective hole size' to represent all the holes in the woven fabric. (He called this Dw., but the Author prefers to rename it Owf when used for a fabric, in order to conform to the convention outlined above.) The concept of an 'effective size' Owf is a reasonably valid approximation, since — as discussed earlier in the book at Figure 4.5 — the variation of hole size within any particular woven fabric is small. Furthermore, it is not unreasonable therefore to substitute and adopt the term O_{50f} instead of Owf, in view of the steep nature of the 'hole distribution diagrams' of woven materials generally. The adoption of O_{50f} allows direct design comparisons with non-woven materials.

Zitscher stated that from practical investigations, the following rules had been determined for the sand tightness of woven fabrics when filtering sands (see Figure 5.15).

Steady state flow $\frac{O_{50m}}{D_{50s}} \leq 1.7$ to 2.7 (See Figure 5.8)

Alternating flow $\frac{O_{50m}}{D_{50s}} \leq 0.5$ to 1.0 (See Figure 5.9)

(For sands with $D_{50s} = 0.1$ to 0.3 mm, and Uniformity Coefficient $U<2$, i.e. uniform poorly graded)

Zitscher does not consider the use of woven materials for cohesive soils. Here, he prefers to use non-woven membranes. To arrive at some design criteria, he considers first granular filters protecting cohesive soils. In this instance, he says that $D_{50f}/D_{50s} = 100$ to 150, i.e. if the soil has $D_{50s} = 0.04$ mm, then the D_{50f} of the filter must be approximately 6 mm. To extend to

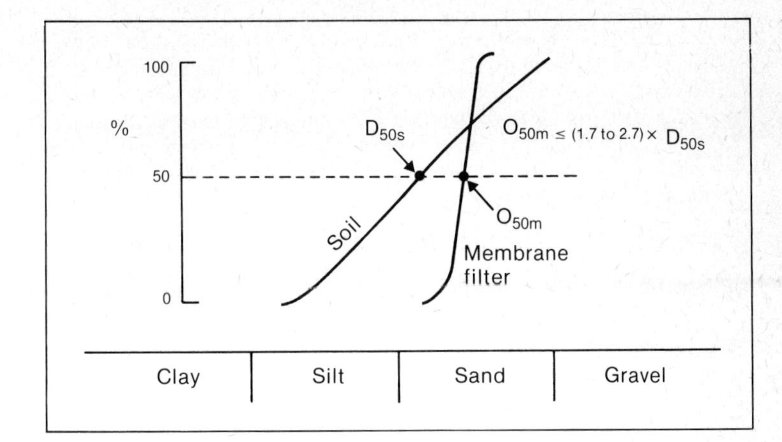

Figure 5.8 Zitscher's criterion for sand tightness of a woven membrane. Steady state flow

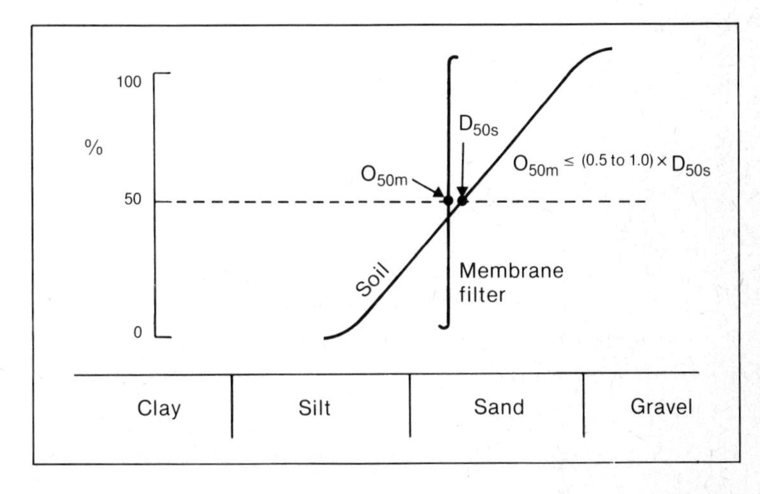

Figure 5.9 Zitscher's criterion for sand tightness of a woven membrane. Alternating flow

fabrics, Zitscher proposes that the Ow (O_{50s}) of a soil is approximately ¼ of its D_{50s}.

Therefore, by substituting O_{50m} for O_{50f},

$$\frac{O_{50m}}{D_{50s}} = 25\text{-}37 \text{ (See Figure 5.15)}$$

(non-woven membrane. Cohesive soil.)

It is interesting to note that ICI Fibres also use a similar quantity conversion, though derived differently, for their non-woven products. On Page 6 of their Manual,[20] the O_{50m} of the design example given is approximately four times bigger than the D_{50s} of the soil to be protected. ICI use the experience of Atterberg[21] who proposed that the average pore size of a soil is equivalent to $0.2 \times D_{10s}$ of the soil mass. In the light of more recent works, ICI have adjusted this concept to a more realistic $O_{50s} = 0.2 \times D_{15s}$. They use this as the basis of their design method, and the following is an extract from their works. (The results are incorporated into Figure 5.15).

Although membranes only contain pores, soils contain both particles and pores, and soil permeability can be defines in terms of either pore distribution (O) or particle distribution (D). It has however been possible to develop the concept of an 'Equivalent Terram Membrane' i.e.; a membrane having the same pore-size distribution characteristics as a particular soil, as shown in the diagram below (Figure 5.10).

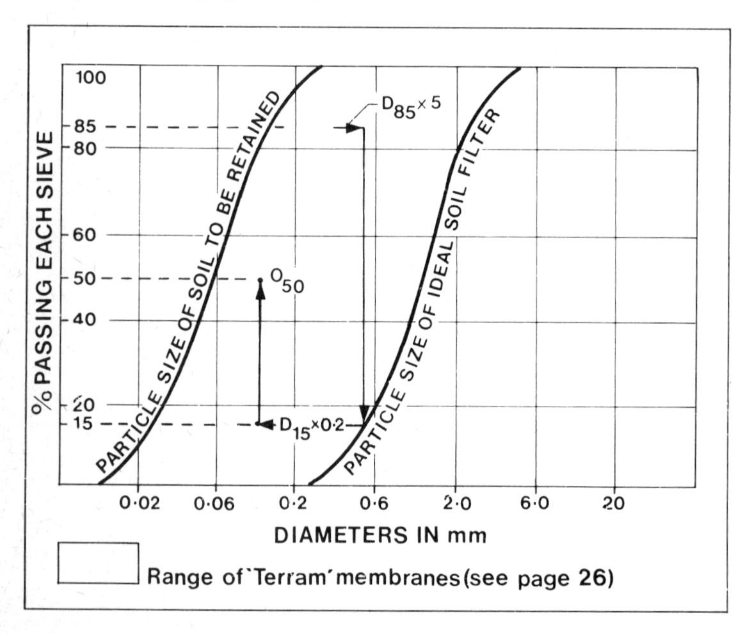

Figure 5.10 Extract from ICI's manual "Designing With Terram"

This is based upon the selection of an 'ideal theoretical soil filter' for which we intend to substitute a particle-less synthetic membrane. The ideal theoretical soil filter is selected by the criteria previously mentioned[2]. Atterberg showed that the average pore size of a soil is equivalent to $0.2 \times D_{15}$ of the soil. This basis has been used for the selection of the 'Equivalent Terram Membrane'.

Consequently, from the diagram, it can be seen that in general terms, if the D_{15} of the theoretical soil filter is $5 \times D_{85}$ of the soil to be retained, and if also the $D_{15}$of the theoretical soil filter is $5 \times O_{50}$ of the equivalent 'Terram' membrane, the soil D_{85} and Terram O_{50} should be the same.

Middle Range Problem Soils (D_{85} between 0.02 mm and 0.25 mm)

From the above discussion, 'Terram' membrane selection has been undertaken by noting the D_{85} size of the soil to be protected, and choosing the 'Terram' product with the O_{50} size exactly the same or slightly smaller.

For example, in the diagram below (Figure 5.11), the soil to be protected is a sandy silt with a D_{85} particle size of 0.18 mm. The pore size distribution diagrams for 'Terram' 700 and 500 are shown, and it can be seen that the O_{50} of 'Terram' 700 is 0.12 mm and the O_{50} of 'Terram' 500 is 0.2 mm. Since the D_{85} of the soil lies between these two, 'Terram' 700 should be chosen to provide minimum piping. On the grounds of economy, the 'Terram' 500 could be chosen, but it would take longer for the filter cake to build up than for the 700.

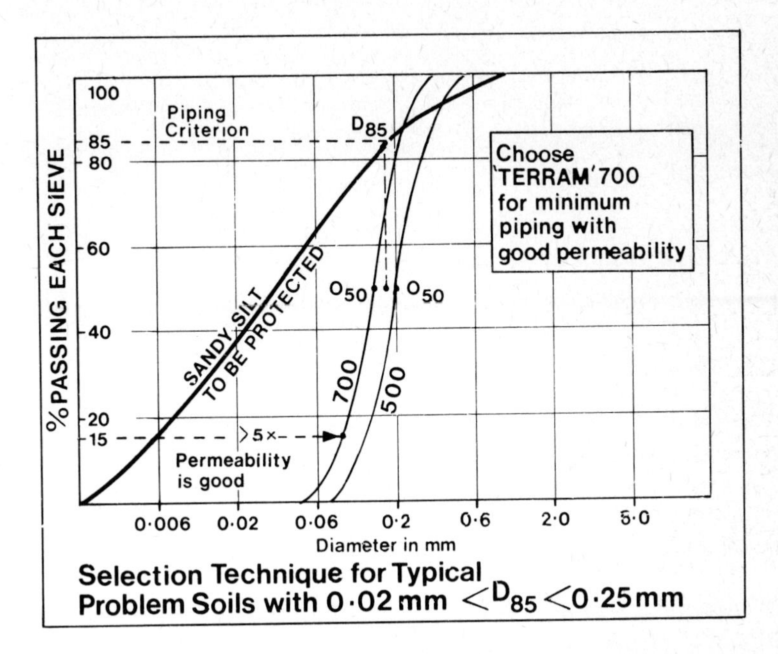

Figure 5.11 Extract from ICI's manual ''Designing With Terram''

Fine Silts and Clays (O_{85} less than 0.02 mm)

Field experience and laboratory investigation have shown that the criterion above does not apply to soils with D_{85} < Approx. 0.02 mm. The graph on page 1 shows that the soils with D_{85} < 0.02mm contain large proportions of clay particles. The cohesive properties of these particles assist in the formation of the filter network, making it possible in conventional drainage practice for filters with O_{50} smaller than 0.02 mm to be unnecessary. Note that this does not apply to conditions of waterflow reversal such as coastal applications which are discussed on Pages 10-13 [of the Manual].

Note:- It is necessary to consider very carefully the *real* environment of the membrane in the ground before choosing the final design. In many soils such as laterites, or laminated clays, permeability and waterflows are much higher than would be expected for the pure soil itself. Similarly, membrane permeability should always be chosen to cope with the waterflows from the most permeable soils present along a structure in order to prevent the build up of water pressures.

This situation of localised waterflows occurs so often that in many of the cases to date, 'Terram' with equivalent pore sizes to 'Terram' 700 has been chosen to ensure effective permeability at a slight loss of piping efficiency. There is no doubt that this has proved a most cost-effective solution in projects where the piping criteria are not as critical as those for permeability.

Sands and Gravels (D_{85} > 0.25 mm and D_{15} > 0.02 mm)

In addition to the smaller particle-size soils, 'Terram' is sometimes used to protect coarser soils such as sand from outwash in a filter structure. In the case of protection of sand soils with D_{85}> 0.25 mm, the critical factor can be seen from Page 6 [of the Manual], this is dependent upon the D_{15} size of the 'Equivalent Terram Membrane', and the following criterion must apply:-

Since from Page 6

D_{15} Min. soil filter $\geq 5 \times D_{15}$ soil to be retained
and D_{15} Min. soil filter $= 5 \times O_{50}$ equivalent 'Terram' Membrane (ETM)
then O_{50} ETM $\geq D_{15}$ of soil to be retained

At present, the most open-textured 'Terram' membrane is 'Terram' 500, therefore since the O_{50} pore size of 'Terram' 500 is 0.2 mm, this limits effectively the size of soil which the 'Terram' range can filter without the build up of pore-water pressures.

If the physical treatment of the membrane is likely to be extremely rigorous — owing to particular site conditions, then it may be necessary to use 'Terram' 1000, which is stronger. In this event a check should be made carefully on required permeabilities.

Ogink,[7] at the Delft Hydraulics Laboratory, studied both wovens and non-wovens in relation to the filtering of sands. He came to the conclusion that the D_{90s} defines the character of a sand most relevant to its behaviour in relation to being filtered by a membrane. He examined the relationship between O_{90m}/D_{90s}. His conclusion was clearly that the relationship O_{90m}/D_{90s} was the most valid for sands with a range of Uniformity Coefficients (U) and a range of O_{90m} sizes. He states:

> It appears that the quantity O_{50}/D_{90} for the mentioned groups differs significantly, depending on the steepness factor O_{50}/O_{98}. The ratio O_{90}/D_{90} is almost equal for these groups. Furthermore, the value O_{90}/D_{90} is independent of the characteristic fabric aperture O_{90}.

For linear steady-state flow for woven materials over sand, Ogink found (see Figure 5.12) that

$$O_{90m}/D_{90s} \doteqdot 1.$$

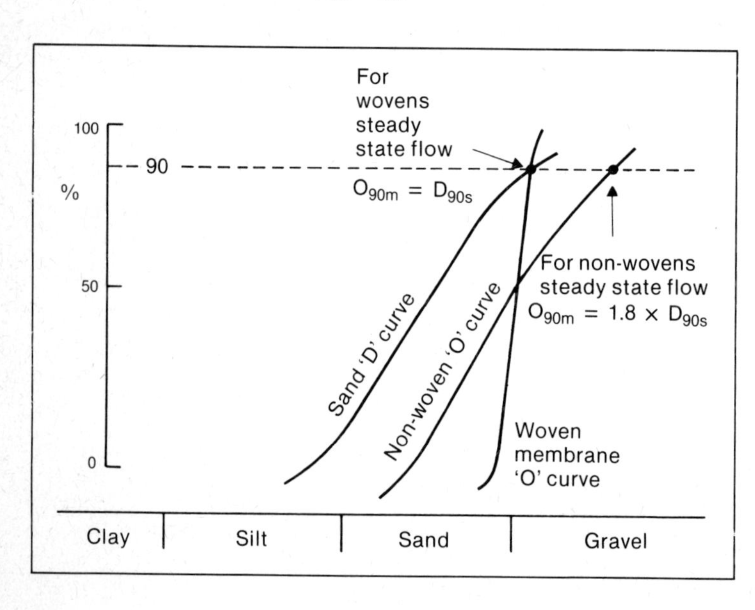

Figure 5.12 Ogink's criteria for sand-tightness of woven and non-woven membranes. Steady state flow

(O_{90m}/D_{90s} can be reduced below 1, but increasing hydrostatic pressures will be generated behind the membrane as it falls below 1).

For *non-woven materials,* Ogink found that

$$O_{90m}/D_{90s} = 1.8$$

This suggests that non-wovens can form a filter more effectively under adverse flow conditions than can woven materials. Ogink said:

> From this the conclusion can be drawn that mats, mesh-nettings, tape and multifilament fabrics fulfil the sand-tightness criterion if $O_{90}/D_{90} \leq 1.0$. For non-wovens this limit is more favourable; here the sand-tightness criterion is obeyed if $O_{90}/D_{90} \leq 1.8$.

Ogink extended his studies into alternating flow conditions with a two metre variation in hydraulic head, e.g. waves. He concluded that for practical conditions:

$$O_{90m}/D_{15s} = 1.0 \text{ if no filter builds up (Figure 5.13)}$$
$$\textit{and } O_{90m}/D_{85s} = 1.0 \text{ if a filter can build up (Figure 5.13)}$$

The above criteria are included in the construction of Figure 5.15.

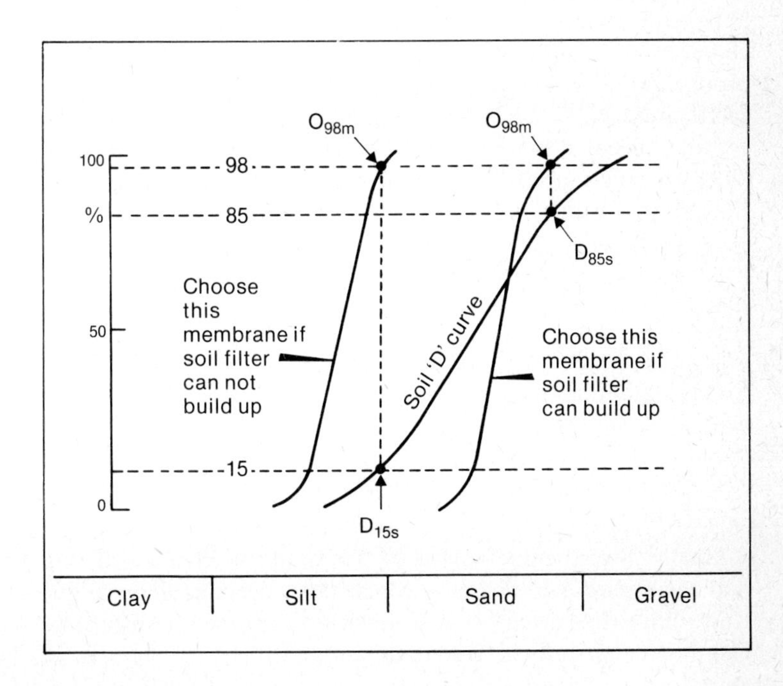

Figure 5.13 Ogink's criteria for sand-tightness of membranes under alternating flow conditions. Note the range of fabrics suggested depending upon the nature of the soil being filtered

Ogink also looked at the problem of 'blocking', and concluded that, although problems of back-pressure are rare with soil hydraulic gradients less than 1, if this is exceeded, then blocking problems can occur if a membrane is used whose O_{90m} is less than the sand diameter protected (especially if the sand is uniform in grain size). He made no comment on 'clogging' other than pointing out a need for research work on that subject. In this respect he echoes the Author's view that generally, more research work into this field — especially on cohesive soils — would be valuable.

Calhoun[5] independently took a very similar approach to that of Zitscher, but undertook tests on the sand-tightness and permeability of woven fabrics only, and opted for the 85 % fraction as being representative of the sand properties. He concluded that:

D_{85s}/EOS ≥ 1 ie. EOS/D_{85s} ≤ 1 (EOS — Effective Opening Size)

This equation is very similar to Zitscher's Owf/D_{50s} ≤ 1.7, but Calhoun is more in agreement with Ogink in choosing D_{85s}. If one follows the previously suggested step of substituting D_{50m} for EOS, then Calhoun's equation becomes:

$D_{50m}/D_{85s} \leq 1$ For granular materials with less than 50% passing a US # sieve (0.07 mm)

This has been incorporated in Figure 5.15.

Calhoun also considered that the Open Area of the fabric was important since it defined the fabric's permeability when not in contact with soil. The % Open Area (OA) represents the proportion of area occupied by the holes in relation to the entire fabric area.

Calhoun stated that always for woven fabrics, both EOS and % OA must be specified to define a filter cloth in relation to its filtering performance. He gave two criteria for choosing fabrics:

Fabric filtering a granular material ($\leq$ 50 % Passing #200 sieve)

D_{85s}/EOS ≤ 1 *and* % OA $\leq$ 40 %

Fabric filtering fine soil with little or no cohesion (i.e. material with > 50 % silt)

EOS < US sieve #70 *and* % OA $\leq$ 10 %

Finally, Sweetland,[8] looking at real granular filters, believed that the Figure 5.14 chart — which takes account of a sand's Uniformity Coefficient — has much to commend it since it is based on empirically derived data.

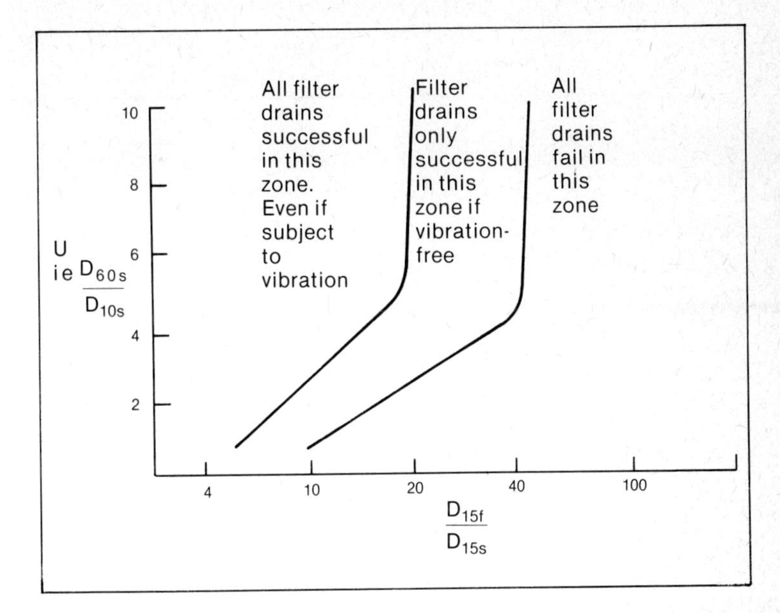

Figure 5.14 U.S.W.E.S. empirically-derived chart showing the performance of many drains with granular filters. (*Not* membrane filters)

In brief, he suggested that the later USWES figures were reasonably valid:

For $U = 1.5$, $D_{15f}/D_{85s} = 6$
For $U = 4$, $D_{15f}/D_{15s} = 40$

Note that these are for granular filters (D_f) protecting granular soils (D_5). It would appear reasonable at this stage to substitute O_m for D_f by the assumption used earlier that $D_{50s} = 4 \times O_{50s}$[19] and allowing for the logarithmic change caused by moving up and down to D_{85s} and D_{15s} respectively, then it is possible to suggest that:

For $U = 1.5$, $O_{15m}/D_{85s} \doteqdot 1$
For $U = 4$, $O_{15m}/D_{15a} \doteqdot 1$

Although this is an approximation, adjusted by the Author into membrane-equivalent terms, nevertheless, it does fit well with fabric experimental data, as can be seen in Figure 5.15.

Most of the above criteria have been summarised in Figure 5.15 which demonstrates the apparent dearth of work on clays and gravels. Little has been done on silts, and most of the work has been done on using membranes to protect sands. The Author could find no published filter design criteria relating to membranes on peat. His personal experience suggests that the rate of water release from peats (whether from simple drains or under load) merits high permeability as being one of the prime requisites, in addition to strength.

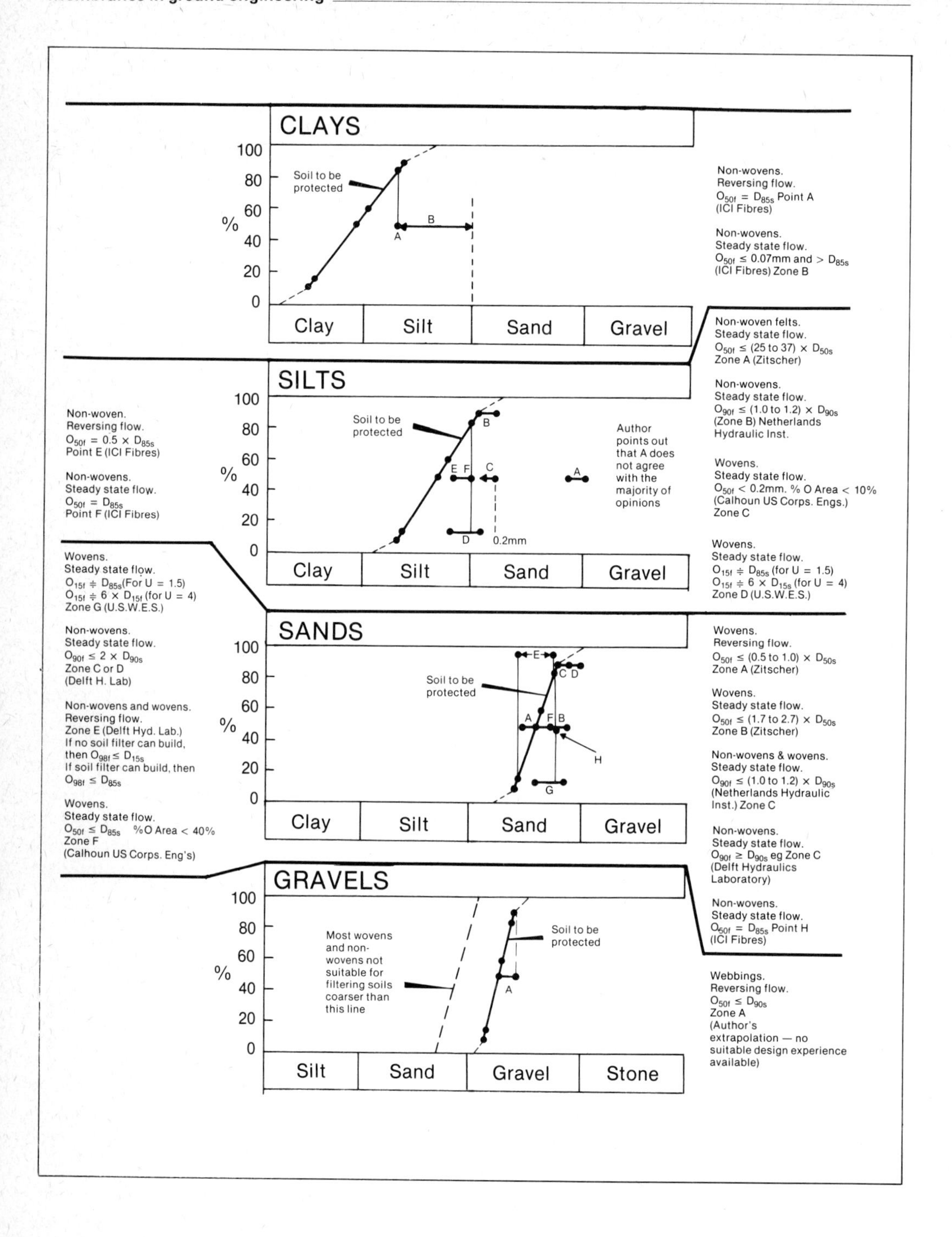

Figure 5.15 Summary of available design works to date on membrane filter criteria

Trench Drains

As a starting point to the design of ground drains, and filters in general, a good understanding of soils and their behaviour under different loading conditions is necessary. Figure 5.16 shows typical permeabilities of different materials in a variety of commonly used units, over a range of different overburden pressures. For example, it can be seen that the permeability of clean gravel hardly varies at all with an overburden pressure varying between 1 and 100 tonnes per square metre. In contrast, the permeability of a typical clay can decrease by ten times as the pressure rises to 100 tonnes per square metre, and an organic clay can have a decrease of permeability of over 30 times at 100 tonnes per square metre overburden pressure.

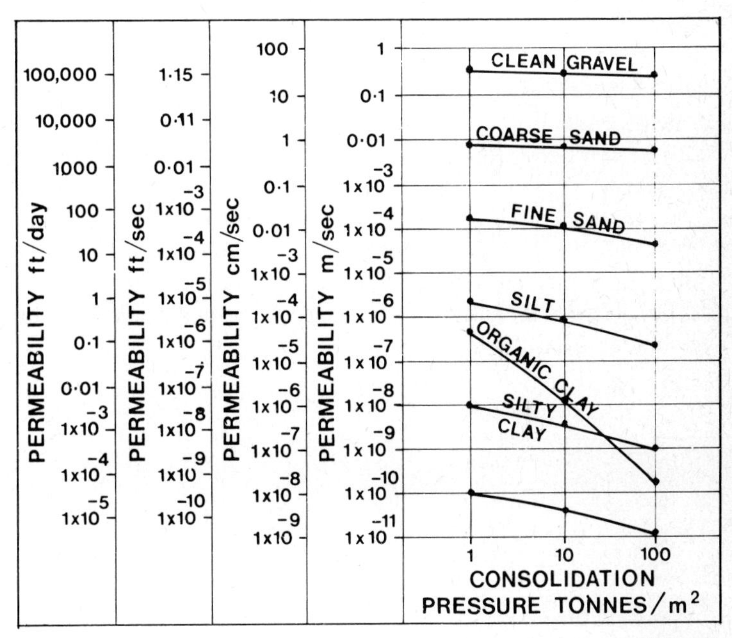

Figure 5.16 Typical soil permeability at different consolidation pressures

Once the membrane has been matched to the soil conditions, then the quantity of water and the flow rates into the drain can be calculated from the permeability of the soil itself, using the gradient equation defined in Figure 2.16 to construct a flow equation known in design terms as Darcy's Law:

$Q = kiA.$

Where Q is the quantity of water flowing through a soil (for example in cubic metres per second):

i = the hydraulic gradient

A = the area (e.g. in square metres) of soil through which the water is flowing at right-angles to the flow, and:
k = the Permeability Constant (dimensioned in metres per second).

Figure 5.16 gives typical values of the permeability constant k for various different typical soil types. It is important to consider the modified situation of natural soils when assessing their permeability under construction conditions. The left-hand side of Figure 5.16, for example, would be typical of ground drainage conditions in surface trenches, but the right-hand side of the figure might have to be considered as being more appropriate beneath the base of a large structure. Naturally, this diagram is only intended as an approximate field guide.

Most natural soils are laminated to some extent, and may contain horizontal partings of coarser soils between thicker layers of clay. The overall mass permeability of such a soil is governed by the small but permeable layers of sandy material sandwiched between the less permeable ones. As a first design principle, therefore, the mode of water egress from the soil body, and the resulting volume and flow rates thereof, should be considered.

If seepage from the soil is to be the sole feeder of water into the drain, then the above Darcy equation can be used directly, otherwise a notional figure for water flow in the drain will have to be arrived at from rainfall figures and the size of the catchment area.

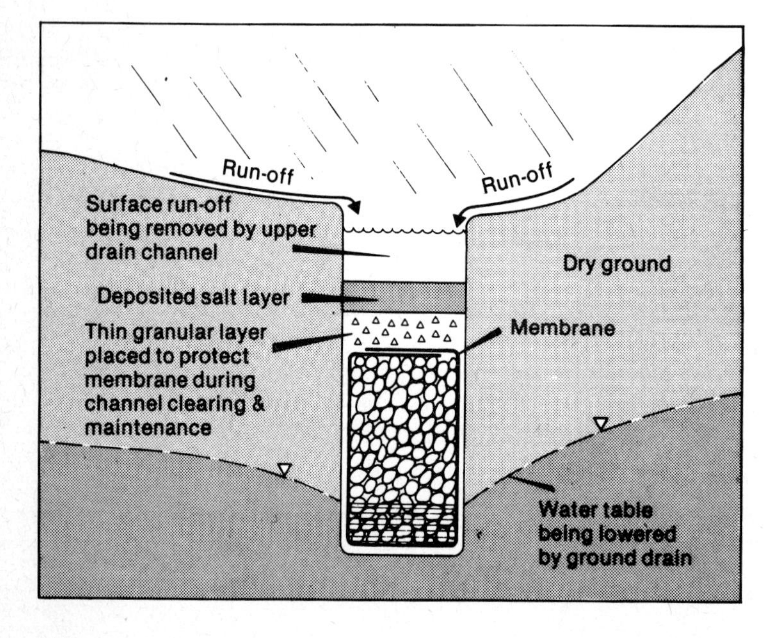

Figure 5.17 2-phase drain handling run-off and ground water simultaneously

In tropical countries, the author has modified the concept of membrane-wrapped drains to cope with tropical storm run-off, and has developed the two-function (or two-phase) drain where the lower section is membrane-wrapped, and takes in percolating groundwater, whilst the upper part is an open trench carrying storm water for short periods during and immediately subsequent to heavy rainfall (Figure 5.17).

Run-off figures for the design of an open trench system can be based on the fact that 10 mm of rain over 1 hectare is equivalent to 100 cubic metres of water (or one inch of rain over one acre is equivalent to 3,630 cubic feet of water). Surface run-off in hilly areas with fairly impervious cover of material accounts for between 45 % and 75 % of the total rainfall in steady conditions. For tropical storms, the author has in the past used 100 % run-off as a safe figure for calculations. For example, a catchment area of two hectares after a storm lasting one hour and producing 25 mm of rain, means that any single drain taking the water from that area will have to cope with a peak flow rate of at least 500 cubic metres of water in one hour (0.14 cubic metres per second).

Figure 5.18 shows some equations which may be used for the approximate calculation of flows in both large and small drainage channels. In large drainage channels, not only does the hydraulic gradient and the size of the channel have a bearing on

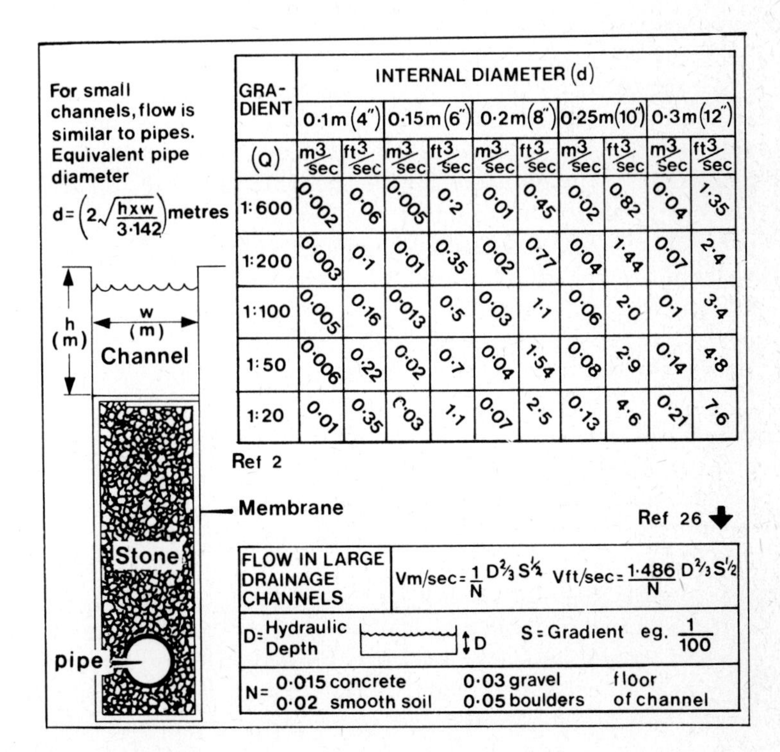

GRADIENT (Q)	INTERNAL DIAMETER (d) 0·1m (4″) m3/sec	ft3/sec	0·15m (6″) m3/sec	ft3/sec	0·2m (8″) m3/sec	ft3/sec	0·25m (10″) m3/sec	ft3/sec	0·3m (12″) m3/sec	ft3/sec
1:600	0·002	0·06	0·005	0·2	0·01	0·45	0·02	0·82	0·04	1·35
1:200	0·003	0·1	0·01	0·35	0·02	0·77	0·04	1·44	0·07	2·4
1:100	0·005	0·16	0·013	0·5	0·03	1·1	0·06	2·0	0·1	3·4
1:50	0·006	0·22	0·02	0·7	0·04	1·54	0·08	2·9	0·14	4·8
1:20	0·01	0·35	C·03	1·1	0·07	2·5	0·13	4·6	0·21	7·6

FLOW IN LARGE DRAINAGE CHANNELS	$V\text{m/sec} = \frac{1}{N} D^{2/3} S^{1/2}$ $V\text{ft/sec} = \frac{1{\cdot}486}{N} D^{2/3} S^{1/2}$
D = Hydraulic Depth	S = Gradient eg. $\frac{1}{100}$
N = 0·015 concrete, 0·02 smooth soil	0·03 gravel, 0·05 boulders — floor of channel

Figure 5.18 Discharge capacities for open channels and circular pipes

the flow rate, but the actual depth of water and the smooth or rough nature of the floor and walls of the channel also affect the velocity of flow.

In the case of small channels, it is suggested that — for ease of calculation — the cross-sectional area of the channel can be equated to that of a cylindrical pipe by the equation in Figure 5.18, and the table in that figure can be used to assess the capability of the channel adequately to permit water flow. For example, if 0.14 cubic metres of water have to be disposed of per second, then from Figure 5.18 it can be seen that a cylindrical pipe of diameter 0.3 metres at a gradient of 1:50 would be needed to carry the flow. By conversion back using the 'effective diameter' equation, the dimensions of a suitable channel can be assessed.

With regard to specifying the dimensions of a membrane-wrapped drain for groundwater flow, the particle size of the stone fill and the gradient of the trench must be taken into account. Figure 5.19 gives a useful indication of the discharge capacity of a 1 m × 0.6 m cross-section, stone-filled drain at a gradient of 1:100. The discharge capacity is, of course, the same whether the drain is membrane-wrapped or not. It can be seen from the chart that the capacity of the drain is proportional to both the cross-section and the gradient. Twice the cross-sectional area will result in the drain having twice the capacity. Half the area will result in half the capacity. A 1:50 gradient will have twice the capacity of a 1:100 gradient, etc.

It is necessary to consider very carefully the real environment of the membrane in the ground before choosing the final design. In many soils such as laterites, or laminated clays, permeability and waterflows are much higher than would be expected from an analysis of the majority of the soil body. This is because the mass permeability of the soil is governed by the smaller more permeable fissures and partings contained with it. Additionally, membrane permeability should always be chosen to cope with the waterflows from the most permeable soils, and under the wettest conditions considered likely to exist within the framework of any particular engineering scheme. These aspects should be observed in order to prevent the build up of water pressure behind any membrane system. Water pressure build-up behind a drain means that the drain is not working properly, and that the groundwater is not being lowered sufficiently. Water pressure build-up beneath the structure such as a dam could cause the overall instability of the dam by decreasing the stresses between the soil particles in the lower layer.

The situation of localised water flow occurs so often that in many of the cases to date, high permeability fabrics are chosen, perhaps at the expense of quality of filter formation. There is no doubt that this has proved a most cost-effective solution in

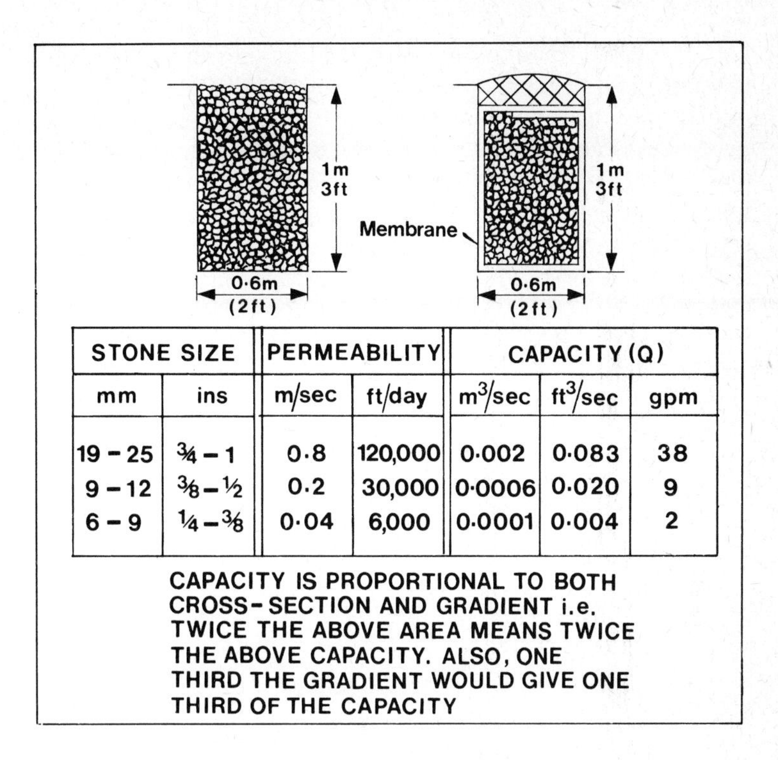

STONE SIZE		PERMEABILITY		CAPACITY (Q)		
mm	ins	m/sec	ft/day	m^3/sec	ft^3/sec	gpm
19 – 25	¾ – 1	0·8	120,000	0·002	0·083	38
9 – 12	⅜ – ½	0·2	30,000	0·0006	0·020	9
6 – 9	¼ – ⅜	0·04	6,000	0·0001	0·004	2

Figure 5.19 Discharge capacities of a 1m × 0.6m (3ft × 2ft) cross section stone filled drain for gradient 1 : 100

projects where piping criteria are not as critical as those of permeability.

In addition to the smaller-particle-size soils, membranes are sometimes used to protect coarser soils such as sand from outwash in a filter structure. In the case of protection of sand soils with a D_{85} greater than 0.25 mm, the critical factor becomes the limit of permeability of the filter membrane. This is dependent upon the D_{15} size of the Soil-Equivalent Membrane in the case of a non-woven fabric. In the case of a woven fabric, it is dependent upon the Open Area exposed for water passage.

Woven membranes have the ability to provide a large Open Area whilst retaining a relatively strong and stable structure, but it is hard to find a non-woven membrane with a very high water permeability. The technical specifications earlier in this book allow the assessment of such factors, and products such as Carthage Mills' Filter-GB (woven), and ICI's Terram 500 (non-woven) come to mind as being among the most permeable.

In the case of woven materials, the previously mentioned Open Area criteria can be applied, whilst keeping a critical eye on the possibility of blocking of the pores under unusual circumstances. In the case of non-wovens, it has been suggested that the following formula might apply: the D_{15} of the soil filter must be greater than or equal to 5 times the D_{15} of the soil to be retained and the minimum D_{15} of the soil filter must be equal to 5 times the O_{50} of the Soil-Equivalent Membrane to be chosen.

In principle, non-woven membranes are limited in the ultimate size of granular materials which they can filter, but woven open-mesh membranes are not. For example, the finest non-woven membrane available on the market at the present time is Terram 500. This has a O_{50} pore size of 0.2 mm. This limits effectively the size of soil which non-woven membranes can filter without the build-up of pore-water pressure, since the Author is not aware of any other non-woven with a higher permeability.

However, the Open Area of woven fabrics may continue to be increased until they effectively become 'meshed'.

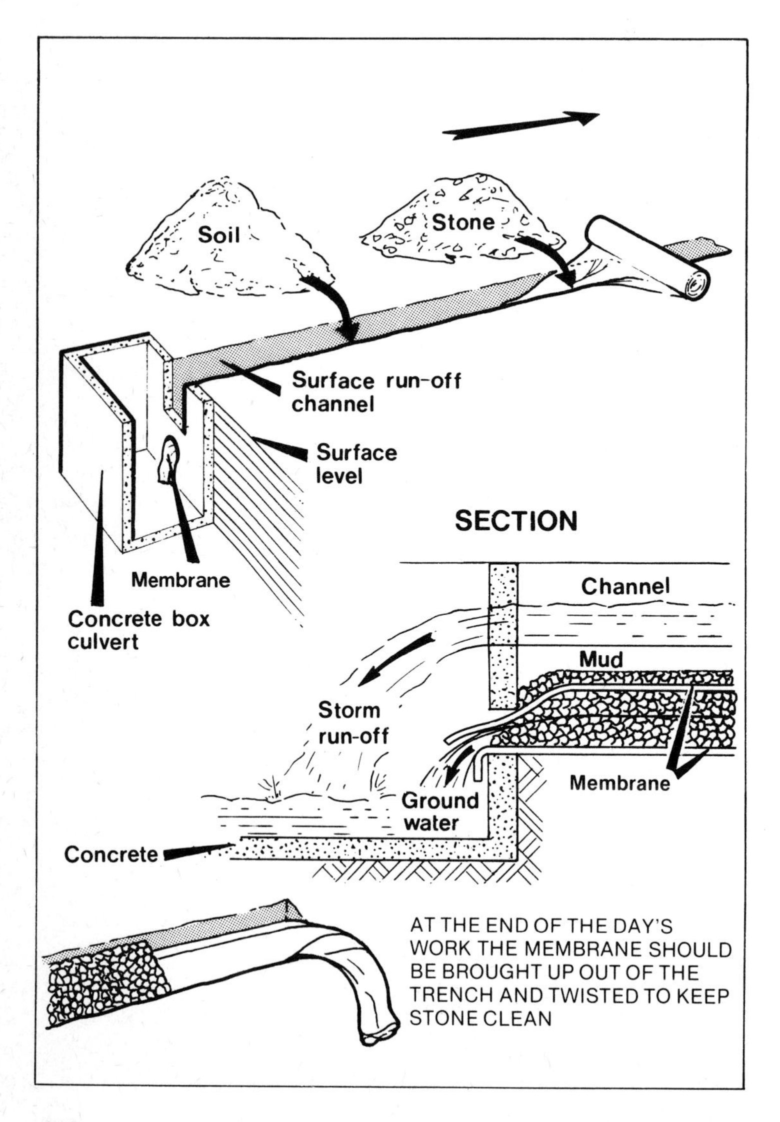

Figure 5.20 Construction technique for membrane-wrapped ground drain. Upper channel is only necessary if storm run-off is to be handled

Area Drainage

There are two fundamental aspects involved in the stabilisation of large areas. First is the removal of excessive water to increase the strength of the existing soils present, and the second is the construction of granular fill layers designed both to convey water laterally in a drainage function and to act as structural units supporting roads, buildings or railways. The fundamentals of this concept are that membrane-wrapped drains are laid at regular intervals in the saturated ground to lower the standing water table level. The detailed construction of a membrane-wrapped trench drain can be seen in Figure 5.20.

The Table in Figure 5.21 suggests suitable drain spacing for lowering the groundwater in conditions where surface saturation is caused by excessive precipitation (where water is moving downward in the ground, and draining away, but too slowly to prevent the soil in its natural condition from being waterlogged).

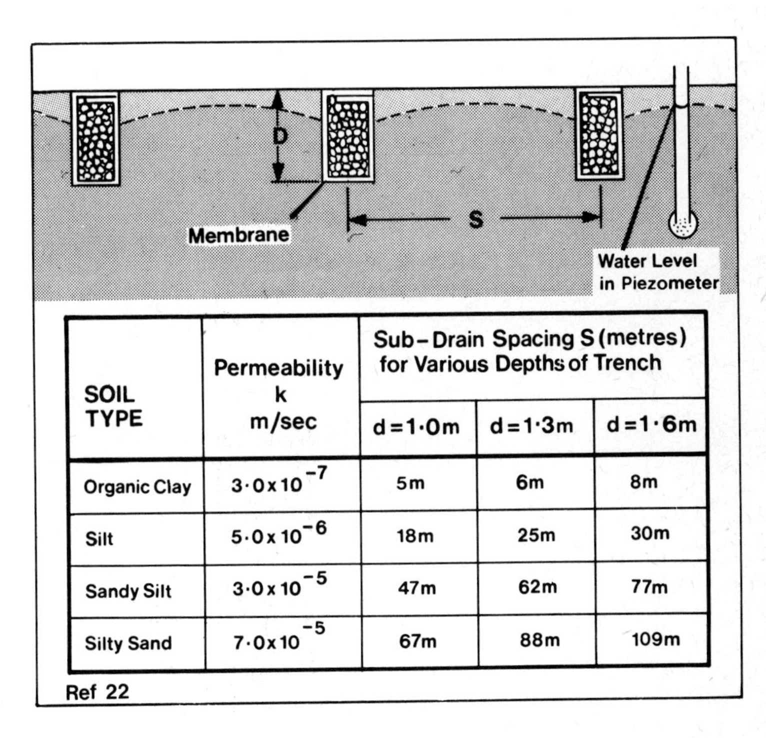

SOIL TYPE	Permeability k m/sec	Sub-Drain Spacing S (metres) for Various Depths of Trench		
		d = 1·0m	d = 1·3m	d = 1·6m
Organic Clay	$3{\cdot}0 \times 10^{-7}$	5m	6m	8m
Silt	$5{\cdot}0 \times 10^{-6}$	18m	25m	30m
Sandy Silt	$3{\cdot}0 \times 10^{-5}$	47m	62m	77m
Silty Sand	$7{\cdot}0 \times 10^{-5}$	67m	88m	109m

Figure 5.21 Ground drain spacing table for lowering ground water

Once a suitable drain spacing has been decided upon for keeping the ground soil well drained, then the size of the drains and the permeability of their stone fill has to be calculated using Figure 5.19. To use these tables it is necessary to assess the quantity of water (Q) which each drain will have to cope with, using the following equation:

$$Q = S \times L \times P$$

S is drain spacing in metres
L is drain length in metres
P is maximum expected rate of precipitation to be dealt with in metres of rain/second and Q is m^3/sec

EXAMPLE: A moderate rainfall would be 0.01 metres falling over a period of 1 hour, i.e. 2.8×10^{-6} m/sec. Assuming a drain spacing S of 15 metres for a 1 m deep trench in silt soil, a trench length (L) of 50 m, then the volume of water to be dealt with per second is:

$$Q = 15 \text{ m} \times 50 \text{ m} \times 2.8 \times 10^{-6} \text{ m/sec}$$
$$= 0.002 \text{ m}^3\text{/sec}$$

Figure 5.19 shows that a 1 m × 0.6 m drain filled with 19-25 mm stone, at a gradient of 1:100 will transport 0.002 m^3/sec.

If surface run-off to the drains is not permissible — as in the case of Sports Playing Fields — then a surface-water drainage blanket must be provided below the surface soil and vegetation. This should be contained between two layers of membrane to prevent contamination. For example, Figure 5.22.

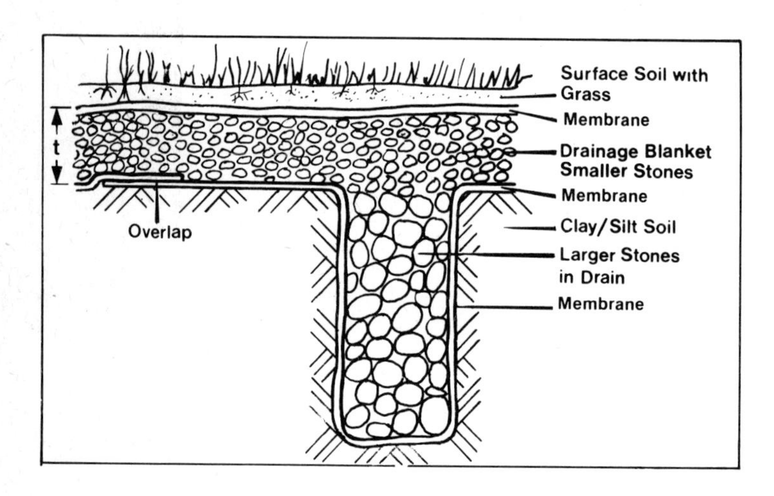

Figure 5.22 Double-membrane ground drainage system

To calculate the thickness and permeability of the drainage blanket, use the equation

$$t = \frac{S}{2}\sqrt{\frac{P}{k}}$$

P = rainfall max in m/sec
S = drain spacing (m)
k = permeability of stone selected (m/sec)
t = thickness (m)

To extend the previous example, where $S = 15$ m and $P = 2.8 \times 10^{-6}$ m/sec, and assuming that it has been decided to use a coarse sand for the drainage blanket, then:

From Figure 5.16 it will be seen that a coarse sand has a permeability of about 0.01 m/sec.

$$t = \frac{15}{2}\sqrt{\frac{2.8 \times 10^{-6}}{0.01}} = 0.13$$

i.e. a 130 mm thick layer of coarse sand is just adequate on a theoretical basis. It would be sensible to increase this thickness by anything up to ten times if only sand is available, to allow for variation in permeability, and for rainfall peaks.

The preferable solution is to use a fine gravel of permeability $k = 1.0$ m/sec at a thickness of say 100 mm. The factor of safety here would now be comfortably high. (Note: a safety factor of at least 10 is advisable in permeability designs.)

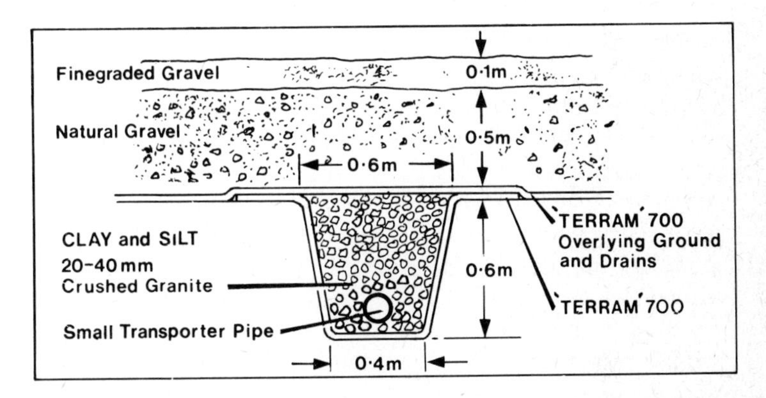

Figure 5.23 Single-membrane ground drainage system

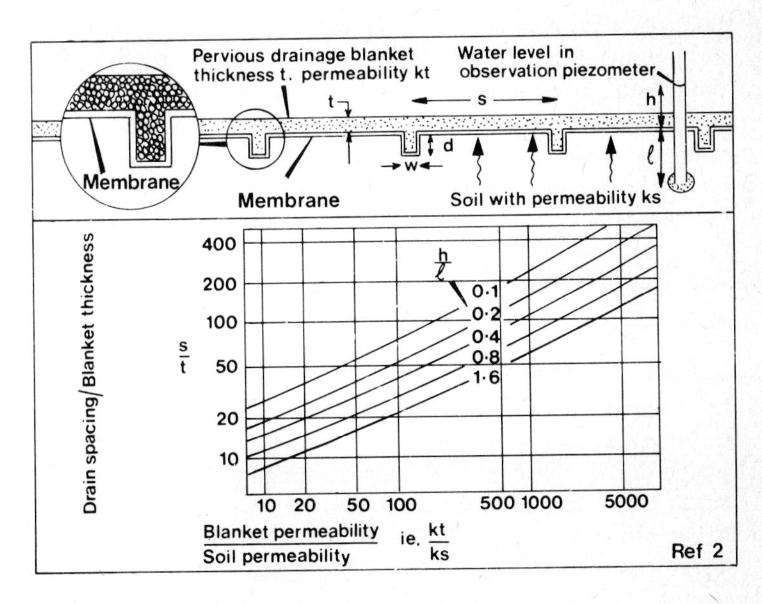

Figure 5.24 Chart for assessing drain spacing and drainage blanket characteristics in saturated ground under artesian pressure. (i.e. rising ground water)

A slight variation of the above technique was used at Gothenburg in Sweden for the construction of a free-draining playing field over an area of 110 × 70 m where the section shown in Figure 5.23[23] was used over silt/clay with a 7 m spacing for the drains. If the groundwater conditions are troublesome as a result of *rising* water, then the drain spacing (*S*) will be determined by the thickness of the over-placed granular fill in accordance with Figure 5.24.

Sloping Filters

There are several instances where sloping filters are required to prevent soil outwash from sloping soil faces with high internal water tables. For example, the sloping faces of road or railway cuttings, or the downstream faces of dams. The problem associated with waterflow from such slopes is that in the case of cuttings, the lowering of the ground level, and in the case of dams, the raising of the water level, has generated an unusually steep localised hydraulic gradient. The hydraulic velocity generated within the soil as a result of this gradient may exceed the critical internal particle transport velocity of the soil. If this is the case, particles may be moved considerable distances through the soil, and may be washed out of the soil on to the sloping face of the embankment, or cutting. As these fine particles are moved out of the soil, so the overall mass permeability of the soil increases with a resulting increase in the velocity of the waterflow. It is apparent that this is a self-aggravating process and eventually waterflows can reach such a point — especially in the case of dams — that the entire structure can collapse.

Once individual piping channels have been established through a soil mass, then water will be coming out of these in similar quantities to pipe flows, and it could be dangerous to cover these with a membrane. However, whilst water is seeping from the soil through the general mass, then it is possible to place a membrane on the soil to act as a filter thus preventing the outward loss of soil particles. The use of membranes in these applications for dams is extremely critical and not so critical on road cuttings. In the case of dams, great consideration must be given to the purpose of the membrane and the choice of membrane consequential upon the establishment of its required function. For example, the sole function of the membrane may be to prevent outwash erosion on the face of the slope. This is a far more common problem than the long distance internal piping of soil particles. The external sloughing of the upper soil layers on an embankment can cause visually unpleasant scars, dangerous blockage of the roadway or railway subsequent to heavy rainfall, and can prevent vegetation establishing itself upon the slope. In such a situation, and in a similar situation on a dam toe, a membrane can be extremely useful, together with an overlying stone layer whose function is to hold the membrane down and to provide a run-off transporter for the water outflow (see Figure 5.25).

The curves in Figure 5.26 can be used to illustrate the relationship between the dimensions and properties of the materials in the sloping filter of Figure 5.25. The membrane eliminates the need for Terzaghi-type sand filters. In constructing the membrane filter, the membrane must be selected to match the soil of the slope according to the criteria specified in Figure 5.15. The membrane must therefore both prevent piping, and yet allow sufficient water outflow to prevent back-pressure from building up.

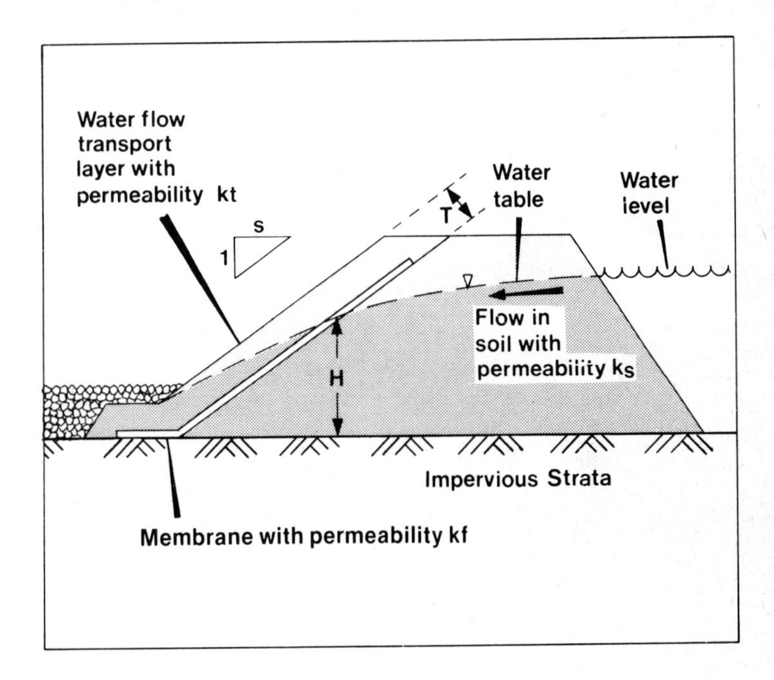

Figure 5.25 Permeable membrane used to prevent slope outwash erosion

Both the permeability k_t and the thickness T of the transporter layer can be calculated from the dimensionless chart in Figure 5.26. It is usually found in practice that k_t is fixed by the local supply of material available, and therefore the chart is used to calculate the thickness T.

Rising Water (Artesian Water) in Road or Railway Cuttings

Figure 5.27 illustrates the situation where a road cutting has been made through an area of relatively high water table. The effective piezometric head is considered to be at the position of the old water table. Consequently, if this were an enclosed

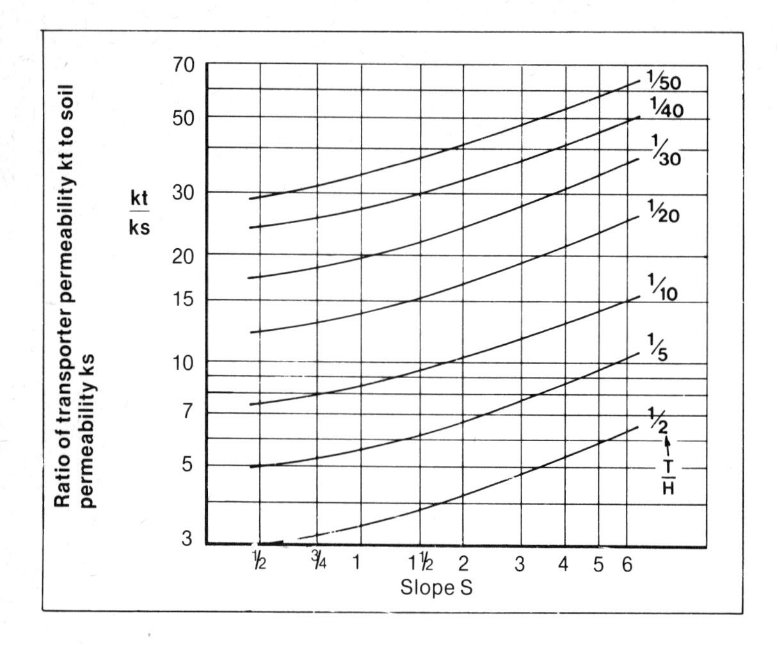

Figure 5.26 Chart for assessing required thickness (T) of slope transport layer in Figure 5.20

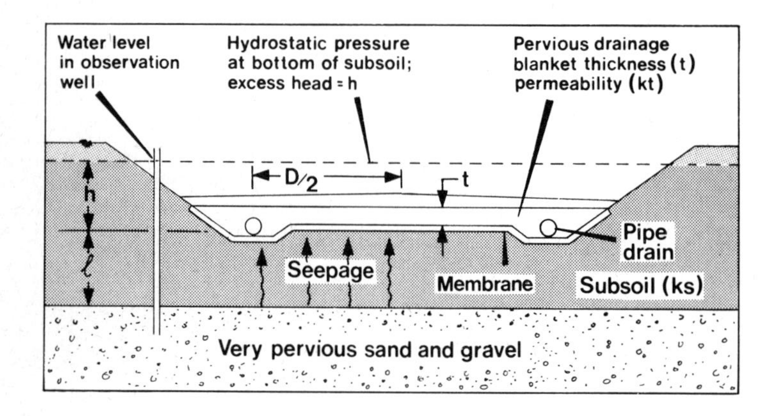

Figure 5.27 Road cutting with artesian water head

excavation, water would run into it and would fill it up to that level and thereby re-establish the old water table. However, being a road cutting, drainage systems must be built in to transport the water away to lower ground. Provided that the water supply within the surrounding country rock is sufficiently large, then the water will continue to flow into the cutting. It will flow in not only through the sides, but — especially if there are subterranean aquifluves such as gravel lenses or beds of sand — then the water will rise up under hydrostatic pressure through the ground and will enter the road structure from beneath. If, inadvertently, the road structure is made relatively impermeable, then the pressure will build up until it reaches a hydrostatic pressure of h, as shown in the diagram. Under practical conditions, long before the pressure reached the value of h, the

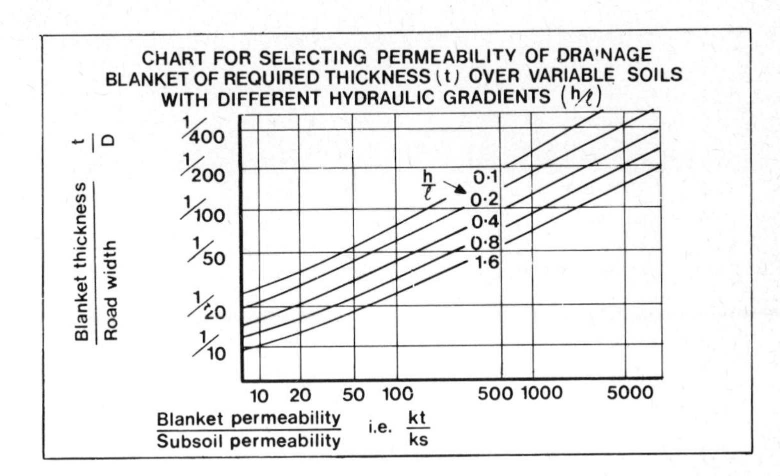

Figure 5.28 Chart for assessing required permeability of horizontal drainage blanket in Figure 5.22

road would have been lifted or partially destroyed by the water. Therefore where such circumstances exist, or are considered possible, it is important to construct a horizontal drainage blanket. The object of this drainage blanket is to intercept the rising water and to act as a lateral transporting drain, taking the water from the road centre out towards the lateral drains. Figure 5.28 is a chart for assessing the required permeability of the horizontal drainage transporter layer in Figure 5.27.

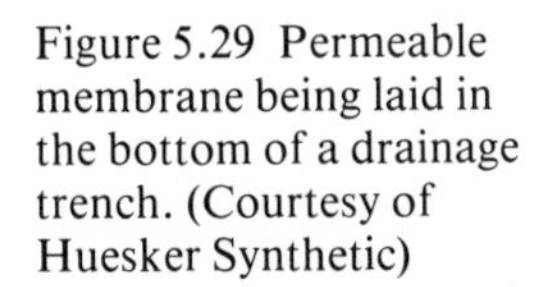

Figure 5.29 Permeable membrane being laid in the bottom of a drainage trench. (Courtesy of Huesker Synthetic)

Figure 5.30 Permeable membrane being used as a vertical support wall for an open ground drain. (Courtesy Nicolon BV)

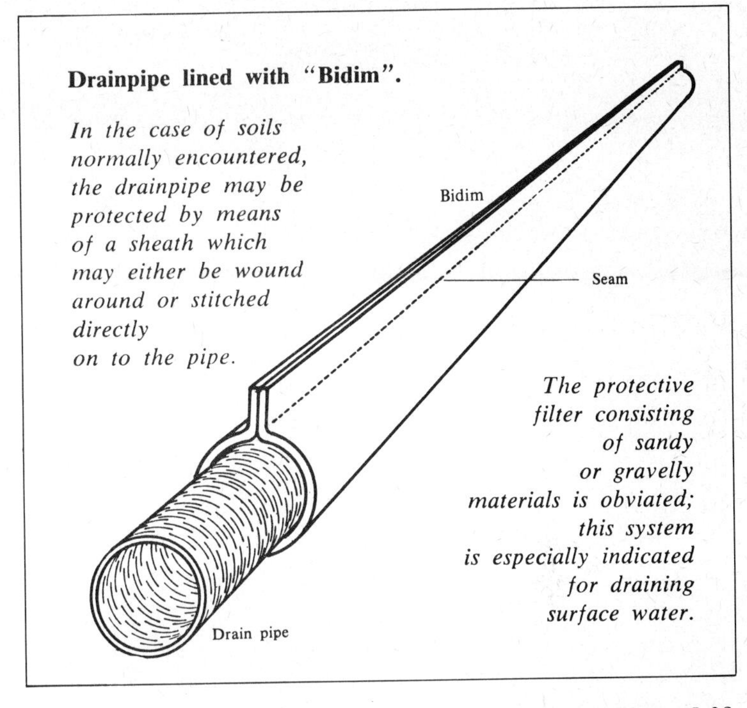

Figure 5.31 Rhone-Poulenc suggest that their Bidim can replace the granular filter around pipes. (However, this membrane does have a varying permeability, with overburden pressure — see Figure 3.19I — which should be considered before use)

The illustrations and photographs in Figures 5.29 to 5.38 show a variety of designs and uses of different types of membranes for one-way filter structures. It is intended that the reader may become familiar with currently accepted designs, notations and detailing techniques.

Figure 5.32 ICI's Terram being laid into a water-filled trench to make a membrane-wrapped drain

Figure 5.33 Membrane filter being used to protect a highway longitudinal drain. (Courtesy ICI Fibres)

Figure 5.34 (Opposite) The automated laying of flexible drain pipe with a permeable membrane cover. (Courtesy ICI Fibres)

Figure 5.35 Filtram drainage material being placed behind a vertical wall structure. (Courtesy ICI Fibres)

Figure 5.36 Filtram being placed in a ground drain. (Courtesy ICI Fibres)

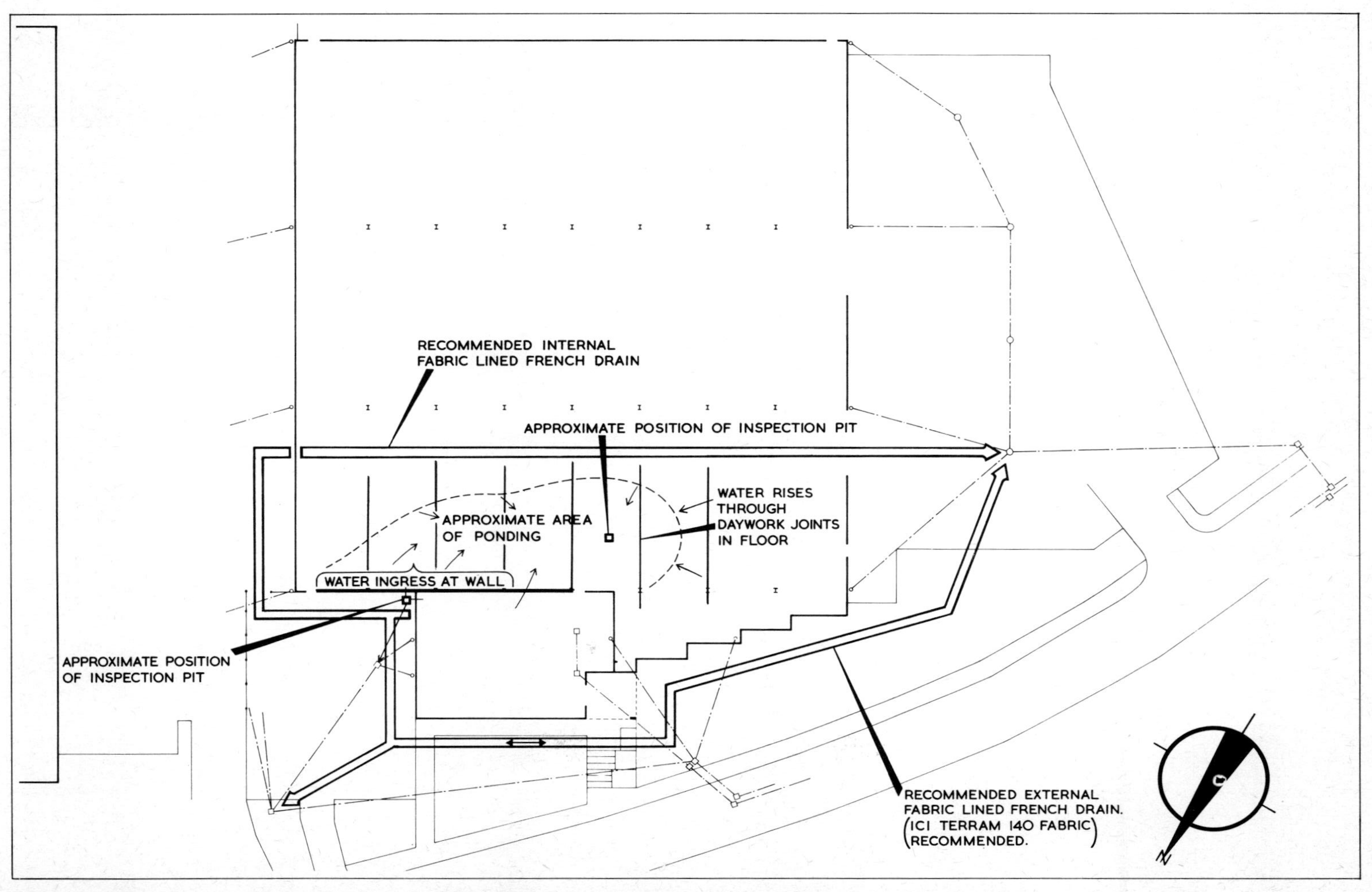

Figure 5.37 Simple drainage layout scheme. (Courtesy Manstock Geotechnical Consultancy Services Ltd.)

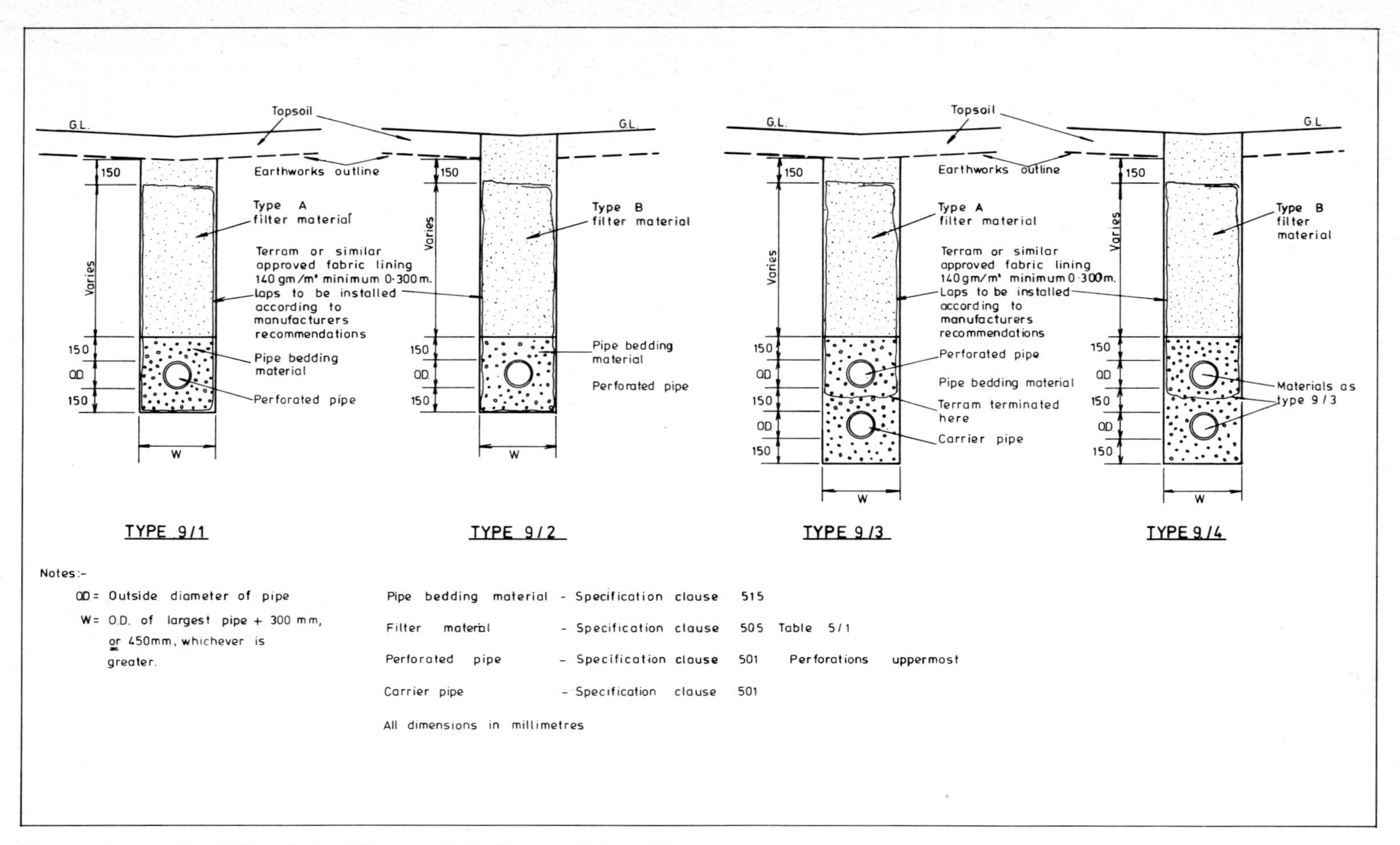

Figure 5.38 Details of filter drain. (Courtesy ICI Fibres and Kent CC)

6

River defences

The predominant problem in the maintenance of rivers is to prevent erosion of the banks on the outer sides of bends. Figure 6.1 shows that, as the water moves round a bend, it travels in a spiral form, cutting the bank on the outer edge and depositing eroded material on the inner side. The outer bank becomes steeper and less stable whilst the inner bank accumulates sediment. The consequence of this is that, as a rule where river defences are required, the working conditions are extremely rigorous. The defence structure has to be placed on a steep bank and under rapidly flowing water. In this respect, the first consideration is to ensure that any construction work is undertaken during the driest season available, when the river is at its lowest level, and running at its slowest speed. The *design* however, must cope with the worst flood conditions expected throughout the life of the structure. This aspect is critically important since the particle-carrying and rolling power of a river increases considerably with only a slight increase in velocity. Figure 6.2 is a chart showing the speed at which particles are picked up, carried and deposited by moving water. Unlike wind, water (because of its relatively high viscosity and density) has the ability to carry particles at a much slower velocity than it requires to pick them up. Once lifted into suspension therefore, soil particles will travel considerable distances before being deposited. In addition, there is the ability of water to move large stones and pebbles by rolling them. This rolling velocity is very much lower than the carrying velocity.

For example, in Figure 6.2 when flow velocity increases to 0.2 m/sec a river would begin to pick up particles of 0.2mm diameter from their natural bed, and erosion commence. If the velocity drops to 0.1 cm/sec, the stream can no longer loosen

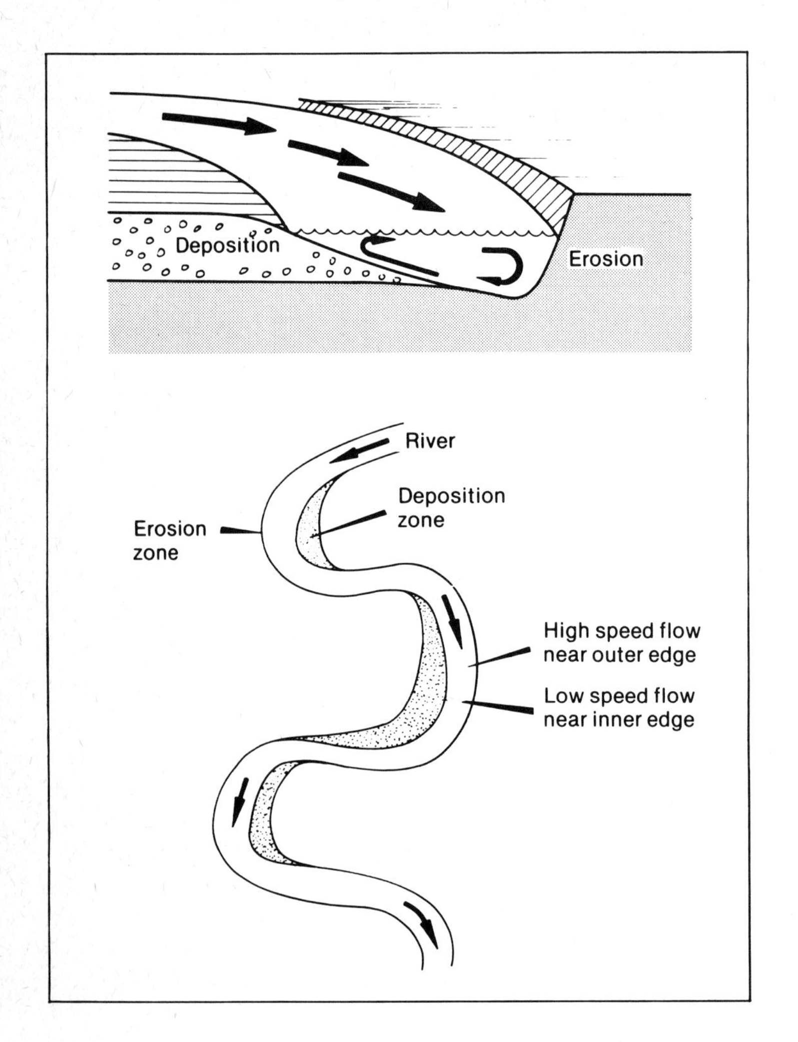

Figure 6.1 Some characteristics of river erosion

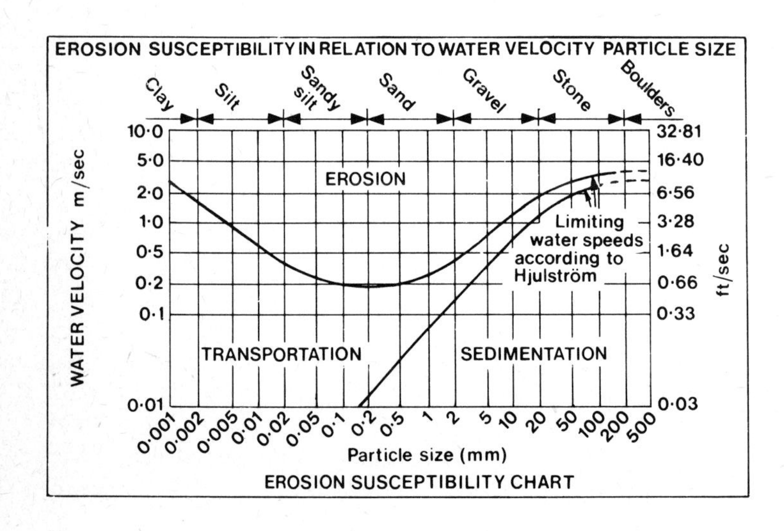

Figure 6.2 Soil erosion susceptibility in relation to water velocity and particle size

these particles, but it will continue to carry them. The velocity would have to fall as low as 0.01 m/sec before the 0.2 particle would eventually be dropped and deposition commence.

The change in river velocity as a result of increased flow caused by the rainy season or major storms, is the agent by which the soils are moved down river, but the erosion on the curved bank and the deposition on the inner curve is caused by the fast moving water at the outer edge picking up soil particles, and then, in following the spiral shown in Figure 6.1, depositing the particle as the velocity drops against the inner edge. It is important to understand this mechanism, since it is the mechanism against which most defences are constructed. However, the overall erosion problem caused by increase in total water flow should not be overlooked (Figure 6.3). A river defence design must anticipate the action of these two erosion types.

A most important aspect to remember is that a river is a dynamic, interactive phenomenon. This means that it is in the nature of a river to change its shape constantly, and therefore stabilisation efforts will be attacked with increasing vigour as time progresses. By being interactive, each part of a river either absorbs energy from the water, or contributes particles to the water which *directly affects* all parts of the river downstream.

In the first instance, if a defence is installed as shown in Figure 6.4A, then the removal of inner bank sediments can cause instability since the supply balance is disturbed, and collapse can result (Figure 6.4B). Deep toe design should be considered as in Figure 6.4C. Secondly, it must be remembered that the introduction of a defence effectively 'robs' the lower stretches of the river of sediment, which would otherwise be obtained from that part of the bank, and thereby increases erosion downstream. No Civil Engineering operation on a river can be considered in isolation (Figure 6.5). In principle therefore it is always worth considering keeping defence works to the minimum necessary.

Assuming that a defence has been decided upon, it can be seen that the insertion of a membrane into moving water is an extremely difficult task. In technical terms, the placing of the finer elements of a graded granular filter structure may be

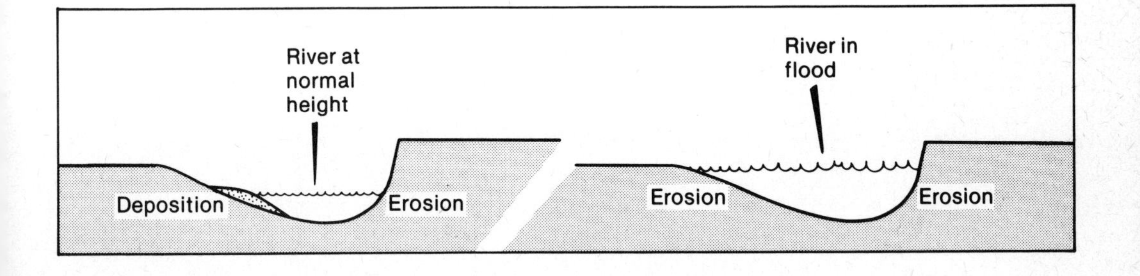

Figure 6.3 During flood flows, cross-deposited sediments are removed and carried downstream

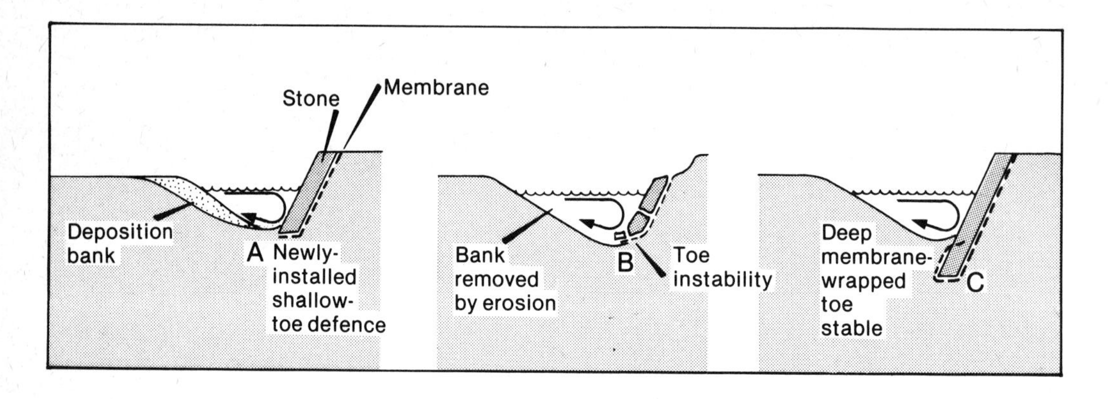

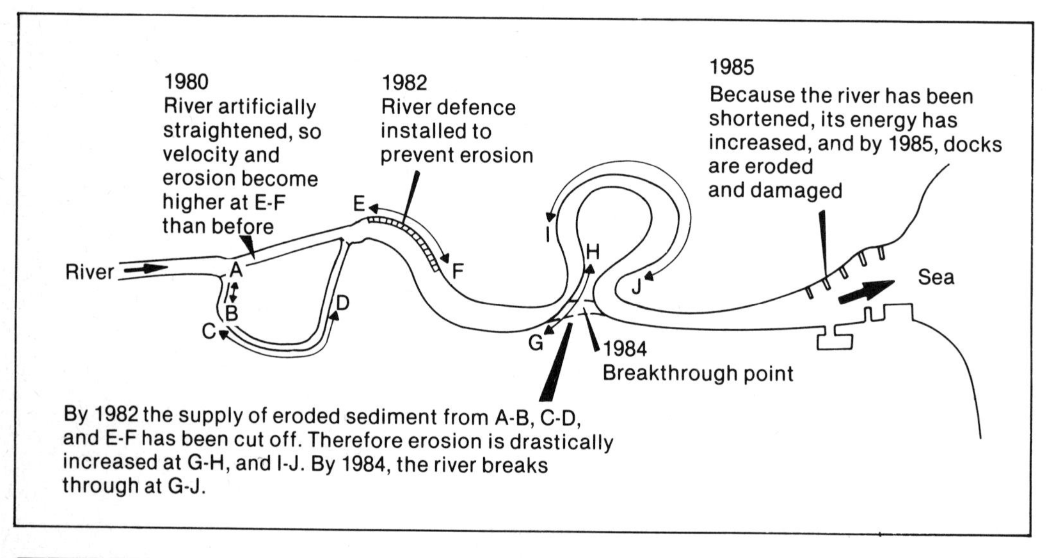

VELOCITY OF FLOW AT THE TIME OF PLACING MEMBRANE		ASSESSMENT OF HANDLING CONDITIONS LIKELY TO BE EXPERIENCED IN PLACING MEMBRANE		
		DIFFICULT	MEDIUM	EASY
Very High > 10m/sec	CLAY SILT SAND	LP.VHS MP.VHS MP.VHS	LP.VHS MP.VHS HP.VHS	LP.HS MP.HS HP.HS
High 5-10m/sec	CLAY SILT SAND	LP.VHS MP.VHS MP.VHS	LP.VHS MP.VHS HP.VHS	LP.HS MP.HS HP.HS
Medium 2-5m/sec	CLAY SILT SAND	LP.VHS MP.VHS HP.VHS	LP.HS MP.HS HP.HS	LP.MS MP.MS HP.MS
Slow 0.2-2m/sec	CLAY SILT SAND	LP.HS MP.HS HP.HS	LP.MS MP.MS HP.MS	LP.LS MP.LS HP.LS
Very Slow < 0.2m/sec	CLAY SILT SAND	LP.HS MP.HS HP.HS	LP.MS MP.MS HP.MS	LP.LS MP.LS HP.LS

VHS = VERY HIGH STRENGTH HP = HIGH PERMEABILITY
LS = LOW STRENGTH LP = LOW PERMEABILITY

Figure 6.4 River bed can become lower after introduction of defence. This can cause collapse

Figure 6.5 A hypothetical 5-year chain of events instigated accidentally by straightening the river between A and E

impossible in a rapidly moving river, and therefore one may have to use a membrane despite the obvious handling difficulties. In the Author's view, high strength of membrane may be of paramount importance in a moving water environment in most cases, but of course other factors can sometimes modify this. For example, consider the case of a large river moving at 2 m/sec, where the membrane is to be placed in six metres depth of water from the steep face of the erosion cliff. Compare this with the case of a river only 30 cm deep moving at 2 m/sec where a shallow ford is to be constructed. The water velocity is the same in both cases, but the handling conditions are far less demanding in the second case.

Assuming that a permeable membrane-lined defence has been decided upon, it may be seen that there are several aspects to be considered in relation to the membrane alone:

1. What are the handling conditions likely to be during placing?
2. What is the water velocity likely to be during placing?
3. What type of soil is to be protected by the membrane?

Figure 6.6 is a table suggesting the type of membrane needed, depending upon the variation in the combinations of the above three factors. After using Figure 6.6 reference can be made to the actual membranes available on the market for the required properties as shown in Section 1 of this book. Figure 5.15 will help the reader to make a more detailed membrane choice with respect to filtering requirements for the particular soil type to be protected.

Having chosen the appropriate membrane for the situation, use Figure 6.7 to select a size of Final Defence stone large enough not to be moved by the river's peak velocity during the

Figure 6.6 (Left) Table indicating strength/permeability properties required for laying membrane in moving water under different conditions of handling difficulty

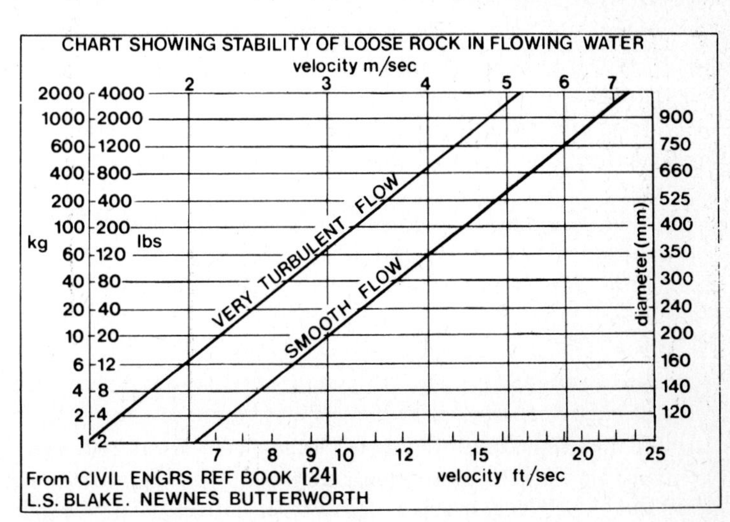

Figure 6.7 (Right) Chart showing the stability of loose rock in flowing water

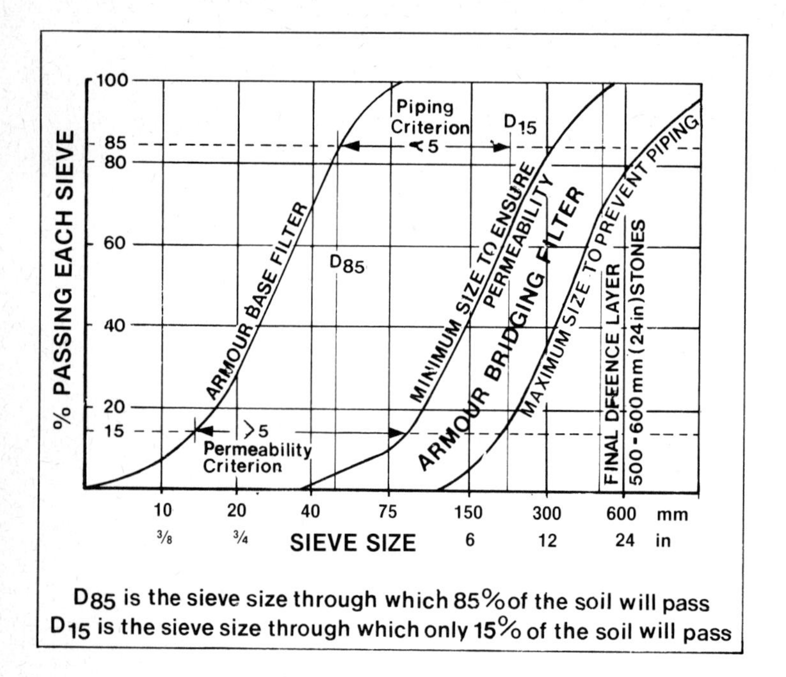

Figure 6.8 Design of a granular Armour Bridging Layer to support a defence stone and protect a granular base filter

rainy season. Then, using the principles shown in Figure 6.8, select the Armour Bridging Filter layer of correct size stones and lay this to the same thickness as the largest stones in the Final Defence Layer.

Empirical Example A river has a maximum flow velocity of 3.5 m/sec. The *in situ* soil is a sand, and it is expected that site handling conditions will be very difficult since the membrane will be laid in deep water on a bend of the river. It is decided to build the defence during the dry season when the flow velocity is expected to be only 2.5 m/sec.,

1. *Choose a membrane.* Using Figure 6.6, it can be seen that a membrane for medium (2.5 m/sec) velocity on sand under difficult handling conditions would have an HP/VHS specification. There are a few membranes that can meet this requirement, and a study of Figures 4.9 to 4.90 would produce — amongst others — ICI's Terram 3015M range, Heusker's H-5/T, and Carthage Mills' Poly-Filter GB.
2. *Choose the Final Defence Stone.* Using Figure 6.7, it can be seen that to resist a flow velocity of 3.5 m/sec, loose placed stones of between 260 and 520 mm diameter (30-200 kg wt) would be needed, depending upon flow type. Using a worst case criterion, 500 mm diameter can be selected.
3. Use the principles in Figure 6.8 to *choose a suitable Armour Bridging Filter layer* to be placed directly on the membrane (which itself may act as an armour base filter). The armour

Figure 6.9 River defence design as discussed in text

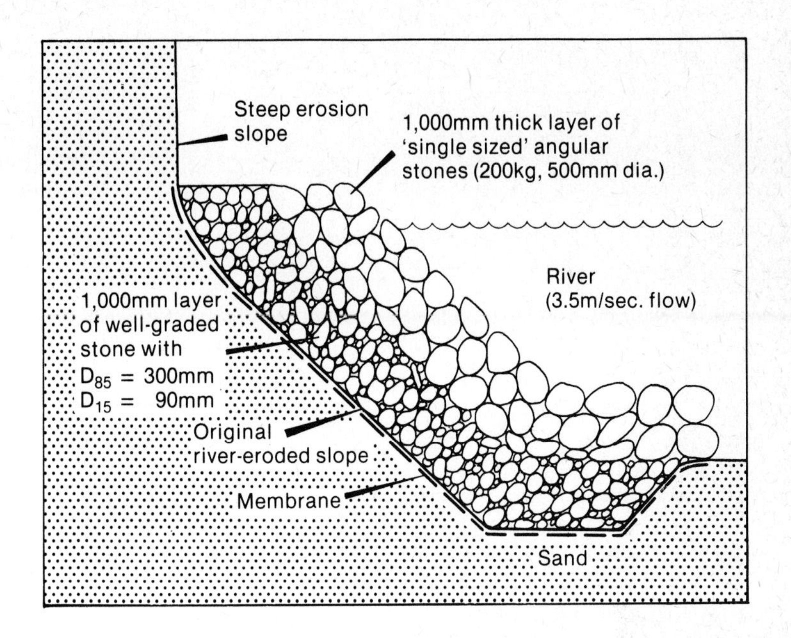

bridging filter will therefore have a D_{85} of 300 mm and a D_{15} of 90 mm.

4. The Armour Bridging Filter is laid to the same thickness as the Final Defence layer, which itself is laid at least twice as thick as the individual stone diameters.

The final design is shown in Figure 6.9, which simple structure replaces the theoretical four or five filter layers of granular material, most of which in practice would be too fine to lay in moving water of 2.5 m/sec. Note that in smooth flow conditions of 2.5 m/sec, the water will not move a 150 mm diameter stone, so the armour bridging filter layer with a D_{40} of 150 mm will be highly stable when laid on the membrane.

Figure 6.10 shows a typical river bend with membrane positioned ready for stonework to be laid. The Author assesses the laying conditions as relatively easy, with the river at a low level, and running smoothly. This has allowed a long stretch of membrane to be fixed without stone cover. Figure 6.11 shows a two-layer stone defence being placed on top of a membrane.

In rough water conditions, the second or underlayer of stones (Armour Bridging Filter) serves to dissipate turbulent energy from the moving water thus preventing erosion. Also, where *in situ* soil is a soft clay, it acts as a load redistribution layer, spreading out the point load stresses that would otherwise be placed on the membrane by the large Final Defence stones. Therefore in quiet water conditions, and where the undersoil is effectively granular, consideration can be given to eliminating the Bridging Filter in the interests of speed and economy, as

Figure 6.10 River embankment with membrane ready for stone laying. (Courtesy of Huesker Synthetic GMBH)

Figure 6.11 River embankment with membrane and two layers of stone for protection as suggested in Figure 6.9. (Courtesy of Huesker Synthetic GMBH)

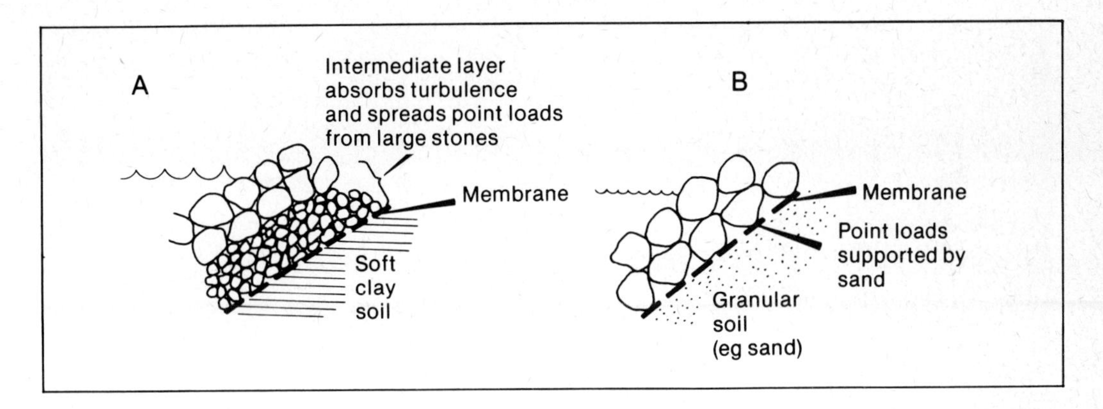

Figure 6.12 The use of an intermediate stone layer can be dispensed-with in quiet water situations on granular materials

illustrated in Figures 6.12 and 6.13. In such a case the membrane selected should have good abrasion resistance since large stones would be bearing directly upon it, and a permeability should be chosen to suit the external/internal waterflow regime. In the case of Figure 6.12A, it is now considered suitable to select a membrane with lower strength and higher extensibility, such as the non-woven types, in order to ensure that the membrane can follow ground movement without rupturing. As mentioned earlier, in reversing water flow conditions, recent research works indicate that a non-woven membrane should cause the development of a more effective filter in the soil structure than a woven one.

Of course, having designed a stable slope, and having chosen a membrane and stone size, there remains the problem of practical field installation. The design principle here is the same, in the Author's view, as for all membrane constructions. The structure is best built in functioning modular units. For example, the Author believes that it is better to build 20 metres of defence with membrane and stone on it, and then to build the next 20 metres, and so on ten times rather than lay out 200 metres of membrane, followed by 200 metres of stone! This latter course of action appeals at first sight because of the superficial ease of project management. However, when a large quantity of membrane is spread out in advance of its cover, then it becomes exposed to sunlight and consequent deterioration, and — more importantly — it becomes exposed to the risk of natural hazard or vandalism. By covering membrane in modular units, a sudden rise in river level for example, would not cause any damage, since each unit would be functional in its own right. Furthermore, the chance of the fabric being stolen or accidently damaged would be avoided.

In terms of detailed fixing, the fabric can be simply rolled down the bank where not underwater, and may be additionally pinned into the ground. However this is rarely necessary — especially if modular construction is used. For underwater

Figure 6.13 Enclosed water lagoon with single stone layer being placed on membrane. (Courtesy of Huesker Synthetic GMBH)

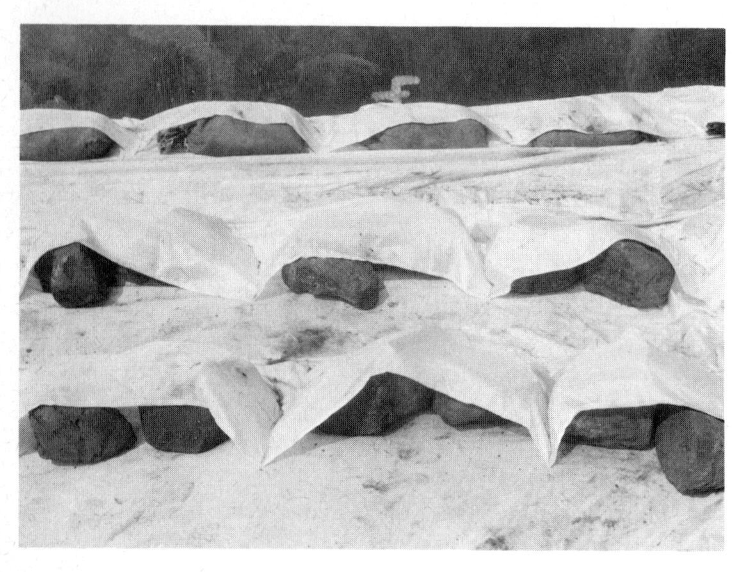

Figure 6.14 Pockets sewn into fabric to weigh it down when being placed underwater. (Courtesy of Huesker Synthetic GMBH)

Figure 6.15 Woven membrane with heavy threads for more-versatile design capability. (Courtesy Huesker Synthetic GMBH)

laying, however, it may be necessary to sew pockets onto the fabric in order to insert stones to weight it down (see Figure 6.14). Alternatively some fabrics can be ordered with heavy threads already sewn in, which can be used to attach to weights, or otherwise improve handling design under difficult laying conditions, as shown in Figure 6.15.

Perhaps one of the most important rules in laying a membrane-backed defence on a slope is to *always lay the toe stones first, and build up the slope!* As illustrated in Figure 6.16, if stones are dumped near the top, they will pull the membrane downslope. This is fairly apparent when a steep slope is considered, but it must also be remembered when work is being carried out on low-angle slopes.

For many minor river erosion problems, the simple 'vertical fence' type of defence is suitable and very cost-effective. Figure 6.17 shows the design concept of this defence, but the problem with most membranes — even if they should prove strong enough for the purpose is that they can suffer from ultra-violet light degradation which limits their life — especially in a ground-supporting situation. Figure 6.18 shows that this problem can be avoided by using a composite material comprising an ultra-violet

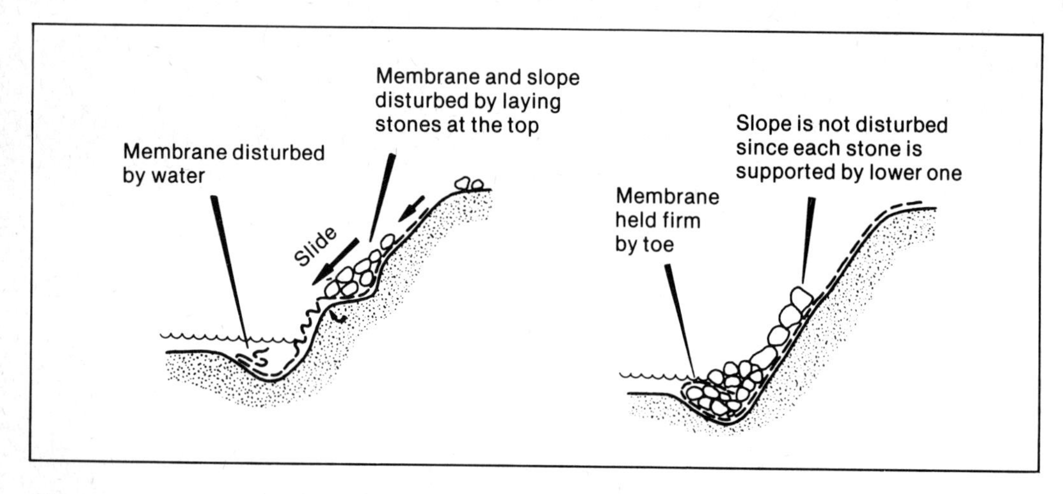

Figure 6.16 River and coastal membrane slope structures should always be constructed from the toe upwards in order to ensure stability

Figure 6.17 Cross-section of membrane vertical revetment

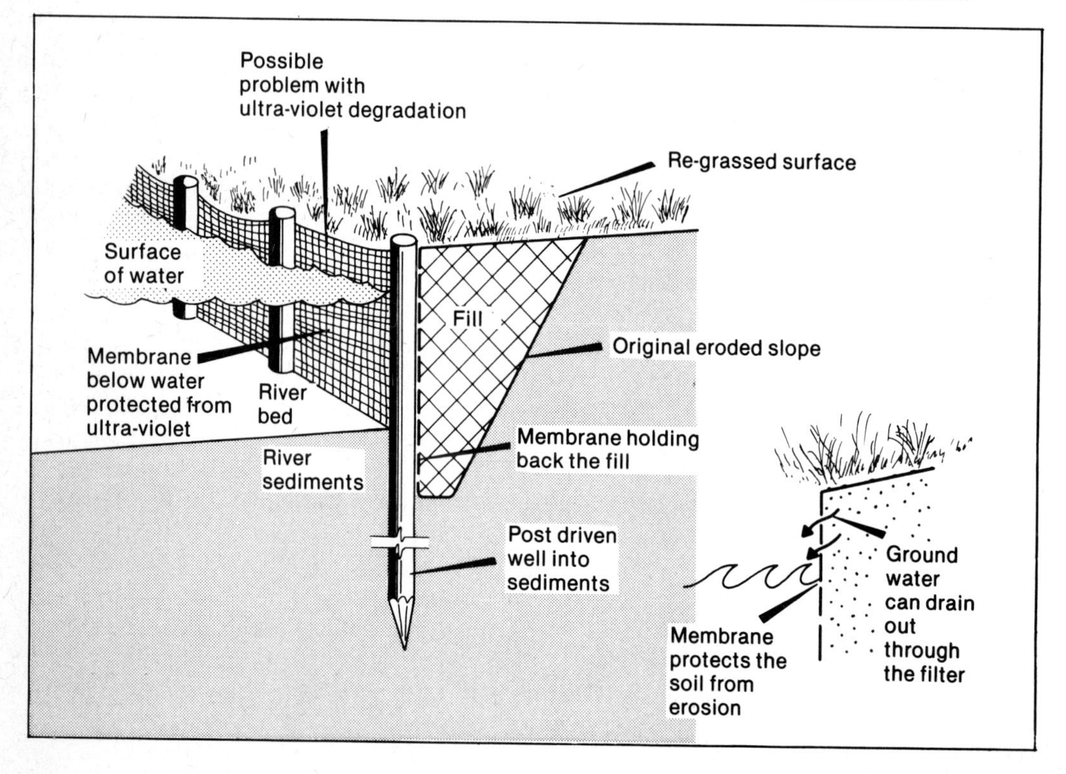

Figure 6.18 Vertical membrane revetment using Paraweb webbing which is unaffected by ultra-violet light. (Courtesy of Linear Composites Ltd.)

proof interwoven webbing to support the soil, whilst a non-woven fabric bonded to the back performs the soil filter function.

Finally, these webbing membranes can be used directly for river defences without any rock cover whatsoever. Owing to the fact that they are ultra-violet proof, they can be used to replace the entire conventional structure (as illustrated in Figures 6.19 and 6.20) in one laying operation.

Figures 6.21-6.28 show a number of either real design works which have been executed, or typical drawn representations of design types, from several different manufacturers, and different parts of the world. These have been unaltered in order that the reader may observe some of the currently acceptable methods of detailing into this type of design work.

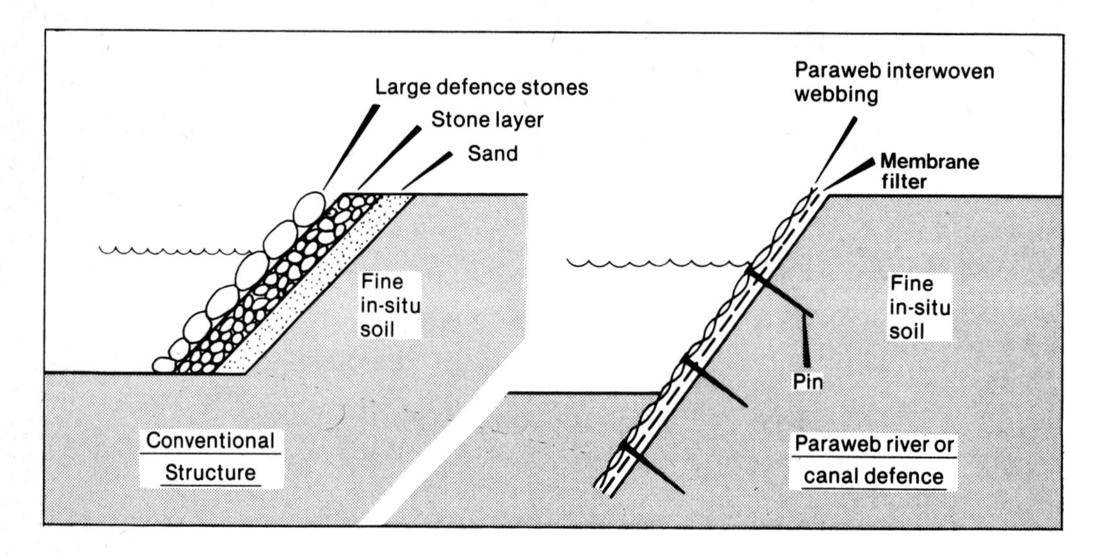

Figure 6.19 Comparison of conventional river defence with modern membrane system

Figure 6.20 River defence with webbing system. (Courtesy of Linear Composites Ltd.)

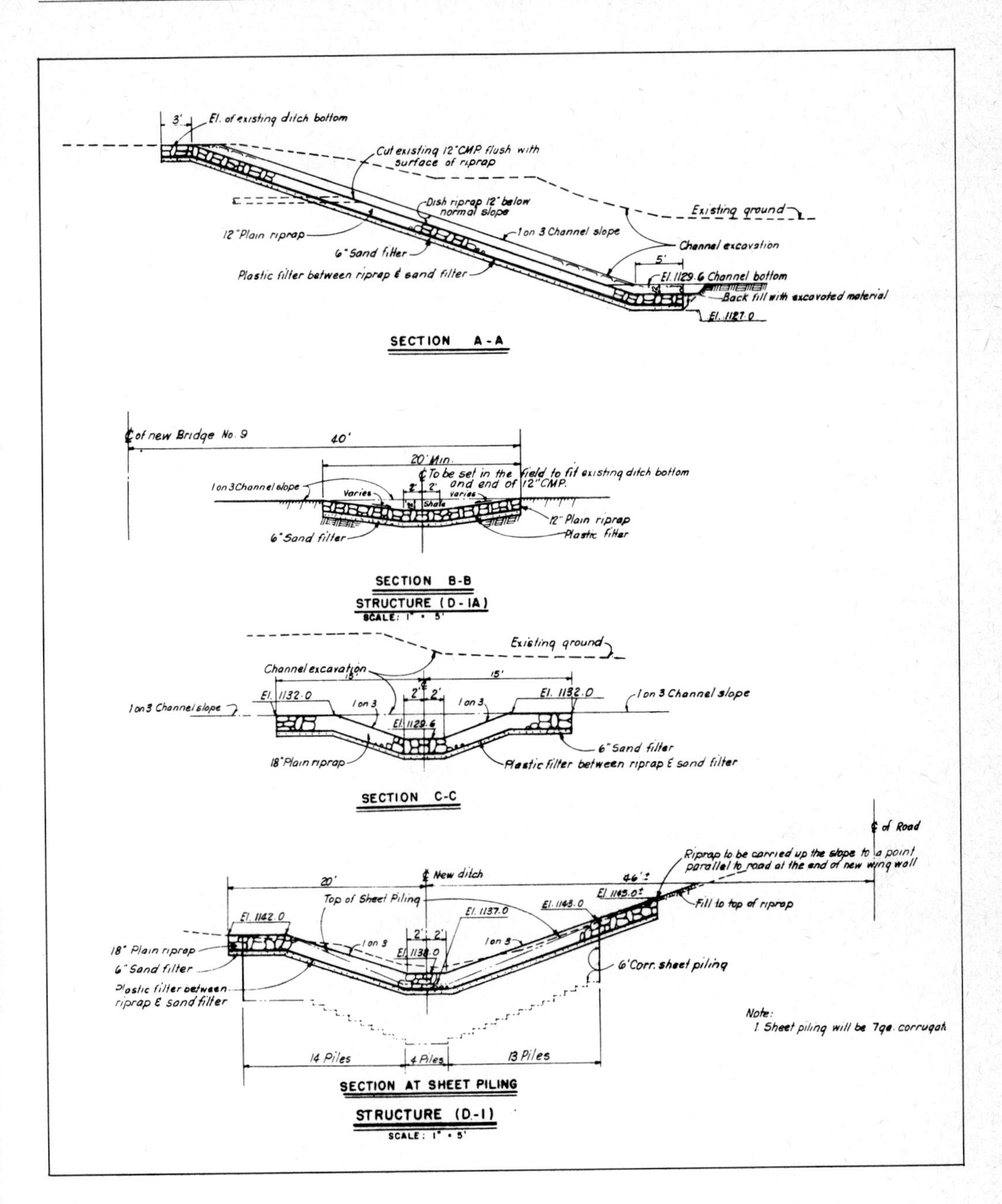

Figure 6.21 Cross-sections of channel linings, Minnesota U.S.A. (Courtesy of Carthage Mills Inc.)

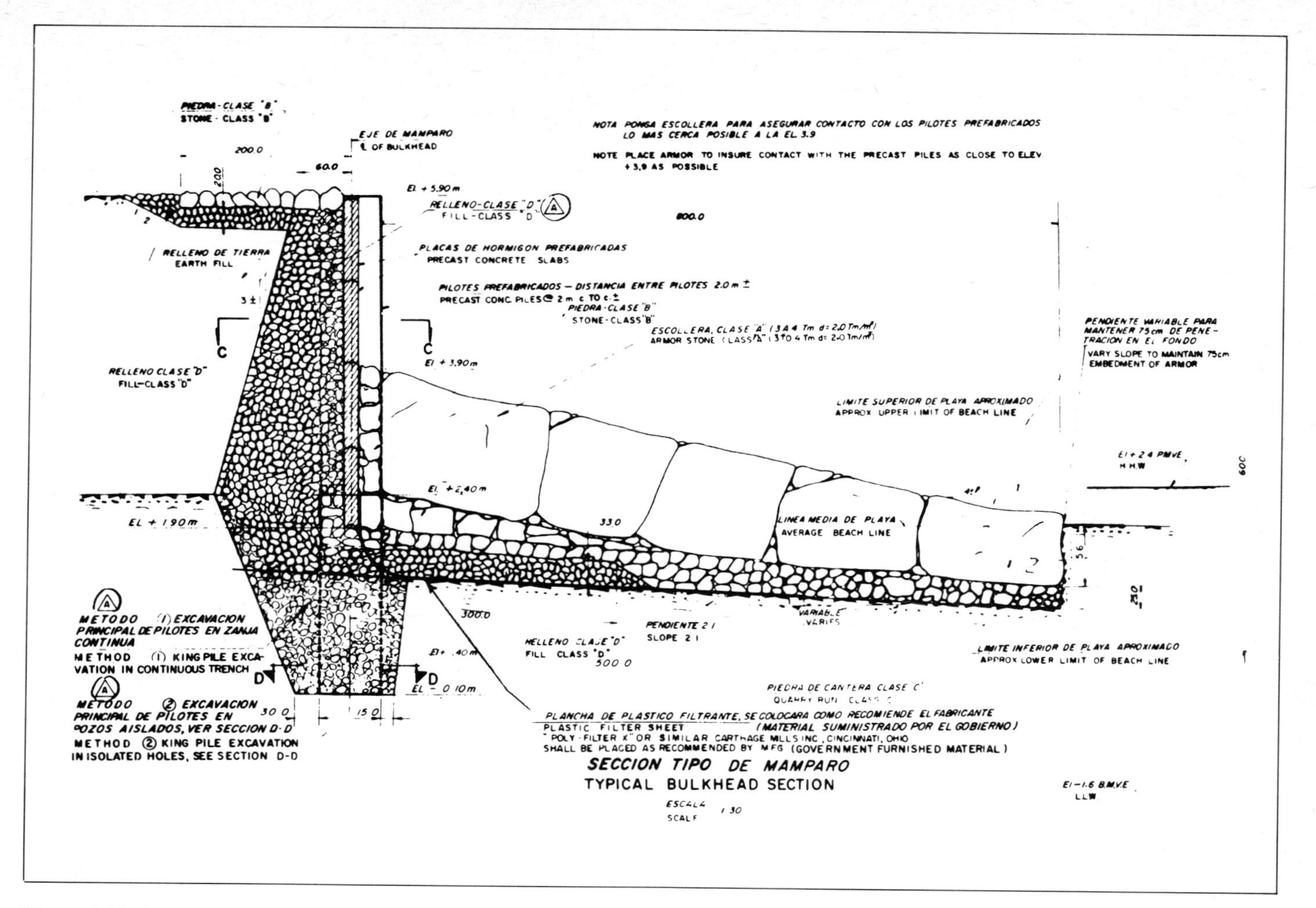

Figure 6.22 Cross section of membrane-lined vertical bulkhead — Madrid, Spain. (Courtesy of Carthage Mills Inc.)

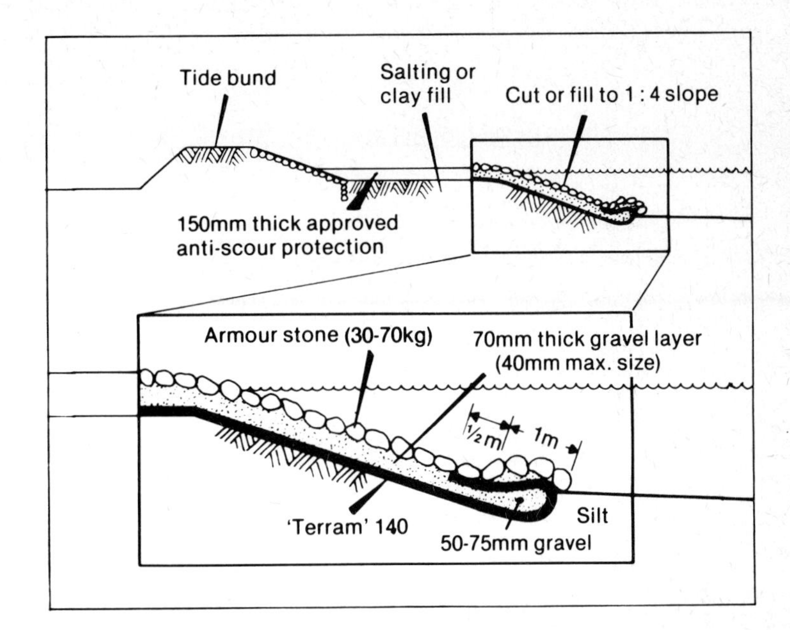

Figure 6.23 Cross section of River Defence London UK (Courtesy of ICI Fibres.)

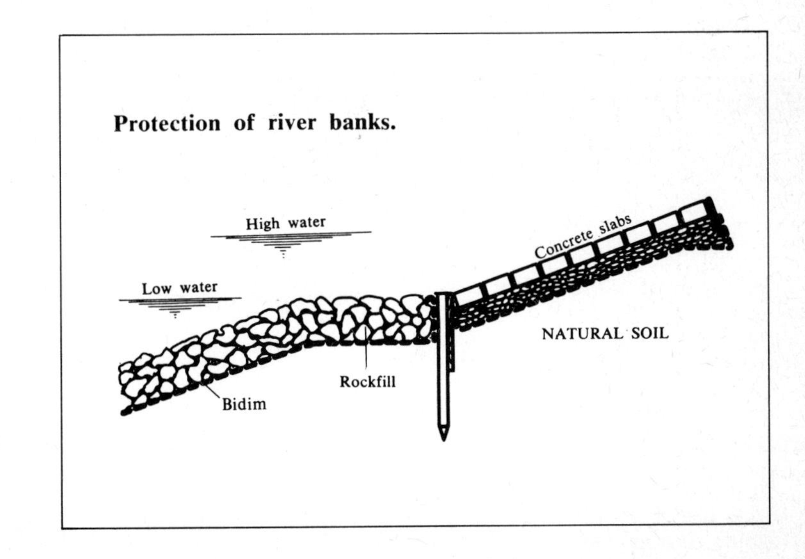

Figure 6.24 A river defence which Rhone-Poulenc Textile Co., suggest is useful for combating the effect of tides and dissipating uplift pressures, but not for absorbing the energy of breaking waves

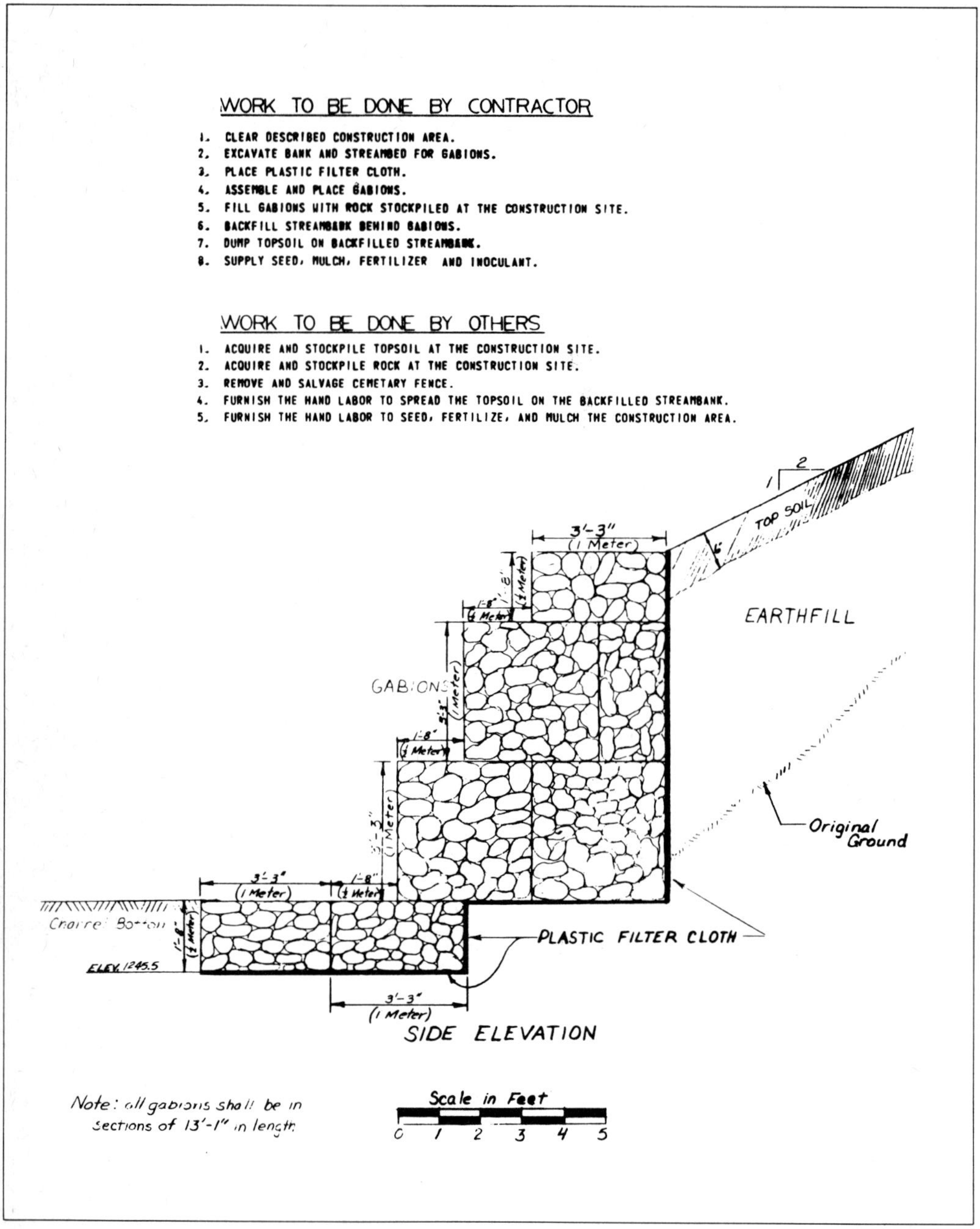

WORK TO BE DONE BY CONTRACTOR

1. CLEAR DESCRIBED CONSTRUCTION AREA.
2. EXCAVATE BANK AND STREAMBED FOR GABIONS.
3. PLACE PLASTIC FILTER CLOTH.
4. ASSEMBLE AND PLACE GABIONS.
5. FILL GABIONS WITH ROCK STOCKPILED AT THE CONSTRUCTION SITE.
6. BACKFILL STREAMBANK BEHIND GABIONS.
7. DUMP TOPSOIL ON BACKFILLED STREAMBANK.
8. SUPPLY SEED, MULCH, FERTILIZER AND INOCULANT.

WORK TO BE DONE BY OTHERS

1. ACQUIRE AND STOCKPILE TOPSOIL AT THE CONSTRUCTION SITE.
2. ACQUIRE AND STOCKPILE ROCK AT THE CONSTRUCTION SITE.
3. REMOVE AND SALVAGE CEMETARY FENCE.
4. FURNISH THE HAND LABOR TO SPREAD THE TOPSOIL ON THE BACKFILLED STREAMBANK.
5. FURNISH THE HAND LABOR TO SEED, FERTILIZE, AND MULCH THE CONSTRUCTION AREA.

TOP SOIL
EARTHFILL
GABIONS
Original Ground
PLASTIC FILTER CLOTH
Channel Bottom
ELEV. 1245.5
3'-3" (1 Meter)
1'-8" (½ Meter)

SIDE ELEVATION

Note: all gabions shall be in sections of 13'-1" in length

Scale in Feet
0 1 2 3 4 5

Figure 6.25 (Above) Cross section of river gabion construction with membrane. (Courtesy of Carthage Mills Inc.)

Figure 6.26 (Opposite) Cross section of channel lining using interlocking concrete block system with membrane filter. (Courtesy Kenross-Naue [Canada] Ltd.)

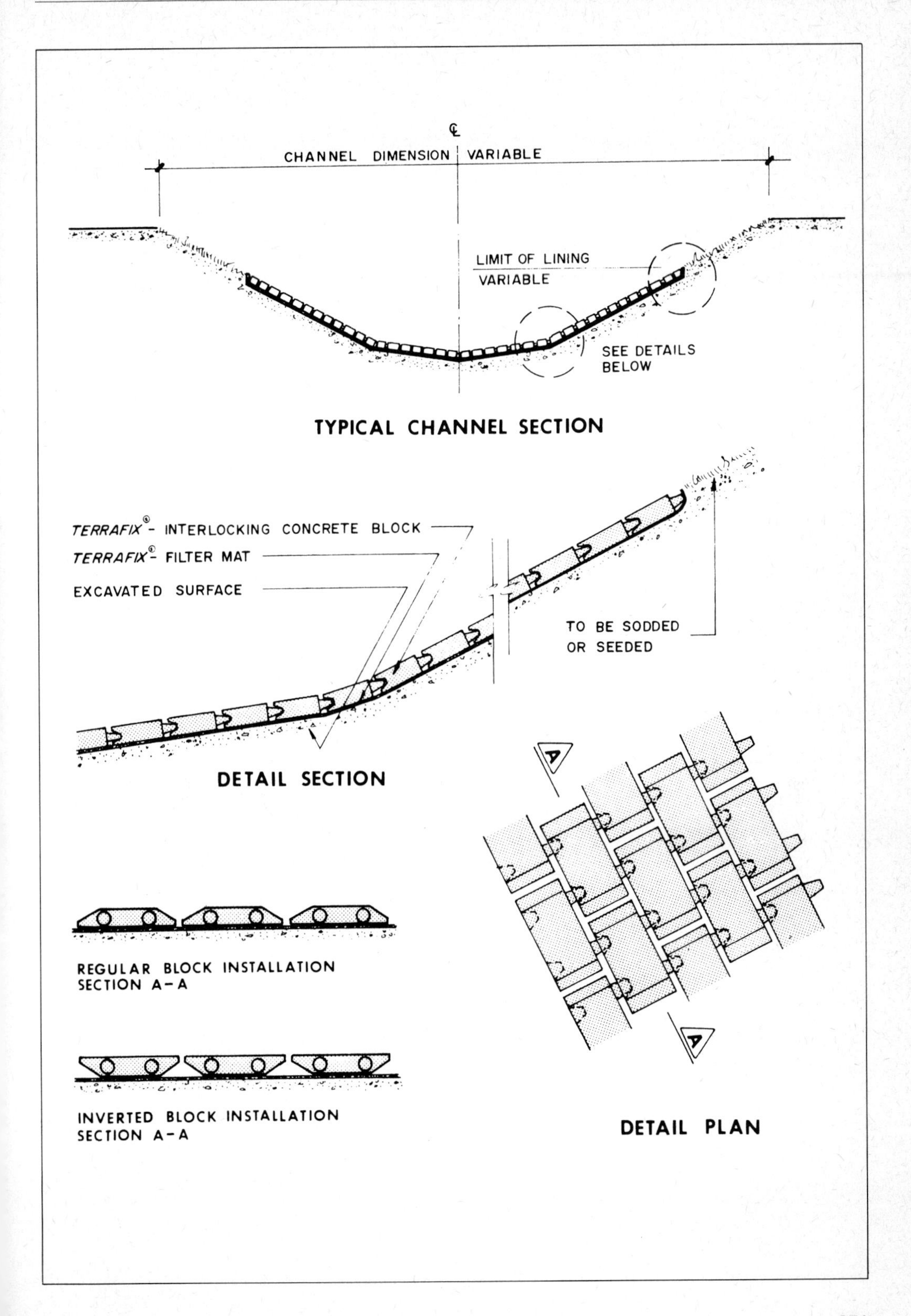
℄
CHANNEL DIMENSION VARIABLE
LIMIT OF LINING
VARIABLE
SEE DETAILS
BELOW
TYPICAL CHANNEL SECTION
TERRAFIX® - INTERLOCKING CONCRETE BLOCK
TERRAFIX® - FILTER MAT
EXCAVATED SURFACE
TO BE SODDED
OR SEEDED
DETAIL SECTION
A
A
REGULAR BLOCK INSTALLATION
SECTION A-A
INVERTED BLOCK INSTALLATION
SECTION A-A
DETAIL PLAN

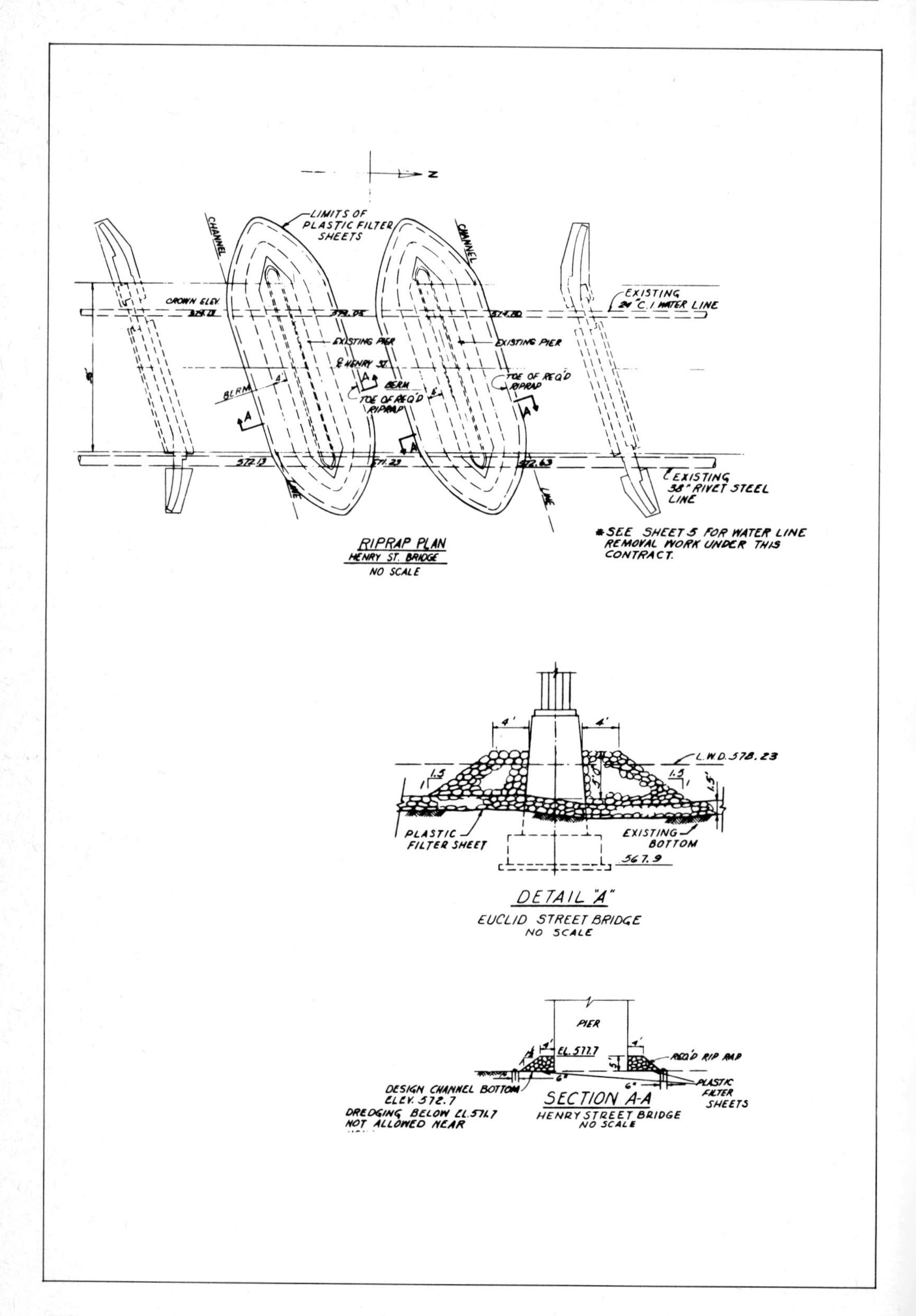
LIMITS OF PLASTIC FILTER SHEETS
CHANNEL
CHANNEL
CROWN ELEV.
EXISTING 24" C.I. WATER LINE
EXISTING PIER
EXISTING PIER
℄ HENRY ST.
BERM
BERM
TOE OF REQ'D RIPRAP
TOE OF REQ'D RIPRAP
LINE
LINE
EXISTING 38" RIVET STEEL LINE
*SEE SHEET 5 FOR WATER LINE REMOVAL WORK UNDER THIS CONTRACT.
RIPRAP PLAN
HENRY ST. BRIDGE
NO SCALE
L.W.D. 578.23
PLASTIC FILTER SHEET
EXISTING BOTTOM
567.9
DETAIL "A"
EUCLID STREET BRIDGE
NO SCALE
PIER
EL. 577.7
REQ'D RIP RAP
PLASTIC FILTER SHEETS
DESIGN CHANNEL BOTTOM ELEV. 572.7
DREDGING BELOW EL.571.7 NOT ALLOWED NEAR
SECTION A-A
HENRY STREET BRIDGE
NO SCALE

Figure 6.28 Coastal estuary defence constructed in Malaysia. (Courtesy ICI Fibres)

Figure 6.27 Detail of Bridge Pier Protection —Michigan, U.S.A. (Courtesy of Carthage Mills Inc.)

7

Marine defences and filters in reversing waterflow situations

Adjacent to the sea, reversing waterflow situations are encountered where wind-generated waves act on the coast producing rapid pressure oscillations, and where tidal water levels change with a lower periodicity to cause reversing waterflows in coastal sediments. Inland, on canals and some navigable rivers, the wash-generated waves from passing river boats and barges cause reversing pressure gradients which tend to break down the soil filter structure of slope defences.

The design of a coastal defence differs from a river defence in one essential aspect. Usually, the river defence has to be laid in moving water on a slope angle already determined by the scour of the river, whereas a coastal defence can often be constructed dry during low tides, and the slope angle is artificially achieved either by excavation or by embankment construction. Therefore, the design operations include the selection of an appropriate slope for the defence cut or embankment. It is important to recognise that this slope must be designed on a standard 'soil mechanics' basis, and must be stable in its own right in terms of potential slip failure. The erosive effect of the sea can naturally be ignored in this calculation since the conclusion of the design will be to place a defence on the slope.

Figure 7.1 is a chart showing a method of assessing the Safety Factor on an existing or proposed slope, as developed by Hoek[24] in 1973. The full method is more complex than shown here, but in order to simplify the analysis, the worst case has been used, being a saturated slope with a tension crack at the top. The Safety Factor in this case is an indication of how likely the slope is to fail in rotational slip failure. The Safety Factor is found on the chart by finding the intersection of two variables X and Y. X is a variable dependent upon the geometry of the slope and the internal water table as follows:

$$X = \alpha - \left[(1.2 - \frac{Hs}{2H}) \times \phi \right]$$

Where α is the slope angle in degrees and Ø is the internal angle of friction of the soil as illustrated in Figure 2.10. If the slope is made of a pure clay with $\phi = 0°$ as illustrated in Figure 2.11, then the part of the equation in square brackets above becomes equal to zero, and $X = \alpha°$, which makes the approximation even safer by indicating a lower Factor of Safety. Y is a variable dependent upon the material properties and the size of the slope.

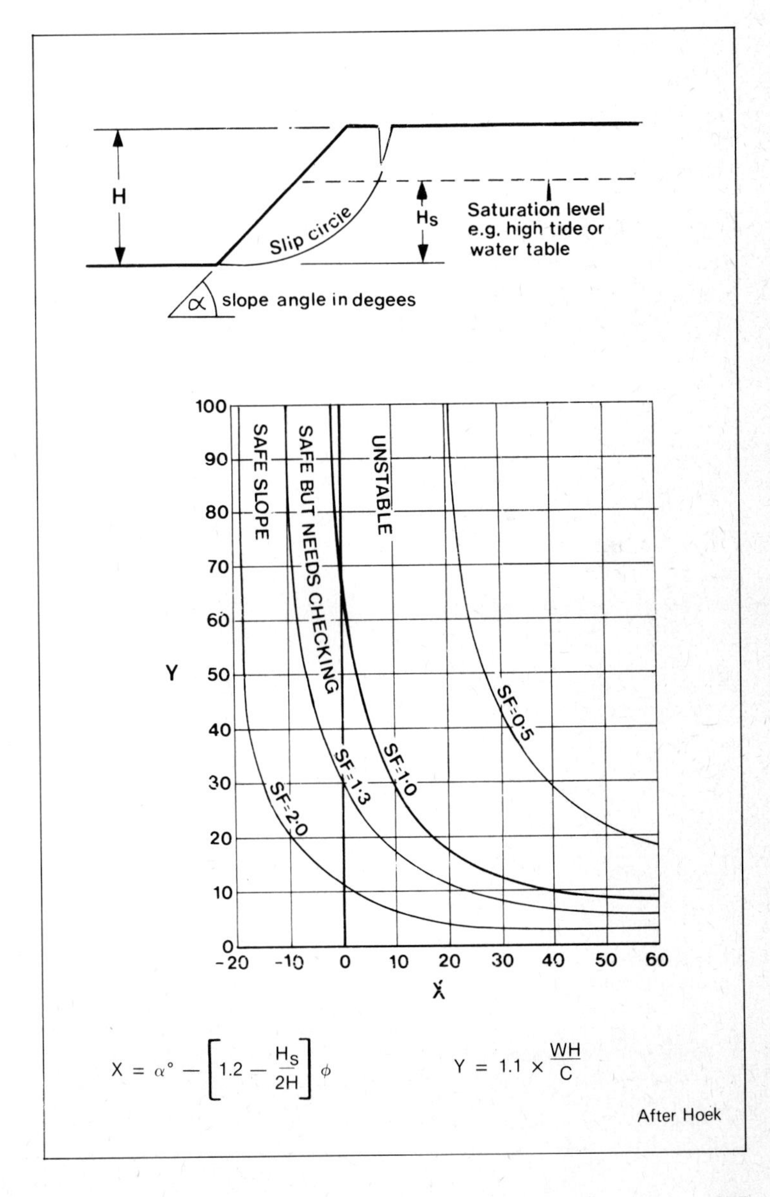

Figure 7.1 Chart for assessing the safety factor of a slope against rotational shear failure

$$Y = 1.1 \times \frac{WH}{C}$$

Where either:—

H is in ft, W is in lb/ft^3, and C is in lb/ft^2

or:

H is in m, W is in Kg/m^3, and C is in Kg/m^2

One of the parameters used in finding the value of Y is the cohesive strength of the material. As discussed earlier in Section 2, if the slope is a pure granular material such as sand then it will not have any cohesive strength. If the slope is a clay, then it will have a cohesive strength which can be determined accurately by laboratory tests, but the table in Figure 7.2 was published by Peck Hanson and Thorburn, and provides a useful field guide for assessing cohesive strength. The Author has added typical soil weights and shear angles as a further guide.

If the Hoek method indicates a Safety Factor below 1.5, then it would be recommended that a proper detailed analysis be undertaken. Suitable methods are well known, and both dry and submerged slopes are dealt with in Reference 25.

Figure 7.2 Chart for assessing the cohesive strength of a clay by use of the hand

Cohesive Soil Type and strength test guide for CLAYS	Cohesive Strength C kg/m² (*lb/ft²*)	kN/m²	Weight kg/m³ (*lb/ft³*)	Shear Angle ϕ
V. Soft Easily penetrated several inches by fist	500 (*100*)	5	1440 (*90*)	0°
Soft Easily penetrated several inches by thumb	1000 (*200*)	10	1600 (*100*)	0°-2°
Med. Can be penetrated several inches by thumb with moderate effort	2500 (*500*)	25	1760 (*110*)	0°-4°
Stiff Readily indented by thumb, but penetrated only with great effort	5000 (*1000*)	50	1920 (*120*)	0°-6°
Very Stiff Readily indented by thumbnail	10,000 (*2000*)	100	2080 (*130*)	0°-10°
Hard Indented with difficulty by thumbnail	20,000 (*4000*)	200	2240 (*140*)	0°-20°

Soils Data of Typical Soils

Loose Sand W = 1200—1600 kg/m^3 (80—100 lb/ft^3)

Sandstone W = 2000—2500 kg/m^3 (128—156 lb/ft^3)

Granite W = 3000 kg/m^3 (200 lb/ft^3)

Although, in practice, different types of marine defence vary widely, the Author considers that there are three degrees of involvement for membranes.

1. Behind a major heavy structure where the membrane is merely acting as a backing separator (see Figure 7.3A).
2. Beneath energy-dissipating stonework where the membrane is replacing a granular filter layer (see Figure 7.3B).
3. Beneath heavy armour where the membrane is replacing both the filter and the smaller energy-dissipating armour material (see Figure 7.3C).

In the first case (A), the membrane is not performing any mechanical function, and therefore its specification will relate more to its permeability than its strength. There is obviously a requirement to dissipate uplift pressures, and relieve groundwater flows without clogging, and possibly a two-dimensional non-woven melded or spun-bonded fabric could be considered best for this application. In the case of Figure 7.3B, the membrane is taking a certain active and passive load, and is acting primarily as a dynamic soil filter. Therefore strength and extensibility would probably balance best in a non-woven needle-punched or strong spun-bonded/melded fabric for soft cohesive *in situ* soils, and in a woven fabric for sandy soils. Lastly, the fabric in Figure 7.3C, is taking direct wave pressure, wave-induced movement of stones, intense point loads from armour stones, and supporting the toe, some minor exposure to ultra-violet light between the larger stones, and acting as a soil filter as well. These qualities are extremely demanding, and to the Author's knowledge, no one membrane could fulfil all these requirements perfectly. Therefore the specification could call for a very strong heavy-weight woven membrane or for a 'composite' product such as Linear Composite's 'Paraweb' interwoven mat which has 'Terram' high permeability membrane welded to the back (see Figure 7.4). The webbing provides high strength, ultra-violet resistance in excess of 30 years, handleability, abrasion resistance and a rough energy-dissipating surface. The welded membrane provides high-permeability filtering to prevent the *in situ* soil being washed out through the webbing. Figure 7.5 shows the case of laying such a membrane onto a 1 : 2.5 clay slope. At this angle, and with construction taking place from the base upwards, the webbing does not need pinning to the slope under normal circumstances, as can be seen in the photograph.

On sloping coastal defences there is the problem of choosing a suitable stone size in order to resist wave action. Figure 7.6 shows the conventionally-accepted formulae for selecting stone

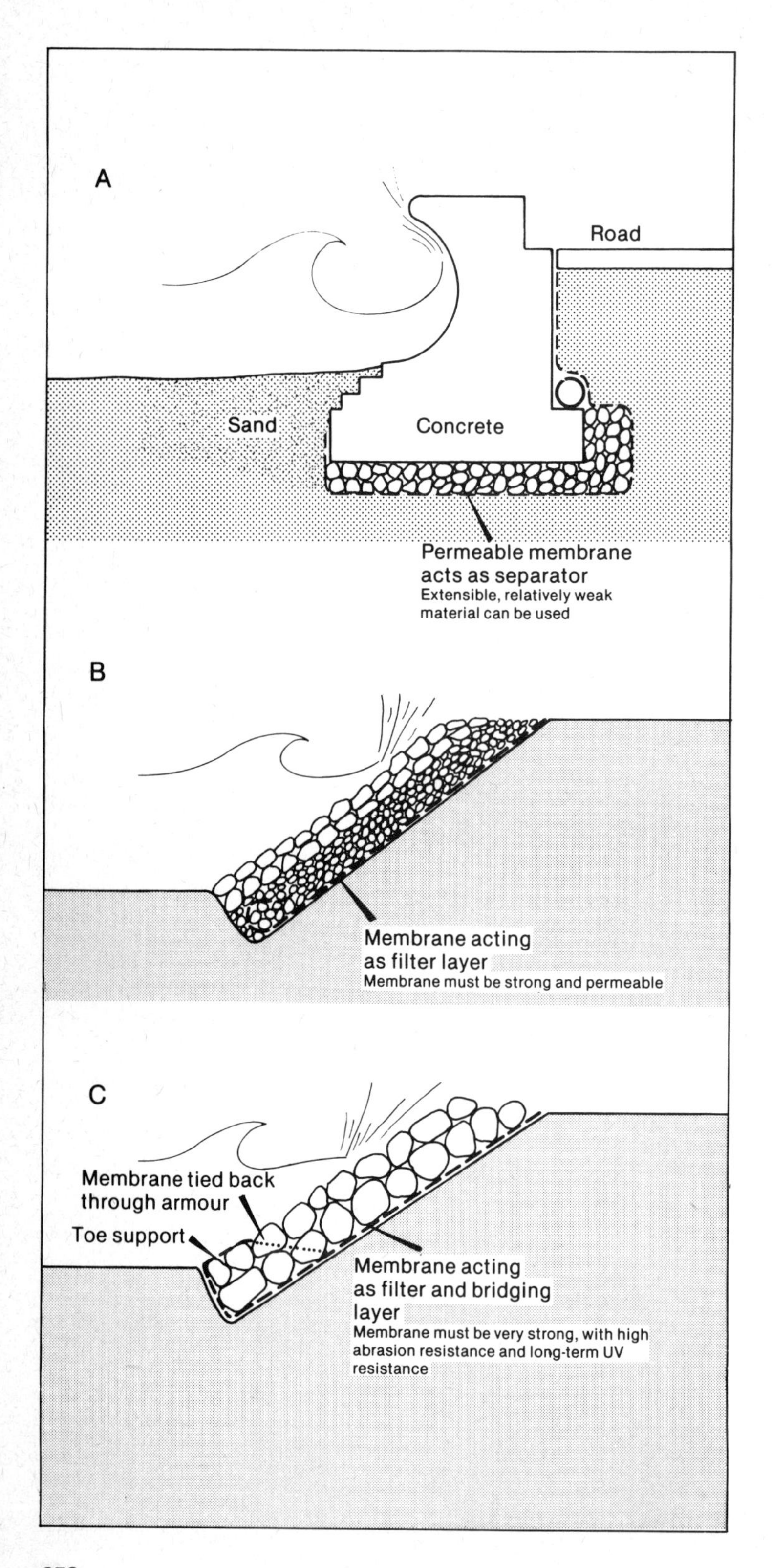

Figure 7.3 Increasing involvement of permeable membrane in defence designed function, necessitates increasingly strong and durable material

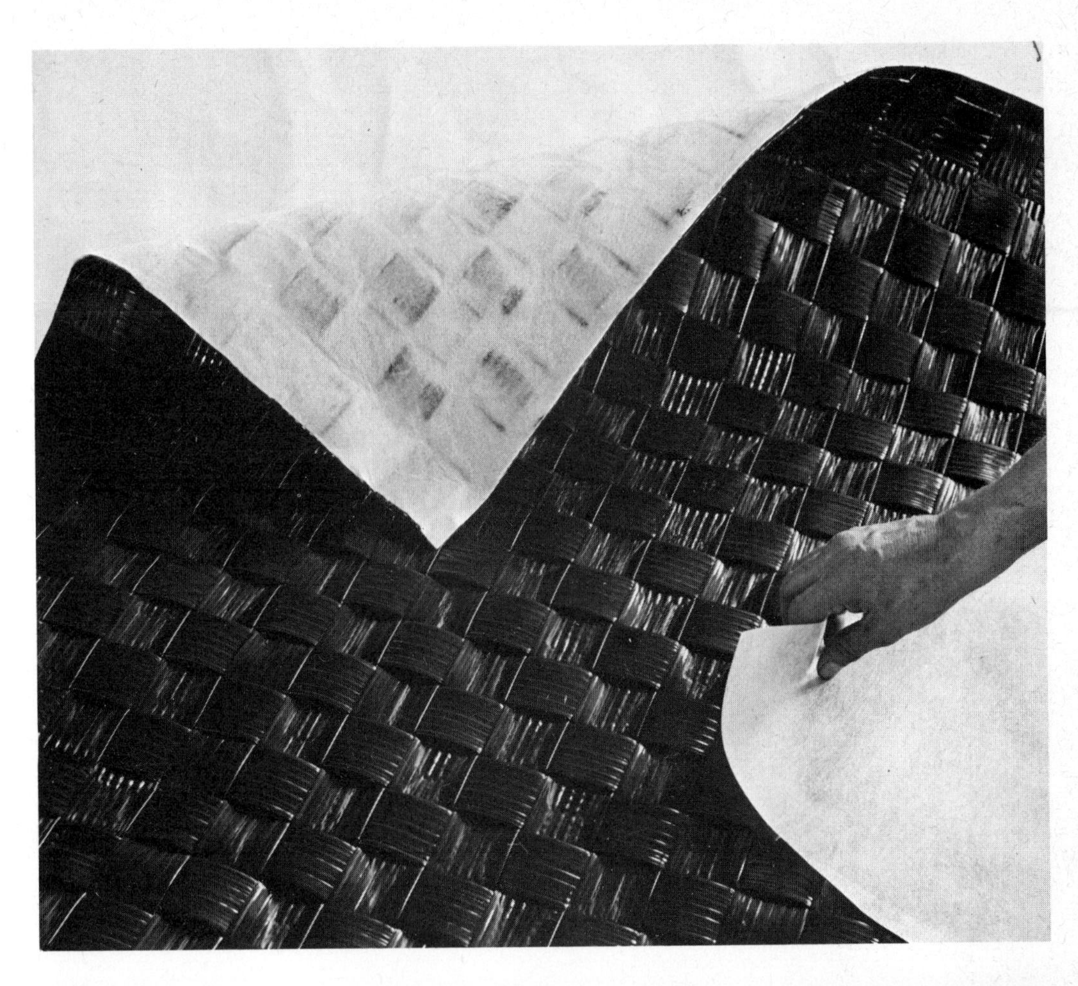

Figure 7.4 Paraweb Interwoven Duplex Mat. A typical specimen of the new generation of composite products which combine the benefits of two different materials. (Courtesy of Linear Composites Ltd.)

Figure 7.5 Interwoven Paraweb webbing and membrane mat being laid directly onto a clay bund in S.E. Asia, and overlain by large stones. Now standing since 1976

size[26], but it should be realised that these sizes are for random-placed or free-dropped stone. Consequently, they are extremely conservative. Furthermore, if bad-weather occasional peak wave heights are used for calculations, then sizes could become prohibitive. In many countries, mechanical equipment and large stone sizes are not available, so in these cases a maximum size of handleable stone (say 60 kg) may be used, and the slope adjusted flatter accordingly. The slope can be re-steepened if (a) instead of considering maximum wave height, the usual wave height is taken for design purposes, and (b) the stones are placed by hand so that they interlock. By this means, a steep slope can be constructed, but regular maintenance will be needed since stones are likely to be dislodged by occasional storms. Hand-set interlocking stones are interdependent, and prompt repair work immediately after storms is highly advisable or serious deterioration can follow — even with normal height waves. For major defence works, free-dropped stones have the long-term advantage of being independent of one another. Each is stable in its own right by virtue of its own mass, therefore if one should be moved the others do not have their support removed.

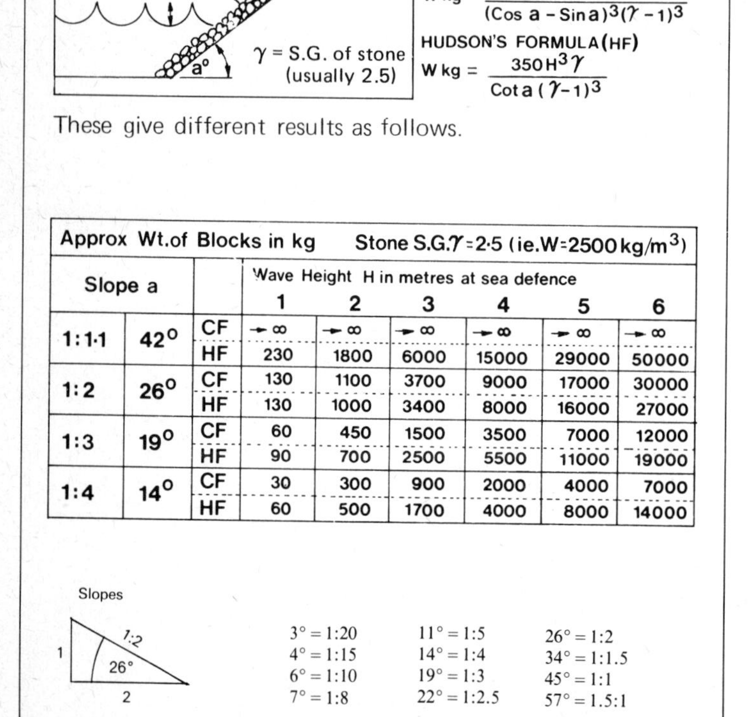

Approx Wt. of Blocks in kg — Stone S.G. γ = 2·5 (ie. W = 2500 kg/m³)

Slope a			Wave Height H in metres at sea defence 1	2	3	4	5	6
1:1·1	42°	CF	→ ∞	→ ∞	→ ∞	→ ∞	→ ∞	→ ∞
		HF	230	1800	6000	15000	29000	50000
1:2	26°	CF	130	1100	3700	9000	17000	30000
		HF	130	1000	3400	8000	16000	27000
1:3	19°	CF	60	450	1500	3500	7000	12000
		HF	90	700	2500	5500	11000	19000
1:4	14°	CF	30	300	900	2000	4000	7000
		HF	60	500	1700	4000	8000	14000

Figure 7.6 Equations for calculating the weight of individual stones for coastal defence design. (From Ref. 26)

The following approach would give a first-order indication of the type of structure needed in any particular situation, prior to full design work being embarked upon:

Information required:—

a. Wave information including heights, and storm occurrence periods.
b. Tide variation and height.
c. ϕ, W kg/m^3, C kg/m^2 of embankment soil.
d. Particle size and permeability order of soil (see Figure 5.16).
e. Longshore or tidal scour velocity.

Then:

1. Add the wave height to the maximum high tide height to establish the minimum height H of the embankment needed to resist waves. Then assess the required angle of slope (α) according to Figures 7.1 and 7.2.
2. Use Figure 7.6 to assess the stone weight needed to resist wave action. Modify this in the light of storm frequency to a smaller size if repair maintenance can be undertaken frequently.
3. If the available stone weight is lighter than than required under (2) above, then use Figure 7.6 to flatten the angle α until the design is satisfactory.
4. Use Figure 6.7 to check that the stone weight used is large enough to resist any longshore or tidal scour.
5. Use Figure 6.7 to convert the stone weight into a dimension — d. The Final Defence layer should be laid to a thickness of at least $2d$.
6. Use the principles in Figure 6.8 to select an Armour Bridging Filter with a D_{85} of $3d/5$. This should be laid to a thickness of $2d$, to absorb turbulence, and to spread point loads over the underlying fabric.
7. Using the design criteria for *alternating flow conditions* as shown in Figures 5.14 and 5.15, and in the context of the *in situ* soil to be filtered, select a suitable membrane.
8. In relation to the final dimensions of the slope and defence, design a modular construction programme allowing for one strip of the defence to be built every tidal cycle. This should include:

   a. Excavate toe trench as tide recedes
   b. At low tide lay membrane
   c. Lay toe stones and toe armour
   d. Lay Armour Bridging Filter and Main Defence upwards as the tide rises

e. At high tide finish off bank top details
f. Re-start cycle as tide goes out

The above sequence a. to e. is shown in Figure 7.7.

Figures 7.8-7.18 are illustrations of field drawings and constructions in order to give the reader some concept of currently accepted practice and actually constructed design detailing.

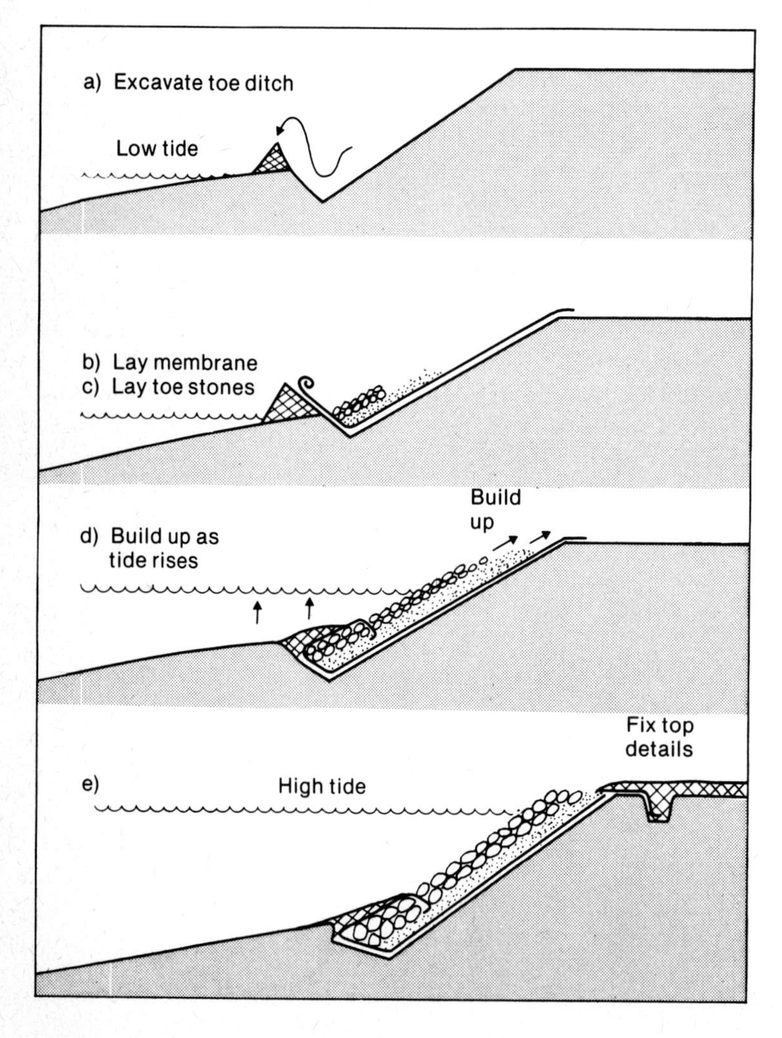

Figure 7.7 Modular construction of a coastal defence using the tidal cycle to construct the defence in unit strip widths

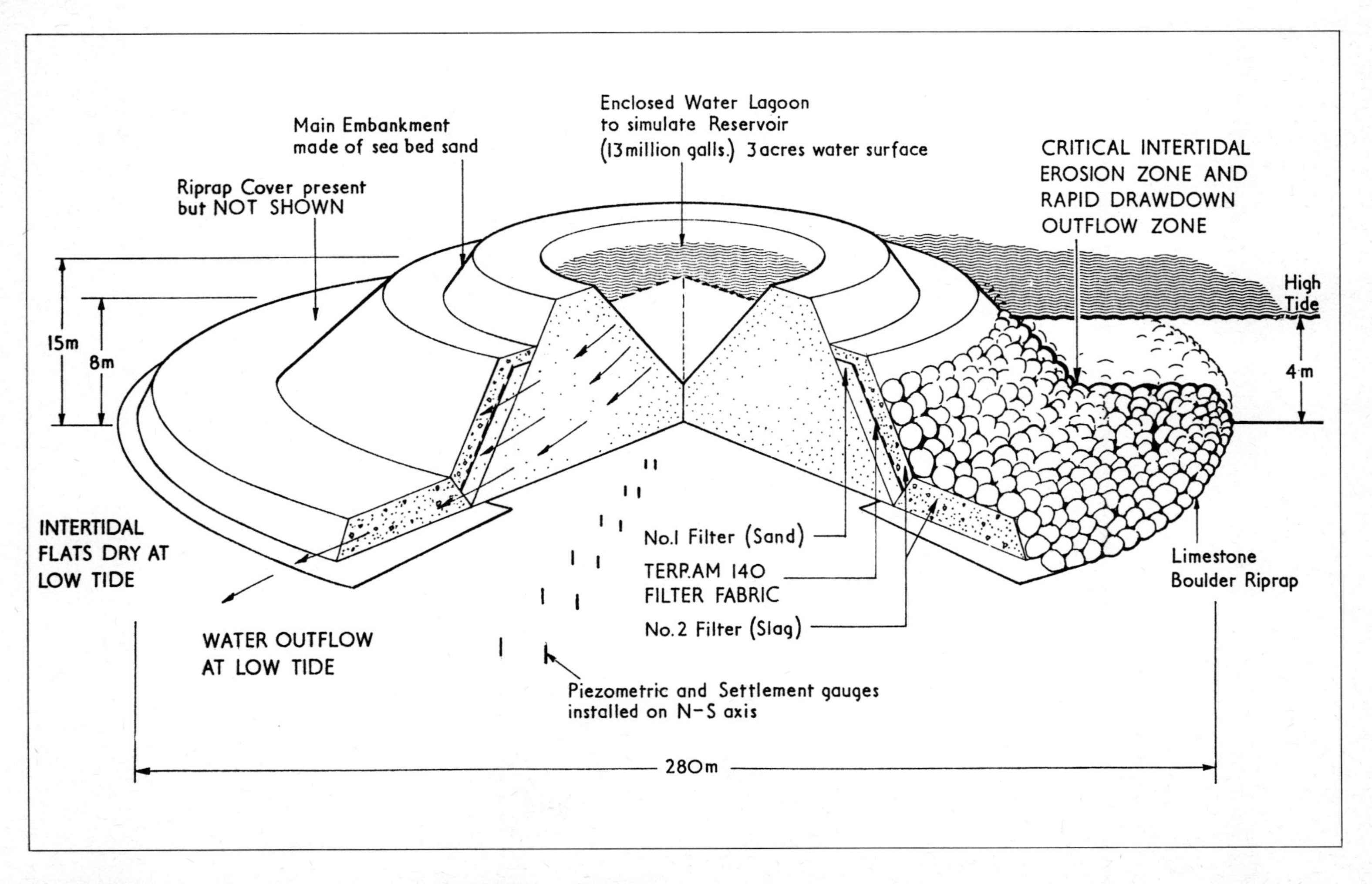

Figure 7.8 Marine structure for evaluating design. (Courtesy ICI FIbres)

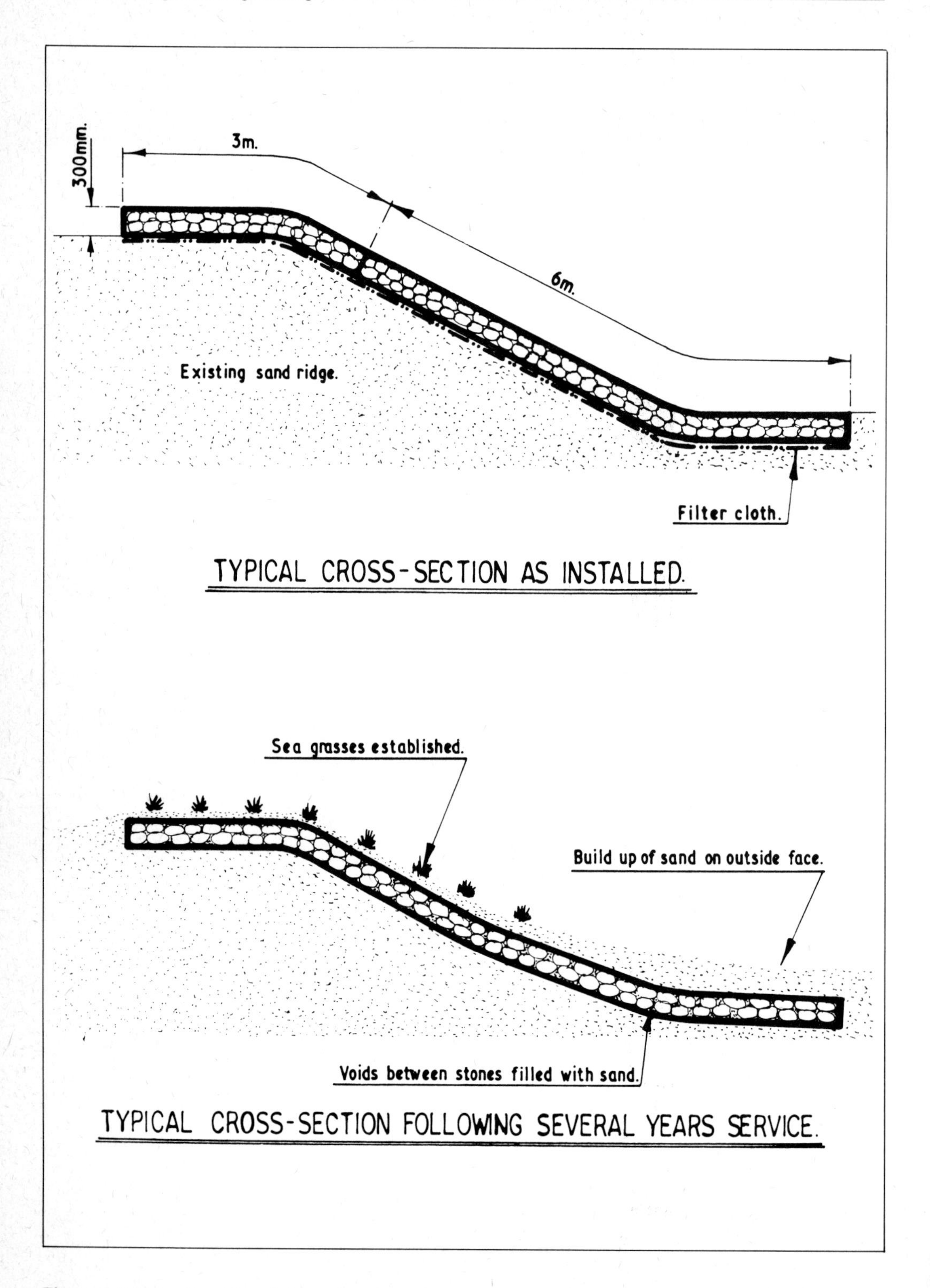

Figure 7.9 Coastal gabion defence drawing. (Courtesy River and Sea Gabions [London] Ltd.)

Figure 7.10 Coastal gabions being installed on sandy coast. Wire mesh prevents smaller stones from being eroded. Membrane beneath prevents sand from being washed out. (Courtesy Fibertex S.A.)

Netlon tubular gabions approx 636mm diameter, of any desired length, erected in batteries of 4, 6 or 8 units, are suspended from a travelling guide framework which spans between two pontoons or barges anchored in position. Broken stone is placed in the gabions initially to provide ballast to sink the gabions in position. The gabions are then filled, progressively, through a hopper head and the guide frame moved in stages as shown towards the bank to allow the stone filled sections of the gabions to settle in position on the river bed/bank surface. As each battery is filled and laid, the pontoons are repositioned and the procedure repeated.

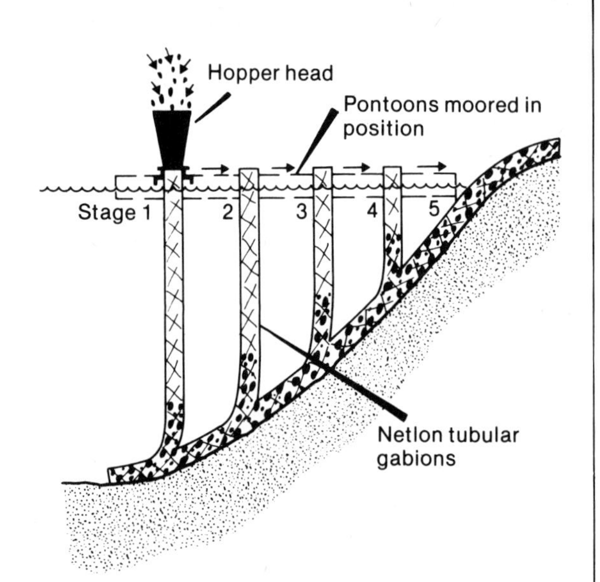

Netlon gabions can be produced in any length, and can be joined on site to give a continuous unit. As the gabions are filled, additional lengths are added and the filled portion allowed to settle on to the bed/bank profile. Locating rings can be used to hold the gabion units firmly in position.

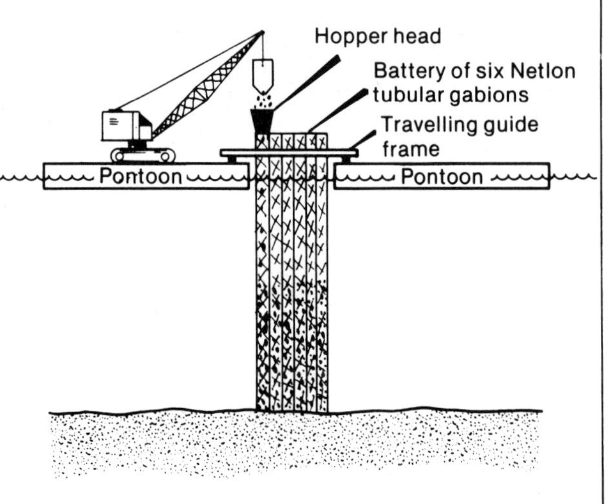

Figure 7.11 (Opposite) Tubular gabion construction for deep water works. (Courtesy Netlon Ltd.)

Figure 7.12 (Top) Non-woven filter separator being laid in a coastal defence. (Courtesy ICI Fibres.)

Figure 7.13 (Bottom) Non-woven filter in a coastal construction. (Courtesy ICI Fibres.)

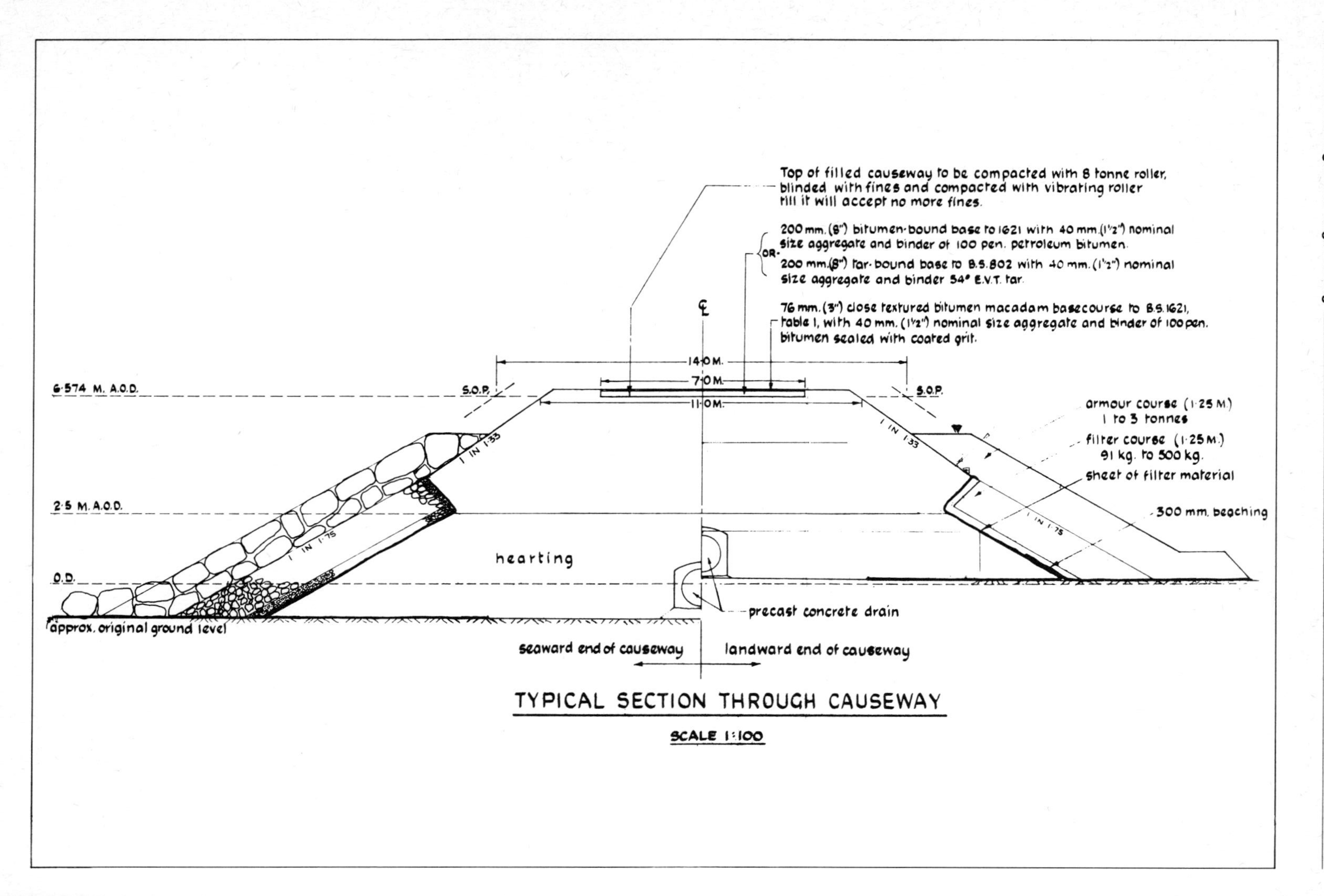

TYPICAL SECTION THROUGH CAUSEWAY

SCALE 1:100

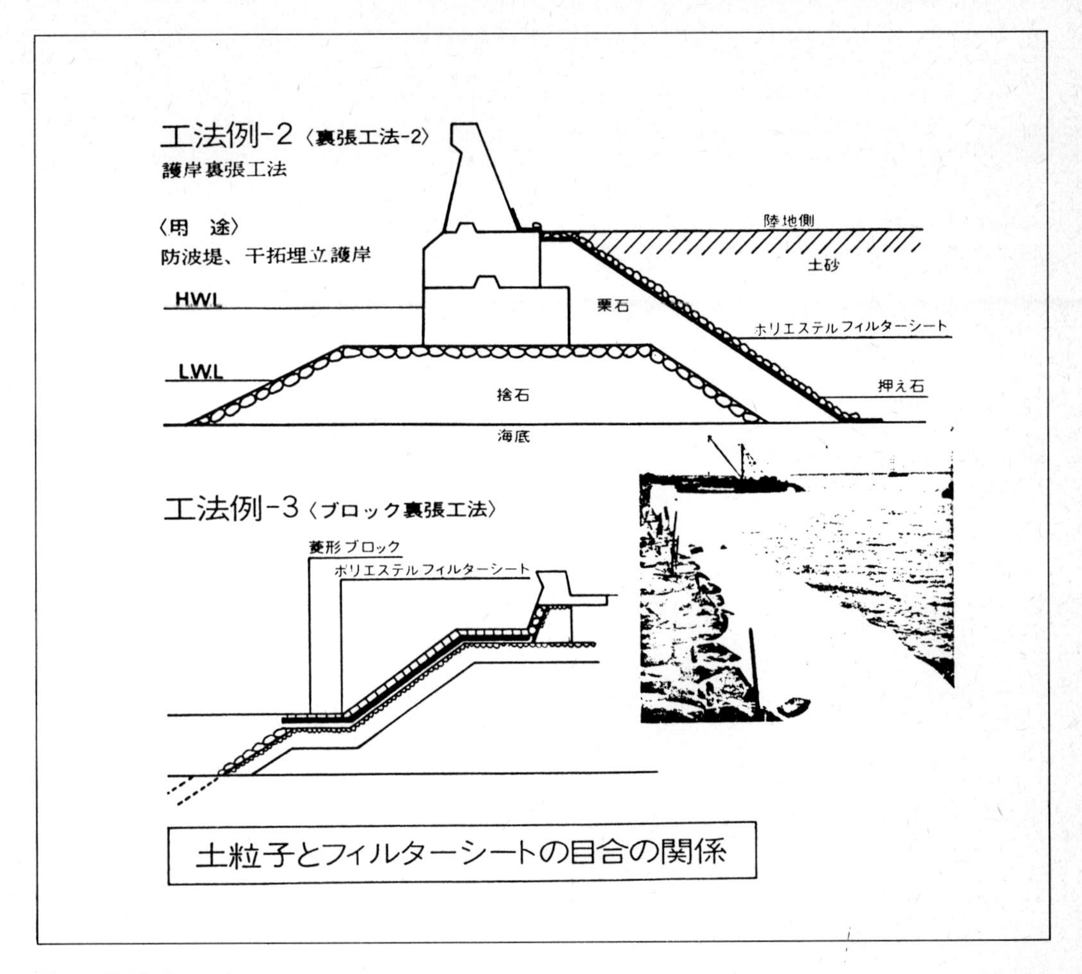

Figure 7.15 Coastal defence designs using Japanese products. (Courtesy Taiyo Co. Ltd.)

Figure 7.14 (Opposite) Cross sectional detail of marine causeway. (Courtesy ICI Fibres.)

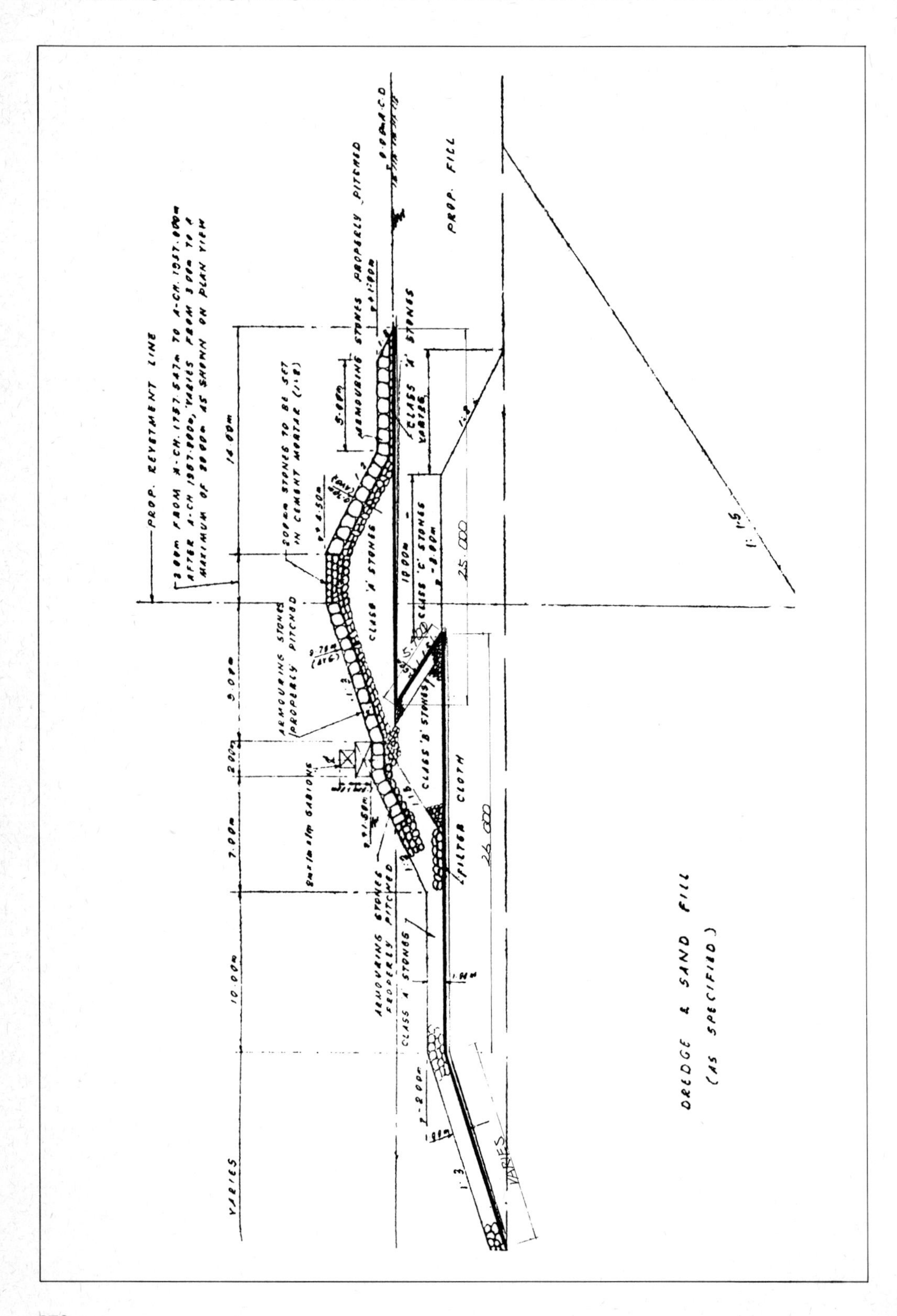
PROP. REVETMENT LINE
14.00m
9.00m
2.00m
7.00m
10.00m
VARIES
PROP. FILL
ARMOURING STONES PROPERLY PITCHED
CLASS 'A' STONES
CLASS 'B' STONES
FILTER CLOTH
25.000
26.000
1:1.5
1:3
DREDGE & SAND FILL
(AS SPECIFIED)

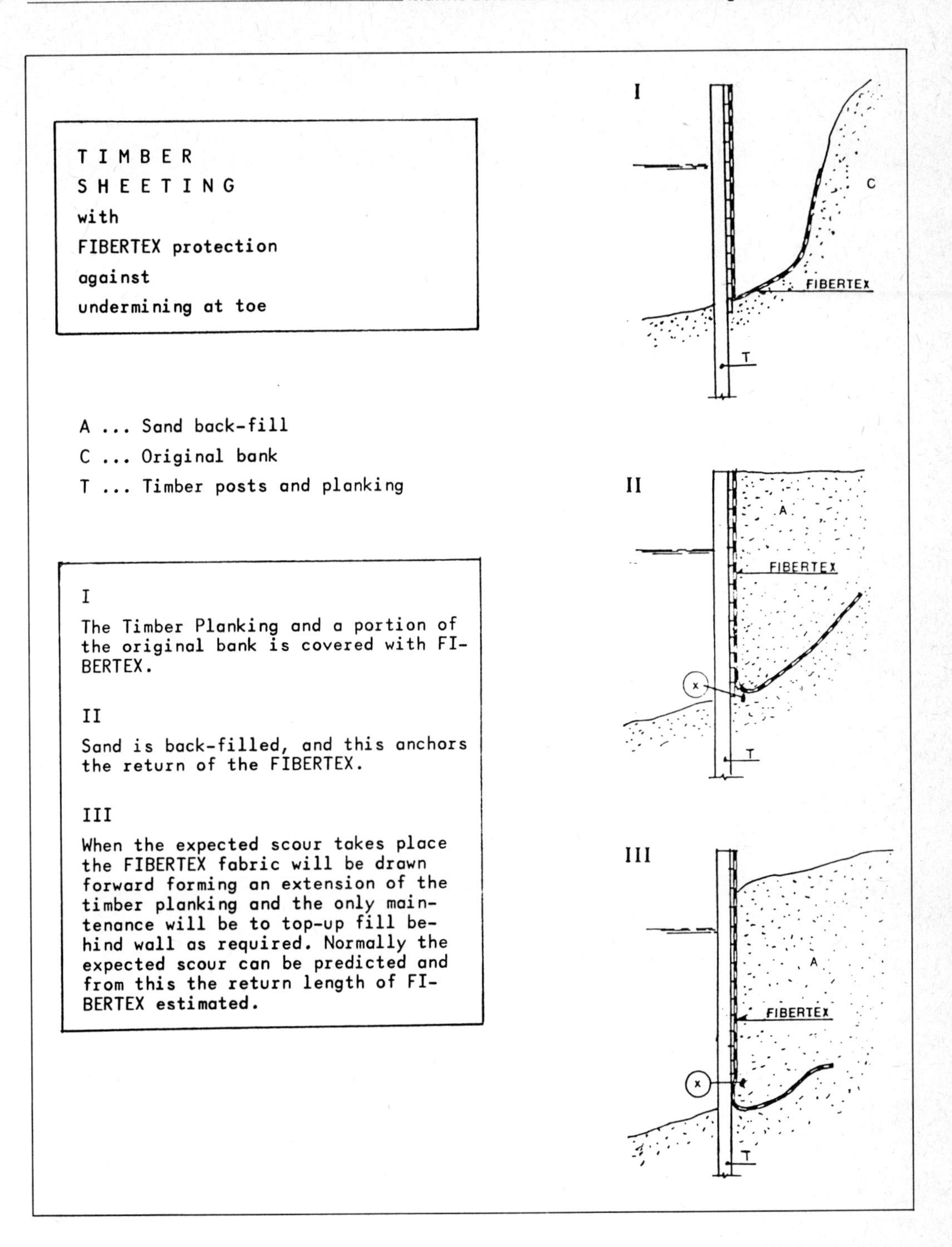

Figure 7.16 (Opposite)
Coastal defence for
land reclamation.
(Courtesy ICI Fibres.)

Figure 7.17 (Above)
Simple vertical structure
suggested by Fibertex to
cope with under-scour.
(Courtesy Fibertex)

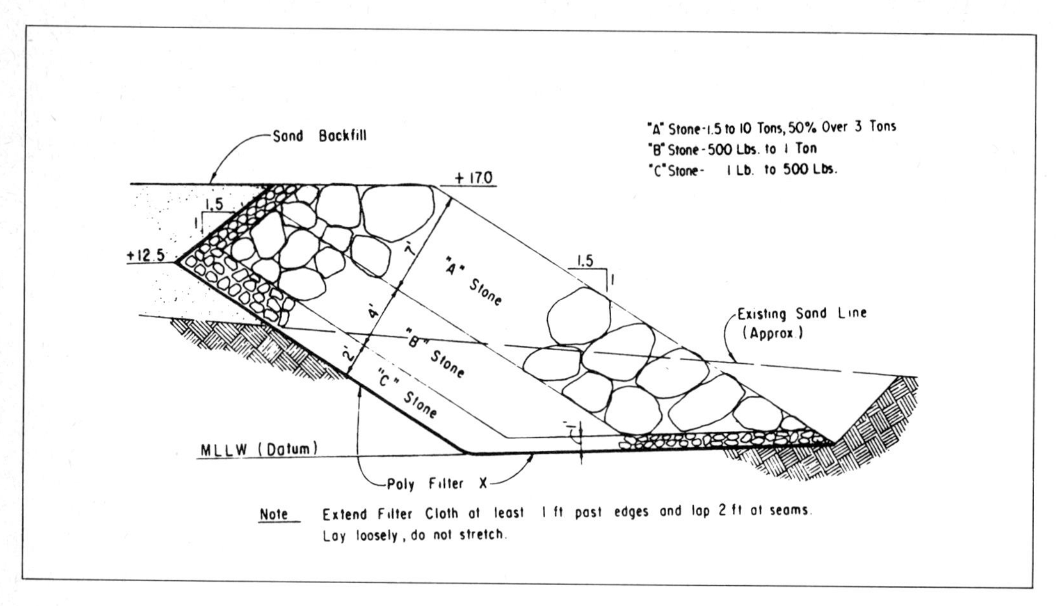

Figure 7.18 American sea wall construction. (Courtesy Carthage Mills Inc.)

8

Land reclamation — inland and coastal

Inland land reclamation

In most cases, this type of reclamation involves the filling in of low-lying waterlogged depressions. Sometimes it includes the rehabilitation of land which has been exploited for industrial purposes such as mining or quarrying.

Low-lying soft soil areas

For the reclamation of poor quality low-lying inland areas, the membrane is rolled out as a single layer directly on the original ground surface if a permeable fill is to be used. If only an impermeable fill is available, then a drainage blanket should be established below to take away groundwater, as shown in Figure 8.1. Special structures should be used to cater for springs and streams. Figure 8.24 is an industrial design showing how such detailing can be drawn (Courtesy of Manstock Geotechnical Consultancy Services Ltd).

If the reclamation is for recreational purposes, then usually, the upper surface of the fill will be covered with a soil layer and grassed. In the case of a non-permeable fill such as clay, then a ground drain system such as shown in Figure 5.22 can be installed. Conversely if a granular fill is used, and especially if the fill is waste material including bricks, broken glass and other potentially dangerous items, then a single layer of membrane can be used across the site to prevent these from 'working up' to the surface (Figure 8.2).

If the reclamation is for constructional purposes, then the fill will usually be of a good quality, well-graded, crushed stone. It can be vibro-compacted in layers to achieve the maximum possible 'beam' strength within the granular material. Buildings

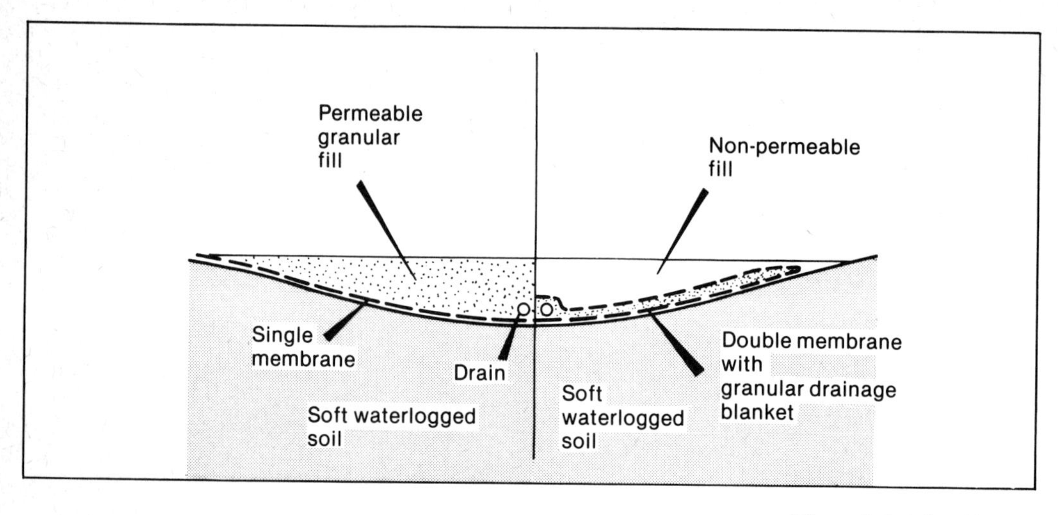

Figure 8.1 Inland land reclamation using membrane to ensure good drainage and prevent uplift pressures establishing

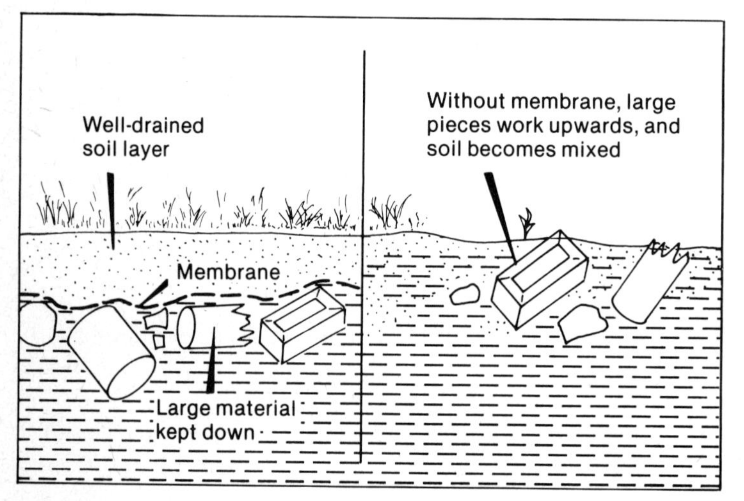

Figure 8.2 Recreational grounds built on domestic fill can be much improved by use of a permeable membrane separator

placed on top of such fill should be constructed with raft foundations at high level.

In very soft ground with high water content, or with standing water, the membrane should be rolled out and covered with a continuous very thin layer of fill. This will have to be either hand laid or laid by very light plant. A second layer can then be placed all over the site, by slightly heavier plant, and the process repeated until the layer is thick enough to support conventional plant. Figure 8.3 illustrates how if this is not considered, then rotational shear failure can take place. Figure 8.4 gives an indication of what should be possible with membrane construction on soft soils providing that multi-layer vibro-compacted construction is used where necessary, and that time is allowed for some pore-water dissipation from the soil if necessary.

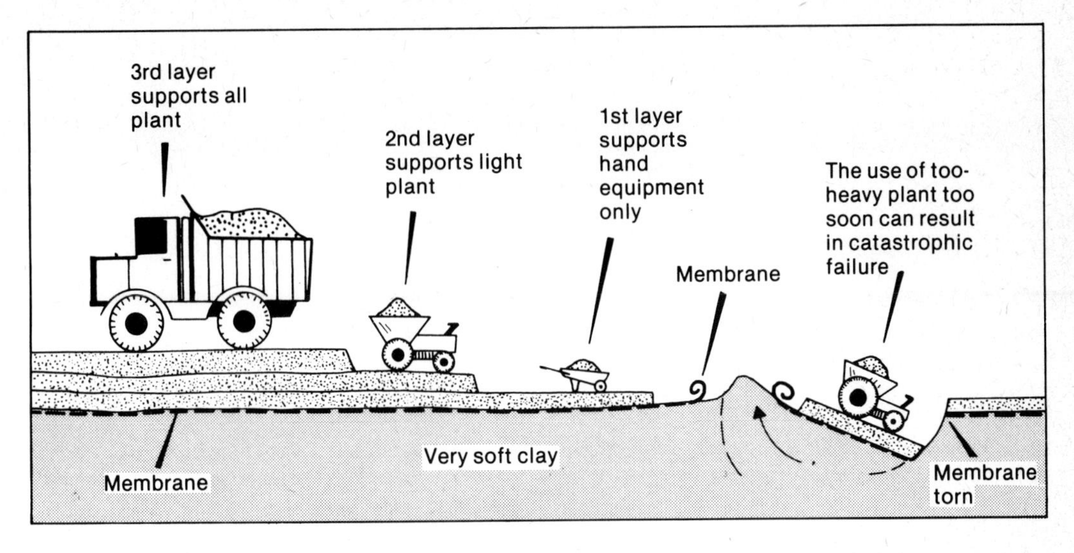

Figure 8.3 Land reclamation on *very soft* soils should be achieved in several thin layers with plant of increasing weight

The curves shown in Figure 8.4 are not based on instrument or laboratory testing, but are *based upon the Author's experience* in land reclamation works on very soft soil and peats both in the UK and South East Asia. Fill materials used have varied from crushed stone, through marine sand to laterites. In the case of laterites, their immediate supporting capability will be less than the graphs given, but with several months hardening, the strength will increase such that final traffic capability may prove to be in excess of that indicated. Because of the intuitive nature of these curves, they should be used as a preliminary guide only, for first order costing estimates and the like. As in all fields of engineering, construction design should be preceded by proper Site Investigation work to provide quantitative data on the condition of the ground to be reclaimed. The shear strength table given in Figure 8.4 is based upon work done by Peck, Hanson and Thorburn.

The curves try to reflect some aspects of practicality that theoretical curves would not. For example there is a practical spreading thickness limitation on virtually any soil area being reclaimed since the surface will be naturally undulating, and the reclaimed surface will be required to be flat. Therefore, if 600 mm is the required thickness, then this should be considered as the *minimum* thickness, with greater thicknesses in any hollows.

Industrially-exploited land

For the reclamation of industrially-exploited land, membranes are often used for fundamentally different purposes than as simple separators. Four examples are given of different types of application for some of the miscellaneous membranes shown in the earlier part of the book (Figures 4.81-4.90).

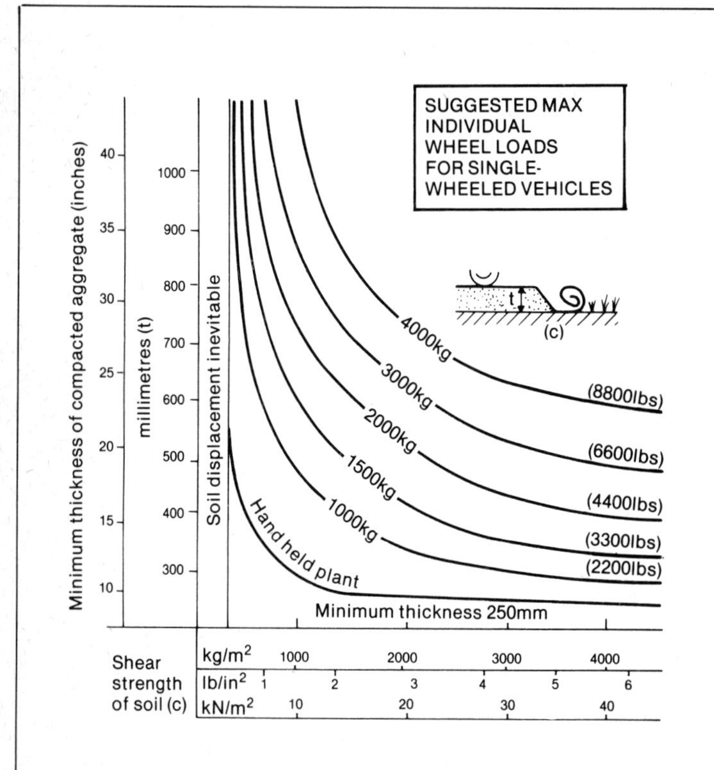

Cohesive soil type and strength test guide for CLAYS	Cohesive strength C (lb/ft²)	kg/m²	kN/m²
V. Soft Easily penetrated several inches by fist	(100)	500	**5**
Soft Easily penetrated several inches by thumb	(200)	1,000	**10**
Med. Can be penetrated several inches by thumb with moderate effort	(500)	2,500	**25**
Stiff Readily indented by thumb, but penetrated only with great effort	(1,000)	5,000	**50**
Very Stiff Readily indented by thumbnail	(2,000)	10,000	**100**
Hard Indented with difficulty by thumbnail	(4,000)	20,000	**200**

Figure 8.4 Suggested thickness requirements for fill material in land reclamation works to support plant wheel loads. Curves are tentative only, and are based on Author's experience only. Assumptions are:
i) Multi-layer vibro compacted construction
ii) Pore-water dissipation allowed if necessary
iii) Suitably strong membrane placed at fill/ground interface.
(Example:
C = 2000 kg/m²
t = 600mm
Fill should support a 4-wheeled vehicle weighing 12,000kg by taking 3,000kg per wheel)

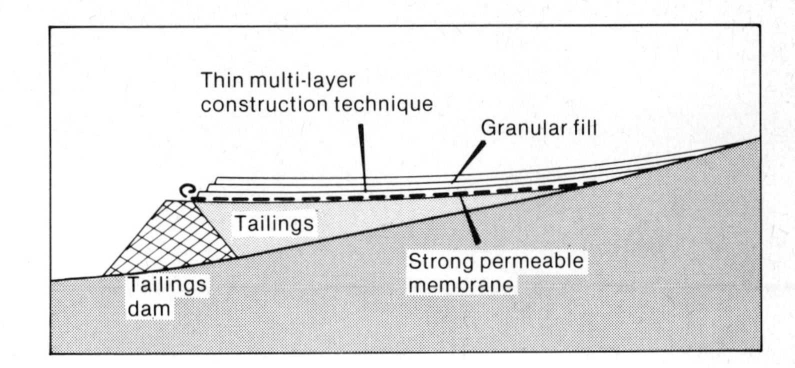

Figure 8.5 Thin multi-layer technique can be used to reclaim tailings lagoons

Figure 8.5 shows how a membrane can be used to reclaim mine or quarry tailings ponds. The principle here is to cover the whole surface with a strong permeable membrane overlain by a *thin* layer of compacted granular fill. By slowly increasing the thickness of granular fill, right across the area, the bearing capacity of the fill can be increased.

Where quarries have been excavated it is increasingly necessary to re-grade slopes with fill material as shown in Figure 8.6. The special paper/polypropylene membrane 'Hold-Gro' shown in Figure 4.81 can be used to cover the fill surface for erosion protection during the grass or vegetation growing period. If normally acceptable slopes can be steepened owing to the protective nature of the membrane then costs can be reduced by importing and laying down less fill. Apart from this, a more rapid vegetation cover is ensured in either wet or dry weather since both the erosive power of rain and the dessicating effect of the sun are reduced. This membrane is specially designed to rot away as the vegetation grows and will eventually disappear altogether. The length of time taken to rot away will depend upon the grade of membrane chosen.

Webbings such as Paraweb Mesh can be used for reclamation purposes in areas subject to localised land settlement such as over mine shafts or over areas of ancient shallow mine workings. In many cases, if industrial development is not to take place immediately over these areas of unpredictable settlement, but is to be undertaken close by, then — as shown in Figure 8.7 — safety meshes made of Paraweb may be placed over them, covered with thick soil, planted with shrubs and trees, and fenced off. Should collapse occur, then the webbing prevents catastrophic failure and loss of life. Reinstatement can be made economically and more safely. The alternative course of action would be to drill into the shaft and mine voids and pressure grout them until filled, which can be a most expensive procedure.

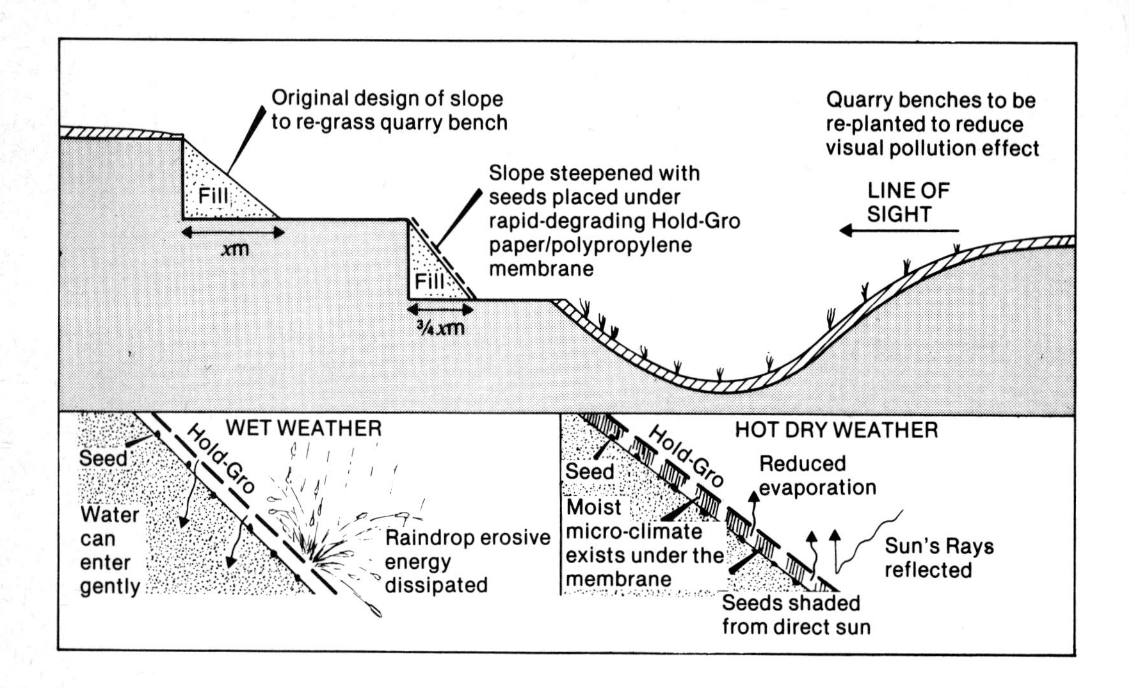

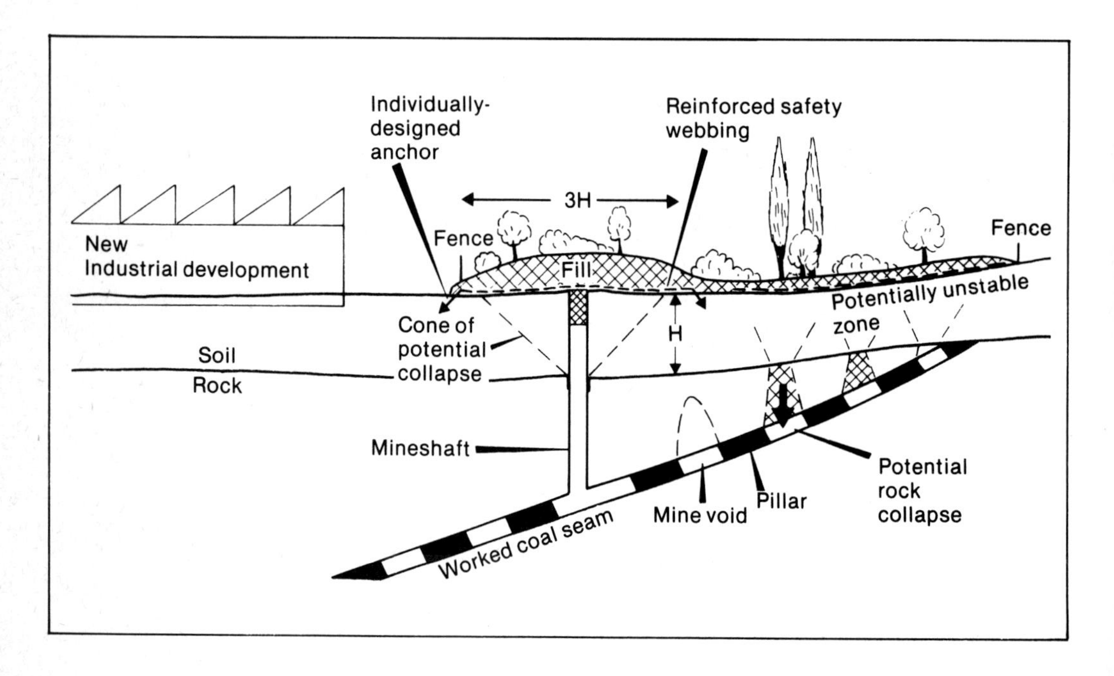

Figure 8.6 (Top) The use of a special degradable textile/paper membrane to promote grass growth in land reclamation works

Figure 8.7 (Bottom) Reinforced safety webbings such as Paraweb can be used to cover over potentially-unstable areas

Marine or coastal land reclamation

There are four main uses for membranes in marine coastal reclamation projects:

1. To protect the sides of a causeway or face of a marine defence from erosion by the sea as shown in Figure 8.8.
2. To act as a filter membrane on the inside of a permeable bund, to retain hydraulic fill (Figure 8.9).
3. To construct drainage elements in hydraulic fill areas to allow more rapid dissipation of included water.
4. To provide basal support for boulder fills placed on soft seabed sediments (Figure 8.10). Without the membrane, large boulders dropped onto the seabed sink in due to their own weight and the pressure of stones above. The seabed becomes sheared and weakened. Further sinking of boulders therefore takes place and material is wasted. If 'diapiric' structures build up, then the overall stability of the fill can be threatened. When membrane is laid prior to boulder placing, they are held at sea-floor level, down-punching is stopped, and a more even load distribution results.

The use of a permeable membrane is particularly beneficial at the edges of intertidal causeways where marine defences are needed on the sides, but complete underneath support is not

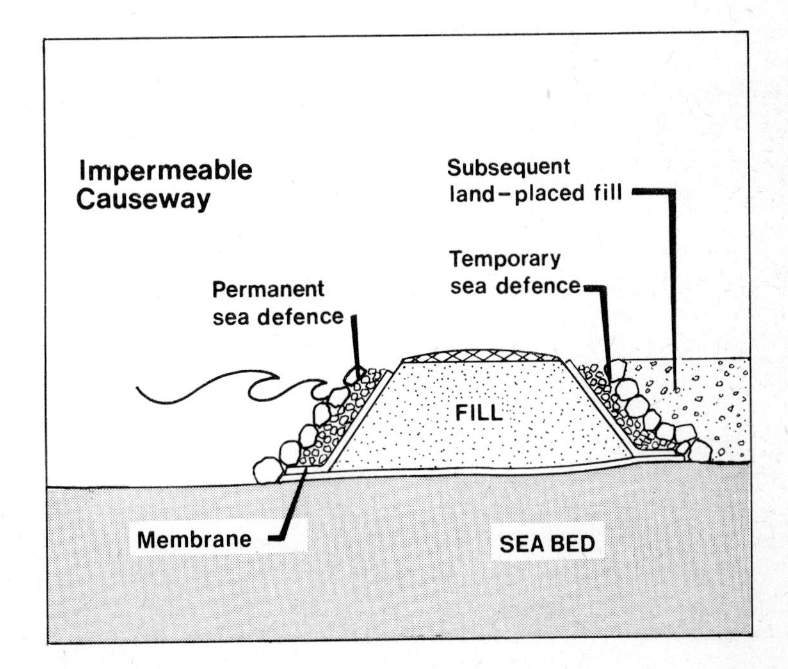

Figure 8.8 Membrane is useful for protecting causeways from erosion by wave and tide action

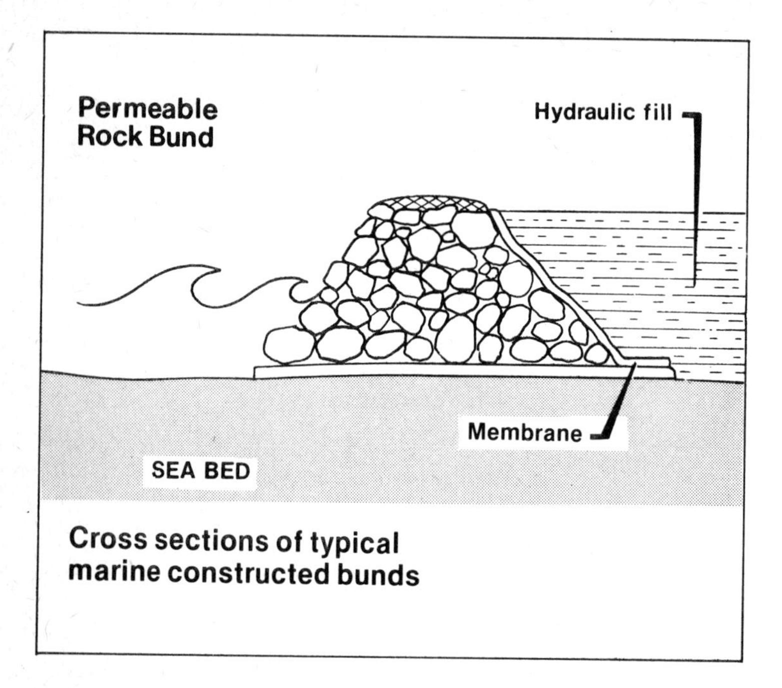

Figure 8.9 A strong permeable membrane can be used to allow water to escape from hydraulic fill whilst retaining the solid particles

necessary. This aspect was utilised, in conjunction with a filter construction on the Hunterston Iron Ore Terminal in Scotland[27] where a causeway was built into the sea, but where it was unnecessary to have the membrane completely across its base. The causeway was constructed on beach-type deposits of sand and laminated clays, and the calculated strengths were considered adequate for the structure in question. The edge design adopted is shown in Figure 8.11.

The choice of membrane will be governed by the criteria described in Chapters 5 and 7, and will particularly take into account the particle distribution characteristics of the fill material and the fact that the fill will be subjected to alternating waterflow erosion.

Where a permeable membrane is used to filter hydraulic fill, care must be taken in selecting the correct membrane for the type of stone against which it is to be laid. If the membrane is to be laid on large stones then the use of a high-extensibility non-woven membrane could prove to be the optimum way to minimise the possibility of fabric tearing. The Author knows of woven fabrics which have developed rips up to 3 metres long when pressed over large boulders by hydraulic sand fill. On the other hand, if a smaller stone layer can be placed beneath the membrane, then the higher permeability/strength ratio of a woven fabric could well be advantageously utilised. Also wovens tend to exhibit good ultra-violet resistance.

Fig. 8.12 shows the principles of hydraulic fill land reclamation. The dimension *'d'* can vary from a few metres (for

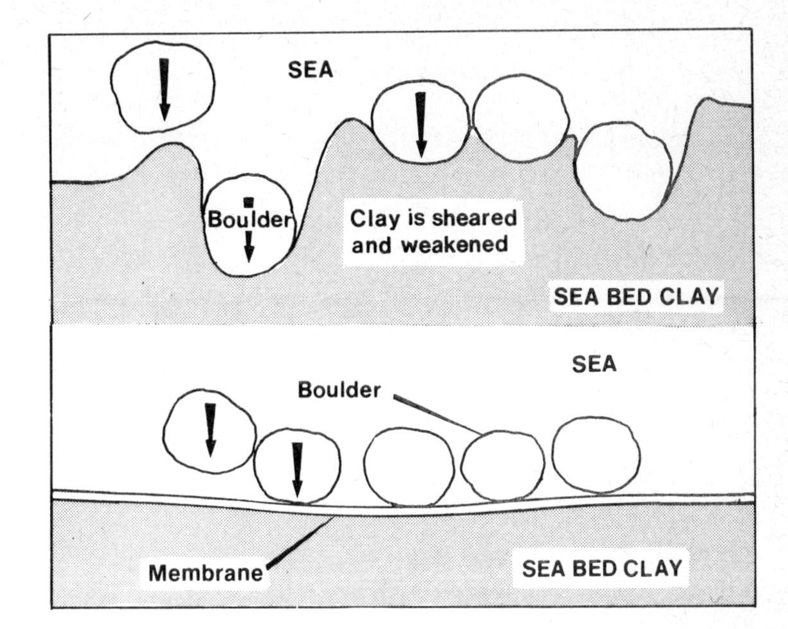

Figure 8.10 The use of a strong membrane on the sea bed restricts basal shearing and saves money by preventing the loss of rocks into the soft sea sediments during the construction of fill areas for reclamation purposes

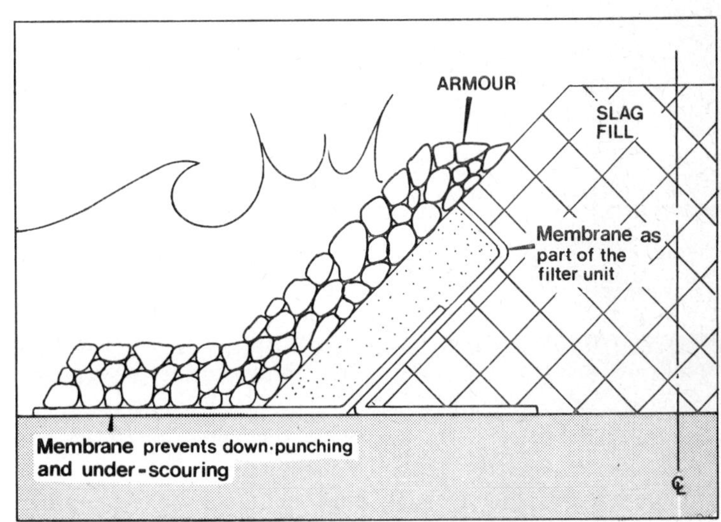

Figure 8.11 Terram membrane incorporated in the design of a marine causeway at the Hunterston Iron Ore Terminal, U.K.

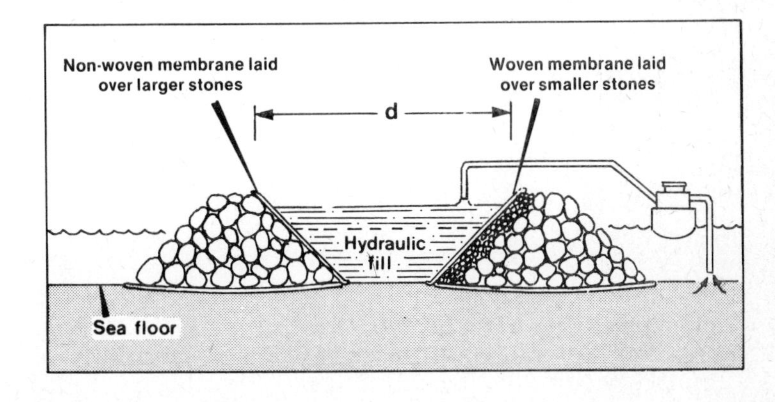

Figure 8.12 Hydraulic-fill land reclamation using rock bunds lined with permeable membrane

causeway construction) through tens of metres (for e.g. coffer-dam construction) to thousands of metres (for large area reclamation).

In calculating the likely rate of water outflow from hydraulic fill through a membrane filter, it should be remembered that a filter-cake will build up on the inside of the membrane. This filter-cake will comprise the smaller particles of the hydraulic fill, and will reduce the water outflow accordingly.

It is useful to establish in one's mind the essential difference between a membrane placed against an *in situ* soil (in which the water is flowing between relatively stable particles) and a membrane filtering particles of soil from a hydraulic suspension.

There is a particular need on large coastal projects to be careful with the specification of the membrane. The Author knows of one case where the rock bund and membrane were laid by one contractor, and the hydraulic fill was the responsibility of a second contractor. Unfortunately, due to some industrial disturbance, the second contractor was unable to cover the laid membrane for about five months. After this period of time exposed to tides and sun, the membrane was unable to be used owing to weathering deterioration. Under the circumstances, neither the first contractor nor the membrane supplier were at fault, since the membrane had met the required strength specification when supplied and laid. Therefore the client had to pay for the membrane to be replaced. The Author includes this example to warn potential users that care must be taken both in the wording of specifications for membranes *and* in the setting out of contractual responsibilities. One client well known to the Author has adopted the following specification which outlines his requirements simply and effectively. It ignores technical details of 'weathering capability' by simply stating the required properties 'prior to being covered'. The supplying contractor is therefore responsible if the membrane has deteriorated prior to being covered. To check this, the client goes on site just before the membrane is about to be covered, and takes random samples from the field-laid material. If these do not meet the required 'spec' then the contractor must replace the membrane, and is charged for the contract delay. The actual specification which was used is reproduced as follows without modification:

"153. The Contractor shall provide the necessary labour, materials and equipment for these tests at his own expense	Pre-cast Concrete Drain Channels, kerbs etc. (Cont'd)
154. The filter cloth shall be made of plastic material that will withstand all the conditions of handling, placing, exposure to salt water	Filter Cloth.

and sunlight, etc. that may be encountered in the construction and service of the Works.
The filter cloth shall have the following properties:

(a) Thickness — 0.5 mm (minimum)
(b) Breaking Strength — 150 kgf/3 cm (minimum)
(c) Pore Size — not larger than 0.2 mm

As most filter cloths deteriorate after exposure to the sun the Contractor shall provide such a material that will have the above minimum strength requirements after exposure encountered in the course of construction of the Works.
Samples will be taken from exposed filter just before the covering up of the filter cloth by the construction. Any filter cloth failing to meet the strength requirements in tests on the portions exposed shall be replaced by new filter cloth at the Contractor's expense. Details of the filter cloth which the Contractor proposes to use in the Works shall be submitted with the Tender.

Joint Filler.

155. The filler to be used at joints in concrete shall be of a bituminous impregnated fibre board type."

As mentioned above, a strong membrane can be used at the base of a permeable sand/rock causeway to support the lower layer at the causeway/seabed interface. Additionally, if the fill is impermeable a less strong membrane can be used in a double layer to construct a drainage blanket for the dissipation of excess pore water pressure generated by the construction of the causeway or embankment (see Figure 8.13).

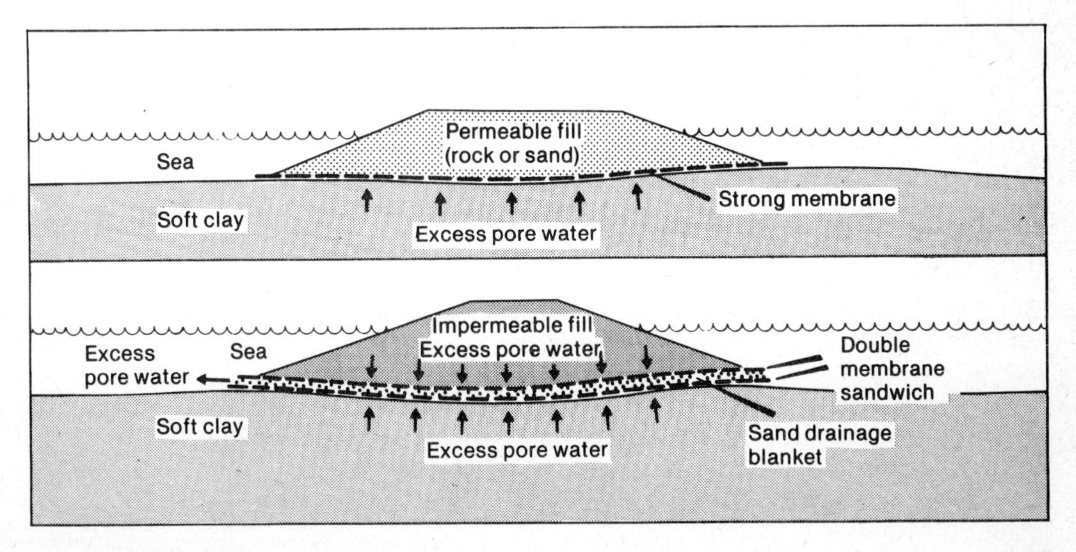

Figure 8.13 Use of permeable membrane at the base of marine embankments and causeways

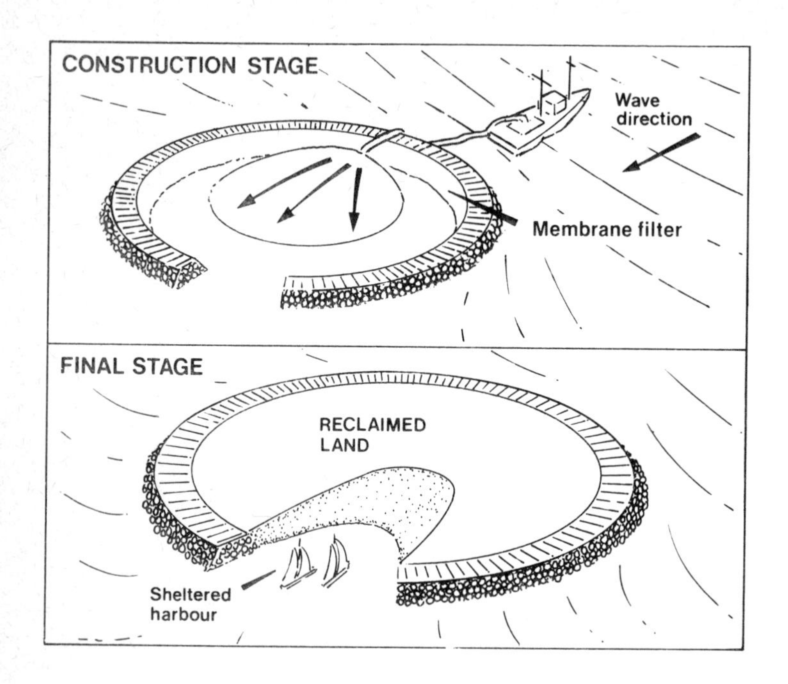

Figure 8.14 Island construction using circular rock bunds lined with membrane

The Author has experience in South East Asia of the successful use of permeable membranes in the construction of new offshore islands. This technique was originally developed in areas of existing coral atolls, where a circular rock bund was built on the coral and lined with membrane as shown in Figure 8.14. Because of the decreased permeability of the membrane when in contact with the fill, in some cases where sand was being pumped, openings were left in the rock bunds in the early stages of reclamation, thus allowing a rapid initial water outflow. In several cases these were left open permanently to form sheltered beach areas ideal for recreational purposes.

In Hong Kong, on the High Island reservoir scheme,[28] the Consultants extended the above concept to construct high-level coffer dams across the seabed from the island to the mainland. The technique developed was to build multiple small bunds in separate 'lifts' in order to save the cost of expensive marine-placed rock material. In this case, an extensible non-woven membrane was used successfully (see Figure 8.15). It is interesting to note that this same membrane was used to cover the sea-floor prior to rock placing, and to do this the Contractor pre-rolled several 4 metre wide rolls onto a long steel bar of such a weight that the pre-wound roll could sink to the seabed and be laid by divers (see Figure 8.16). Both the innovative construction technique and the membrane laying technique are to be considered as excellent examples of original thinking of the type which the availability should be inspiring in all branches of ground engineering.

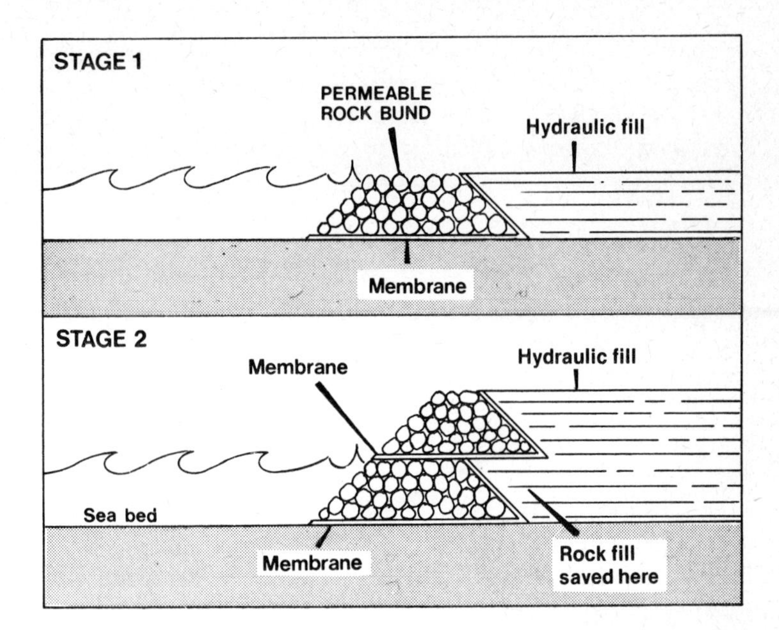

Figure 8.15 The use of a permeable membrane in a novel design to construct a high coffer dam

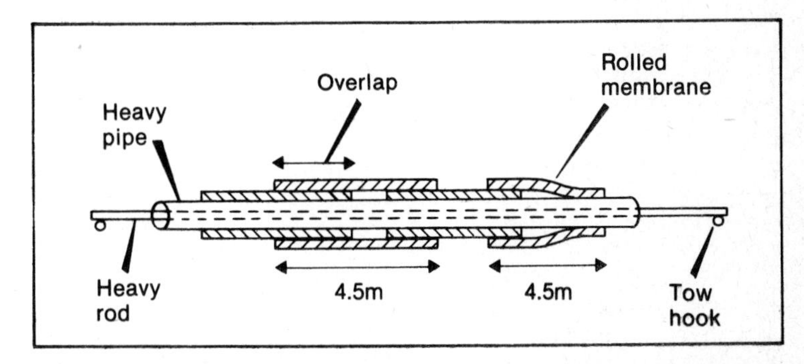

Figure 8.16 Pre-rolling several widths of membrane onto a heavy steel rod and pipe for submarine laying.

Another example of new products leading to novel solutions is the advent of self-contained internal-flow drainage products such as 'Filtram' (shown in Figure 4.83). This product is constructed with a central sheet of Netlon plastic mesh and Terram filter membrane welded to each side. As shown in Figures 8.17 and 8.18, if the sides of this product are sealed with tape, then the membrane can act as a composite 'blanket drain', without allowing the ingress of mud or silt. This is especially useful where the hydraulic fill is marine clay, and where dissipation of excess pore water pressure can be speeded by the order of years on major projects. The cost of using such systems must be weighed against the economic income of say one year's earlier rental for new industrial land. The layout in Figure 8.18 is simply the Author's suggestion for a novel approach to an old problem. It is in the reader's interest to try to use modern membrane technology to arrive at novel solutions rather than simply placing new materials into conventional solutions. No doubt the latter approach will work, but it is in the former that the best technical advances and cost savings can be made.

Figures 8.19-8.25 show some typical installations and engineering drawings, in order to familiarise the reader with currently acceptable designs and design detailing.

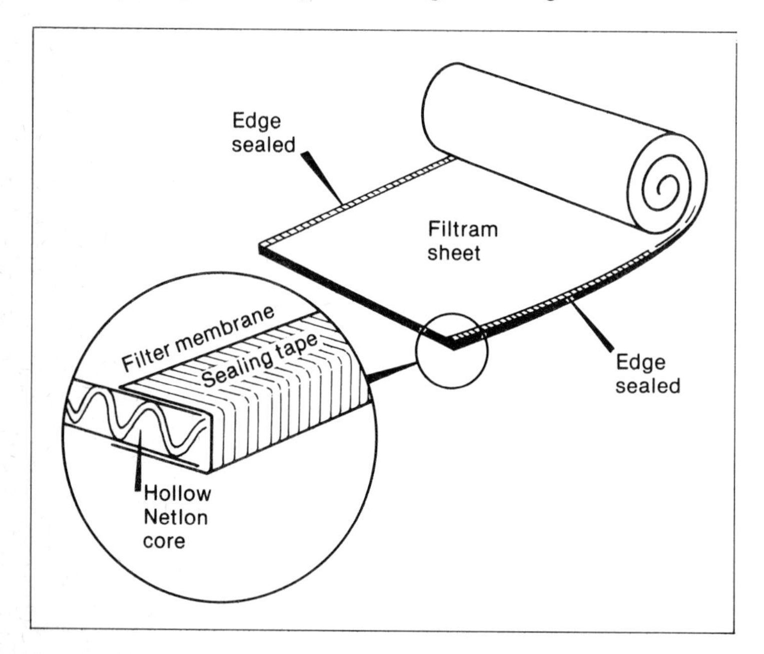

Figure 8.17 Showing how Filtram material can be sealed with tape for use as a pore-water dissipation blanket

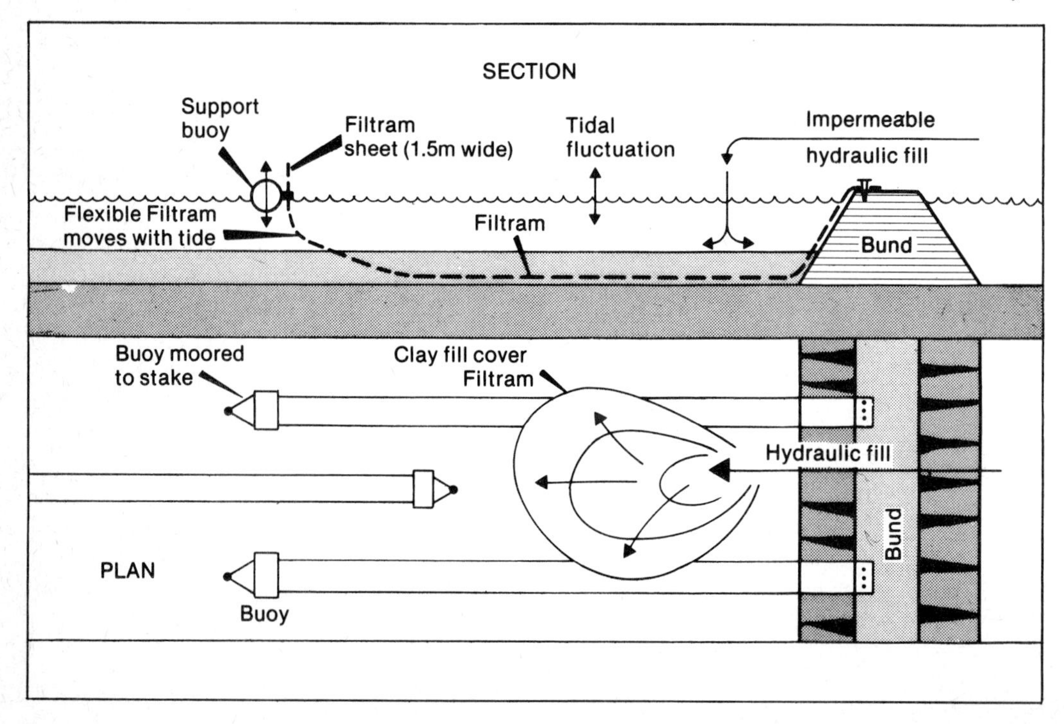

Figure 8.18 Author's suggestion for possible use of Filtram material to relieve excess pore water pressure and speed consolidation during the laying of relatively impermeable hydraulic fills such as silts or clays

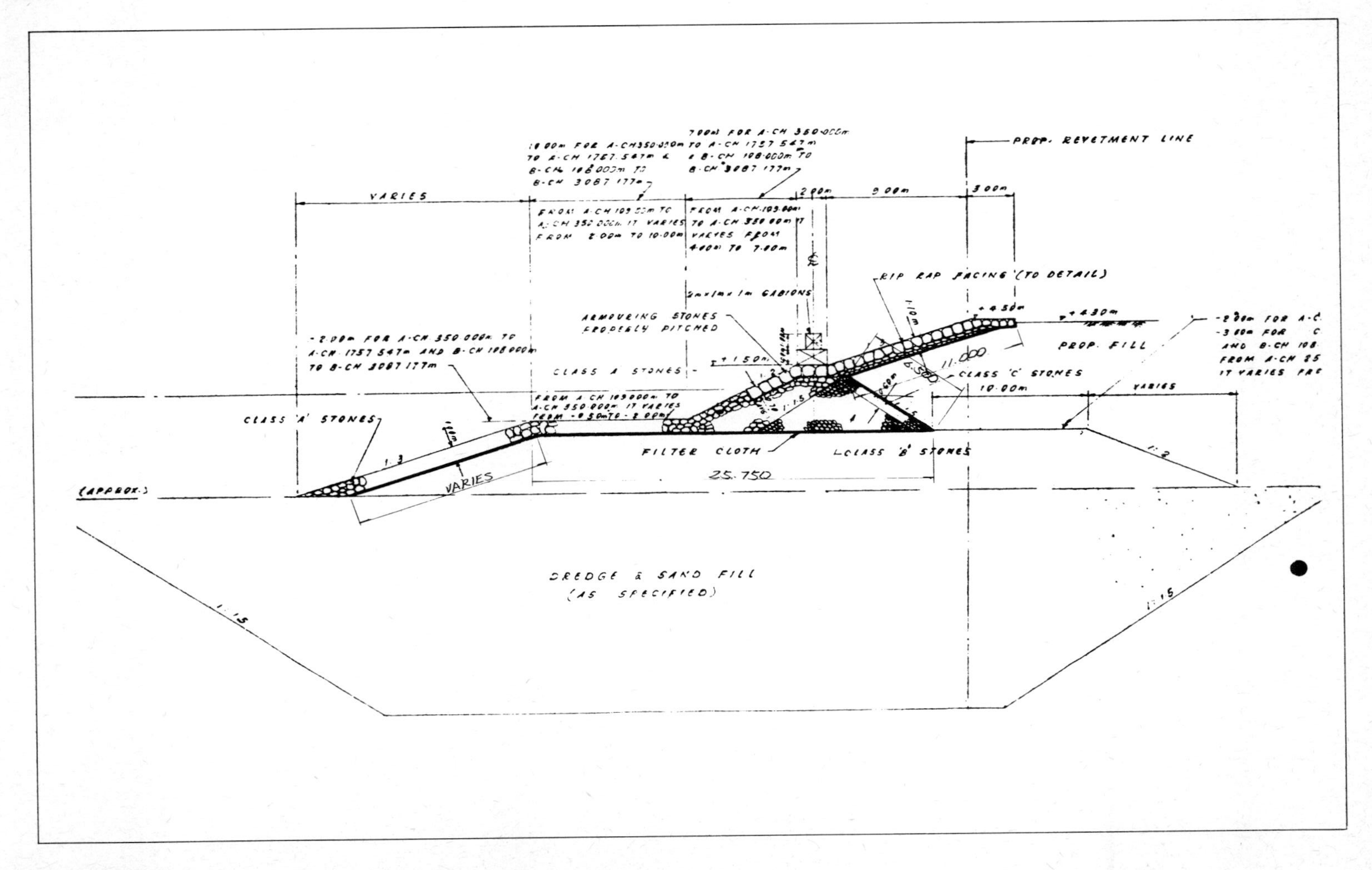

Figure 8.19 Coastal defence for land reclamation

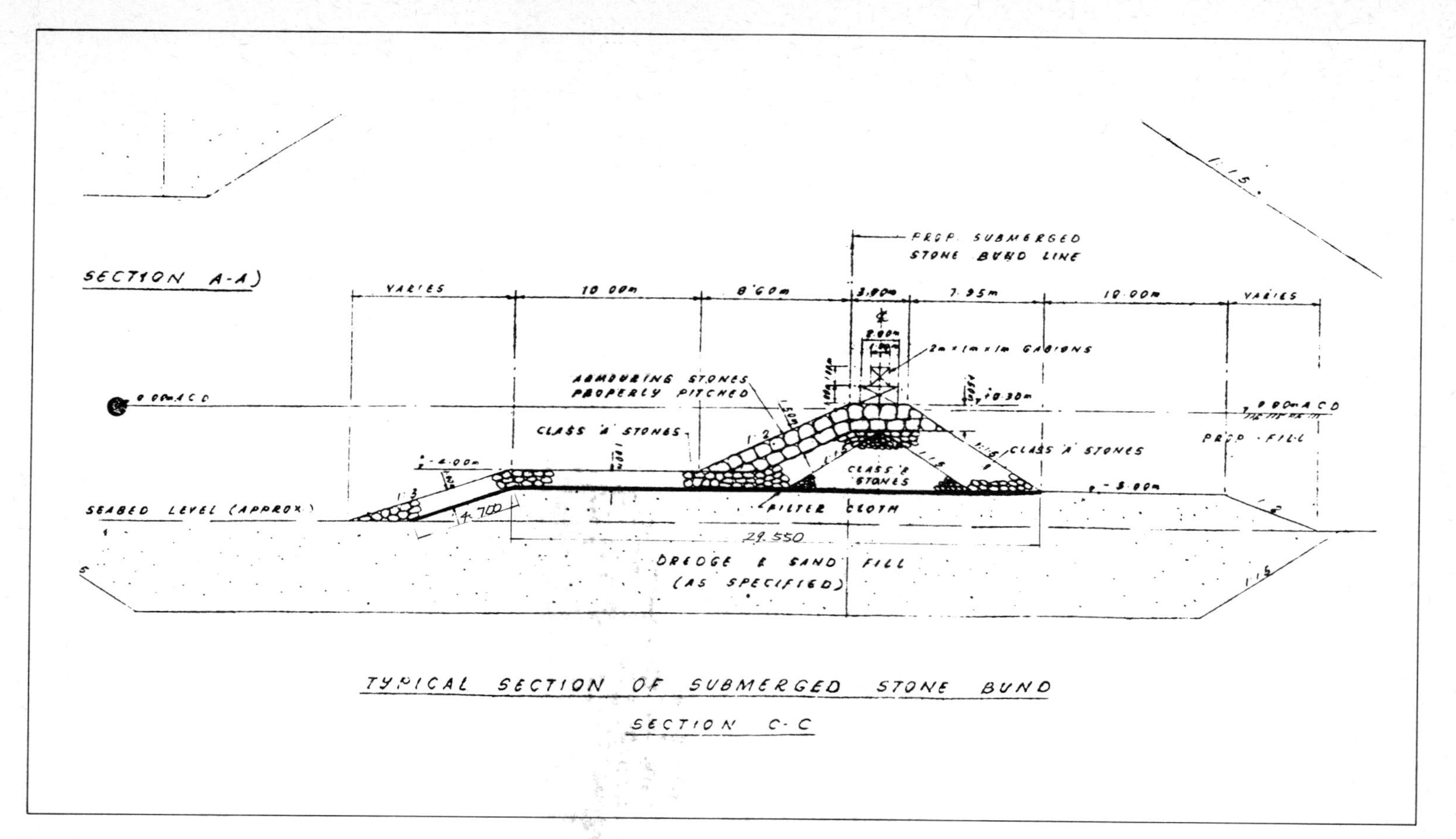

Figure 8.20 Coastal defence for land reclamation

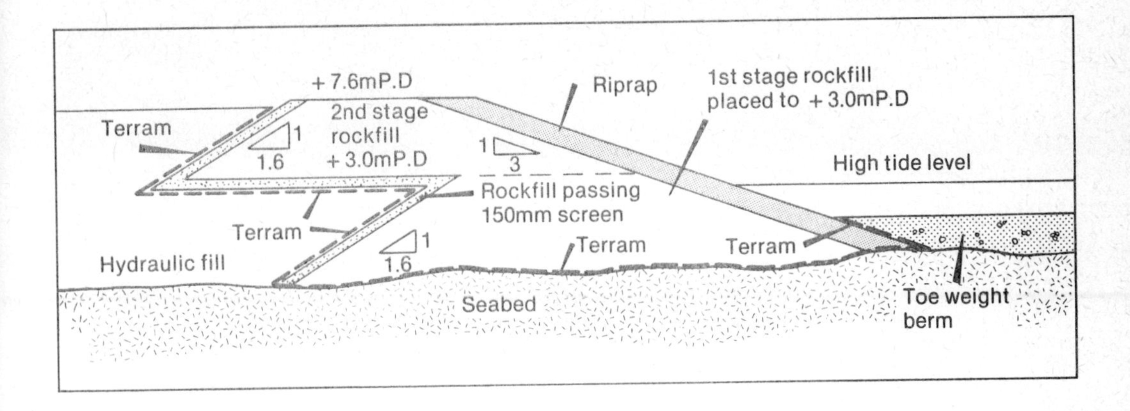

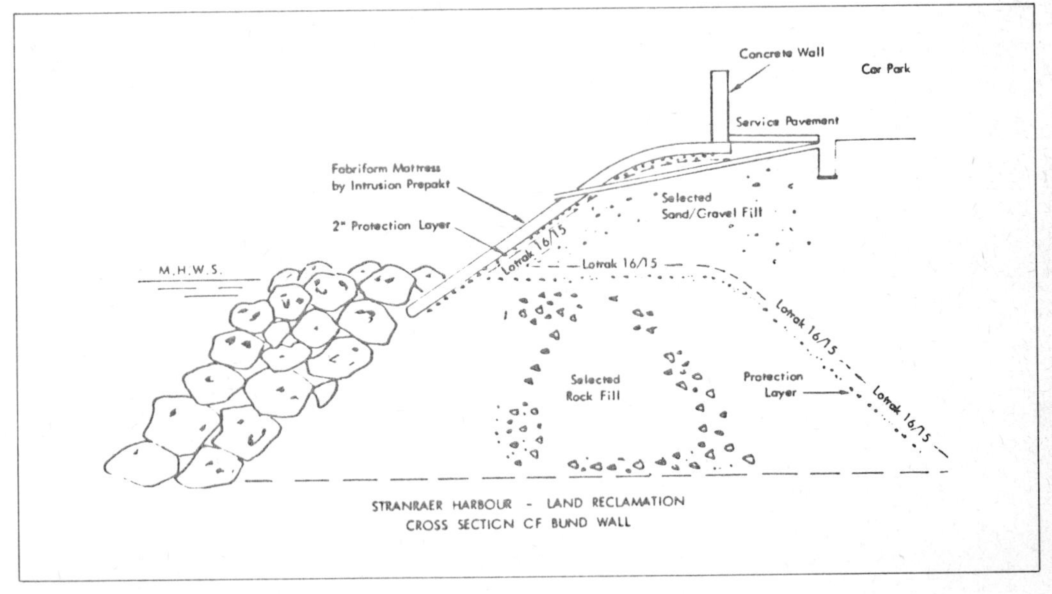

Figure 8.21 (Top) Use of Terram in the construction of a marine defence for land reclamation using hydraulic fill to build the defence in two stages

Figure 8.22 (Bottom) Defence for land reclamation. (Courtesy Lowe Brothers)

Figure 8.23 Land reclamation using permeable membrane separator (Courtesy ICI Fibres)

Figure 8.24 Membrane design for housing development scheme. (Courtesy Manstock Geotechnical Consultancy Services Ltd.)

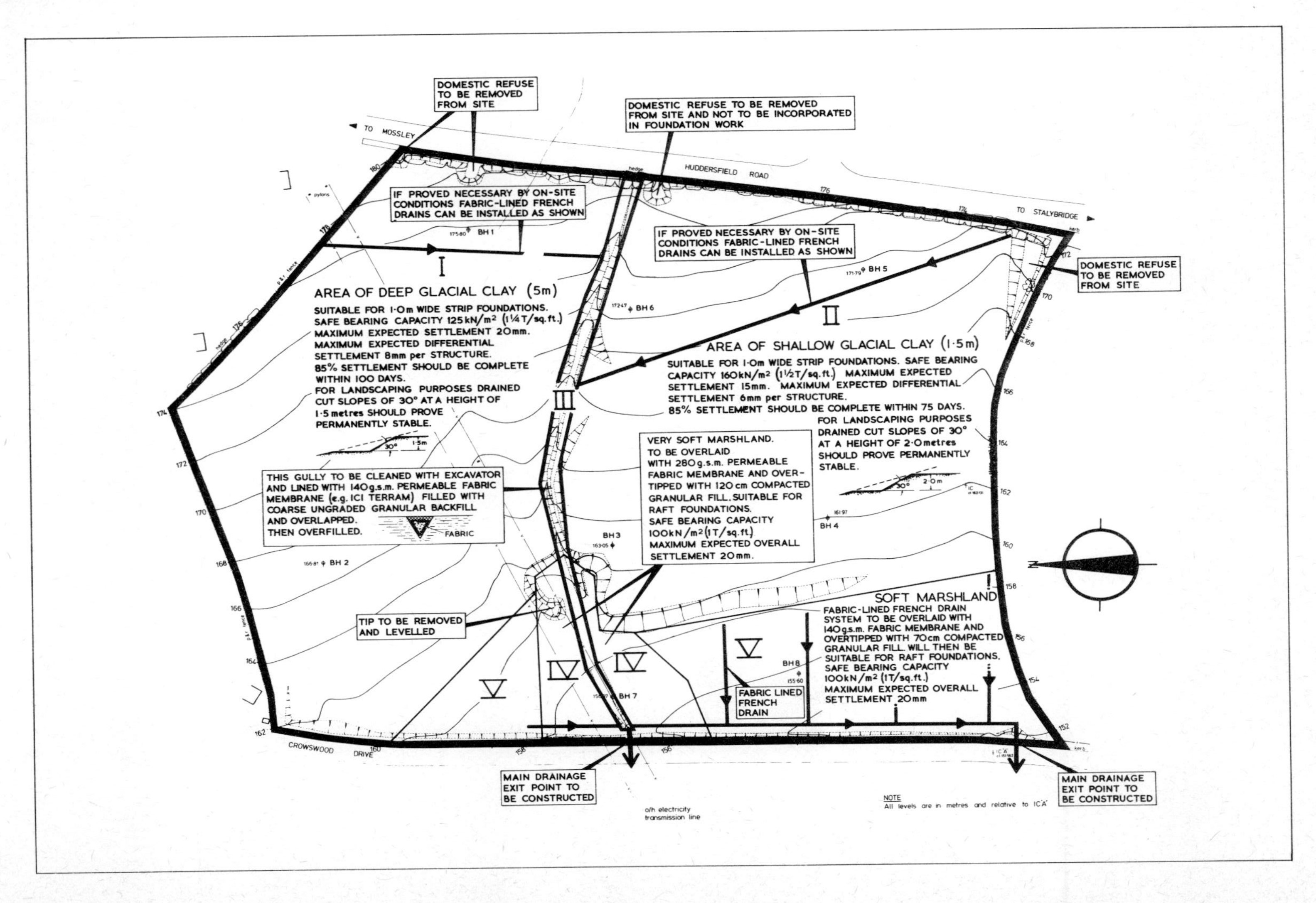
DOMESTIC REFUSE TO BE REMOVED FROM SITE
DOMESTIC REFUSE TO BE REMOVED FROM SITE AND NOT TO BE INCORPORATED IN FOUNDATION WORK
TO MOSSLEY
HUDDERSFIELD ROAD
TO STALYBRIDGE
DOMESTIC REFUSE TO BE REMOVED FROM SITE
IF PROVED NECESSARY BY ON-SITE CONDITIONS FABRIC-LINED FRENCH DRAINS CAN BE INSTALLED AS SHOWN
IF PROVED NECESSARY BY ON-SITE CONDITIONS FABRIC-LINED FRENCH DRAINS CAN BE INSTALLED AS SHOWN
BH 1
BH 5
BH 6
I
II
III
AREA OF DEEP GLACIAL CLAY (5m)
SUITABLE FOR 1·0m WIDE STRIP FOUNDATIONS. SAFE BEARING CAPACITY 125 kN/m² (1¼ T/sq. ft.) MAXIMUM EXPECTED SETTLEMENT 20mm. MAXIMUM EXPECTED DIFFERENTIAL SETTLEMENT 8mm per STRUCTURE. 85% SETTLEMENT SHOULD BE COMPLETE WITHIN 100 DAYS. FOR LANDSCAPING PURPOSES DRAINED CUT SLOPES OF 30° AT A HEIGHT OF 1·5 metres SHOULD PROVE PERMANENTLY STABLE.
AREA OF SHALLOW GLACIAL CLAY (1·5m)
SUITABLE FOR 1·0m WIDE STRIP FOUNDATIONS. SAFE BEARING CAPACITY 160kN/m² (1½T/sq.ft.) MAXIMUM EXPECTED SETTLEMENT 15mm. MAXIMUM EXPECTED DIFFERENTIAL SETTLEMENT 6mm per STRUCTURE. 85% SETTLEMENT SHOULD BE COMPLETE WITHIN 75 DAYS. FOR LANDSCAPING PURPOSES DRAINED CUT SLOPES OF 30° AT A HEIGHT OF 2·0 metres SHOULD PROVE PERMANENTLY STABLE.
30°
1·5m
2·0m
THIS GULLY TO BE CLEANED WITH EXCAVATOR AND LINED WITH 140g.s.m. PERMEABLE FABRIC MEMBRANE (e.g. ICI TERRAM) FILLED WITH COARSE UNGRADED GRANULAR BACKFILL AND OVERLAPPED. THEN OVERFILLED.
FABRIC
VERY SOFT MARSHLAND. TO BE OVERLAID WITH 280g.s.m. PERMEABLE FABRIC MEMBRANE AND OVER-TIPPED WITH 120 cm COMPACTED GRANULAR FILL. SUITABLE FOR RAFT FOUNDATIONS. SAFE BEARING CAPACITY 100kN/m² (1T/sq.ft.) MAXIMUM EXPECTED OVERALL SETTLEMENT 20mm.
BH 2
BH 3
BH 4
BH 7
BH 8
TIP TO BE REMOVED AND LEVELLED
SOFT MARSHLAND
FABRIC-LINED FRENCH DRAIN SYSTEM TO BE OVERLAID WITH 140g.s.m. FABRIC MEMBRANE AND OVERTIPPED WITH 70cm COMPACTED GRANULAR FILL. WILL THEN BE SUITABLE FOR RAFT FOUNDATIONS. SAFE BEARING CAPACITY 100kN/m² (1T/sq.ft.) MAXIMUM EXPECTED OVERALL SETTLEMENT 20mm
IV
IV
V
V
FABRIC LINED FRENCH DRAIN
CROWSWOOD DRIVE
MAIN DRAINAGE EXIT POINT TO BE CONSTRUCTED
MAIN DRAINAGE EXIT POINT TO BE CONSTRUCTED
o/h electricity transmission line
NOTE
All levels are in metres and relative to ICA'
N

Figure 8.25 Drainage blanket being constructed under the base of an embankment to dissipate pore water pressure. (Courtesy B.S.P. International Foundation Ltd. — Fibertex)

9

Permanent and temporary road construction

There are three major uses for permeable membranes in the field of road construction.

1. Water drainage schemes associated with road construction — already dealt with in earlier chapters.
2. As a sub-base/sub-grade separator in 'temporary road' construction. Such uses include quarry and mine haul roads, site access roads, farm roads, etc.
3. Use as a separator either at the sub-base/sub-grade interface or at another interface. This can include the use of permeable membrane just below the wearing course of a black-top road to provide an element of reinforcement (see Figures 3.17 and 3.18).

Drainage in relation to roads

Although the principles of the use of permeable membranes in general drainage applications have been dealt with earlier in Figures 5.1-5.24, the specific application areas of membranes in roadway construction are summarised in Figure 9.1.

Use of permeable membranes as sub-base/sub-grade separators in temporary roads

This application is extremely well known and is perhaps the one most widely advertised by manufacturers. Figure 9.2 shows a typical access road being constructed across a field.

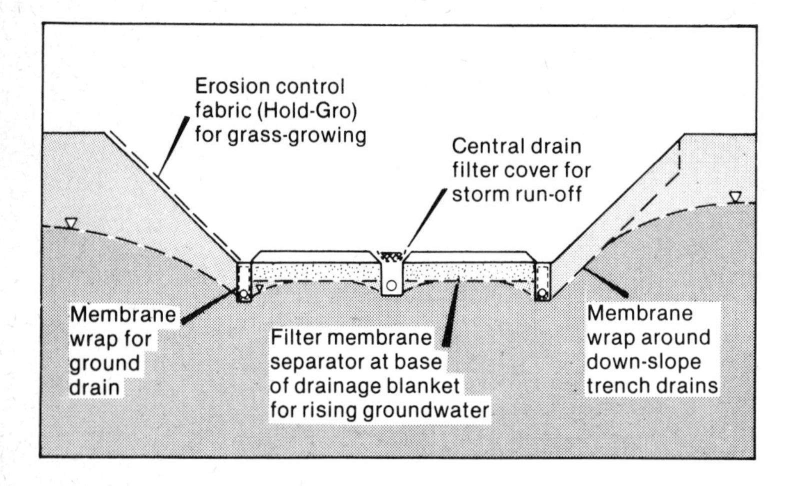

Figure 9.1 General uses for membranes in road construction

Figure 9.2 Photograph of a temporary access road being constructed across a field. (Courtesy Enka Glanzstoff)

The construction principles are easy in so much as the membrane is rolled out ahead of the granular fill, and can be rolled directly onto the existing vegetation surface. Naturally, if large tree stumps or boulders are present in the ground, then it is necessary to remove these prior to laying the road. In practice, temporary access roads are usually only of the order of four to five metres wide, and they can — in most cases — be taken along such a route as to avoid major obstructions. Therefore it is common practice to lay the membrane directly on the surface vegetation and follow this with the dumping of granular fill. Vehicles should not be allowed to travel directly on top of the

Figure 9.3 Lorry dumping stone onto compacted material during temporary road construction. (Courtesy ICI Fibres)

membrane, and lorries carrying granular fill normally reverse along the road dumping their load on top of already-compacted material (see Figure 9.3).

At first it is difficult to see exactly what function the membrane is performing in these circumstances, until one tries to examine the fundamental engineering principles behind the concept of laying granular material onto a soft cohesive sub-grade in order to provide vehicular trafficking capability.

The most important part of understanding the behaviour of a permeable membrane at the sub-base/sub-grade interface of a road structure is to learn that the membrane is performing many small functions which are combining to produce a substantial and observable effect. (Some of these functions are illustrated in Figures 3.4 and 3.11.)

One object of laying granular material on top of a wet cohesive sub-grade is that the granular layer should spread the point load of a vehicle tyre over a large surface of the sub-grade by frictional energy dissipation (Figure 9.4). By this means the sub-grade material can support the weight of the vehicle. The

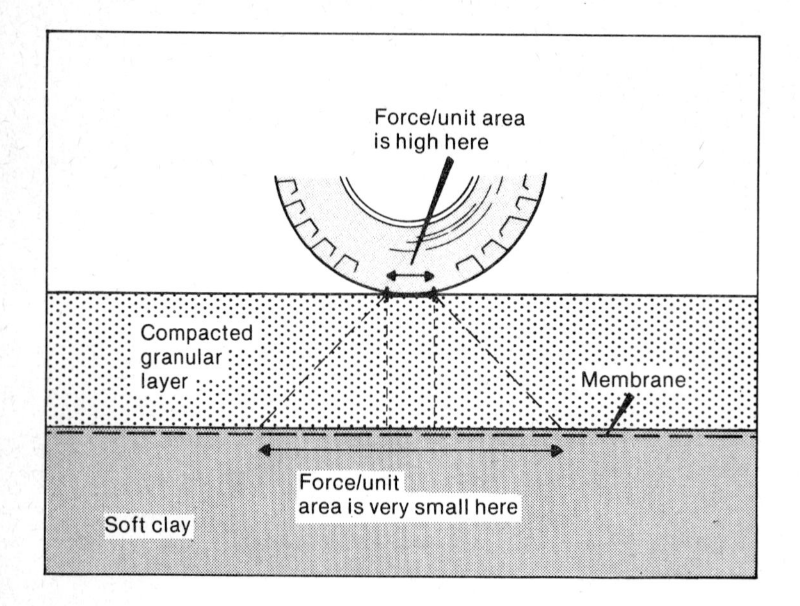

Figure 9.4 Load being spread as a force/unit area with increasing depth in a granular material

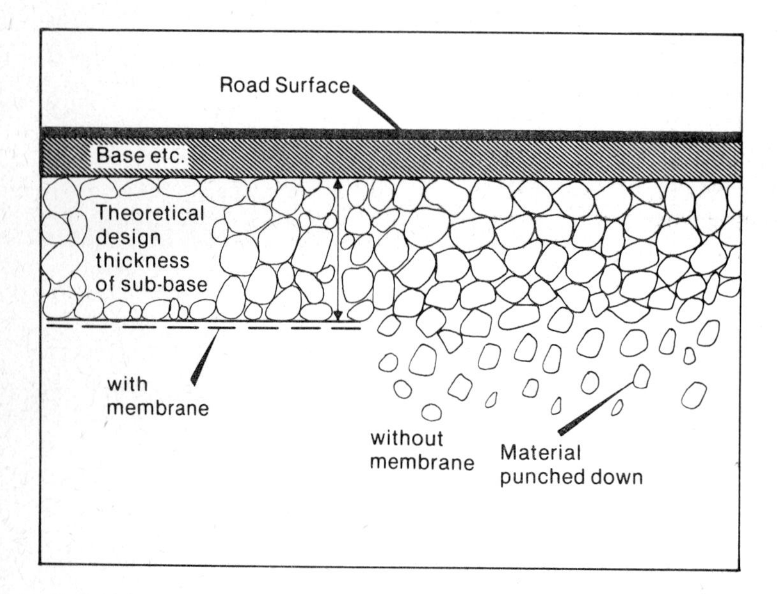

Figure 9.5 Where sub-base is laid without membrane, stone is always lost below the designed thickness level

practical problems connected with this theoretically-admirable solution are concerned with the fact that, in practice, without a membrane it is difficult to construct, and the structure rapidly deteriorates with use as is illustrated in Figure 3.4. It may be considered questionable that a simple haul road should be difficult to construct, but the Author particularly means 'hard to construct *in accordance with the theoretical drawn design'*. In particular, whenever a road is to be built — and this applies much more to permanent roads than to temporary roads — a design thickness has been specified by an engineer, in order that the road may carry the required traffic for the required number of vehicle passes. Figure 9.5 shows that without a membrane present at the sub-base/sub-grade interface, this design can never be practically achieved. The first stones dumped upon the cohesive soil are pushed into it by the weight of stones above and become separated from the main body of the granular material; their value is therefore completely lost. In effect this is money, used in transport and for materials, that has been totally wasted.

Since the strength of a granular material in bearing a surface load is found in the frictional contact between the granular particles, the maximum frictional force can only be utilised if (1) the particles are dry, (2) the particles are clean, (3) the particles are in firm contact with one another, (4) the particles are angular in shape, and (5) the particles are well graded in size. (I.e. there are many particles of different sizes to provide a large frictional surface contact area.) Looking at the above requirements, it can be seen why, if some granular particles become separated from the main mass of material by downward punching into the underlying sub-grade, they can no longer perform any useful function. Similarly, if by trafficking, mud is pumped upwards into the road as illustrated in Figure 3.4, then those layers of road sub-base which have been penetrated by mud will no longer be performing a useful function.

A list of the multiple functions of a membrane in a simple road structure is given below:

1. A membrane can be rolled over vegetation and therefore there is no cost for stripping and preparing the ground.
2. Without the membrane, individual stones can punch through the vegetation or the upper layers of soil, disturbing the cohesive material and — even though only slightly — by shearing it, these stones can reduce its shear strength.
3. The agitation of these stones penetrating and mixing the sub-grade causes any surface water to be mixed in with the upper layer of cohesive material thus reducing its strength.

(The loss of stones in this way is a direct economic loss since the money is wasted and the stones are no longer functional. In addition, the weakening of the upper layer of soil caused by this disturbance generates an additional loss of stone into the sub-grade which cyclically goes on to weaken the sub-grade and cause further loss of stone. This damaging of the sub-grade by the action of placing granular materials upon it, is almost self-defeating and takes large quantities of stone to overcome. Conversely, when a membrane is placed on the sub-grade material, the loss of the first stones into the sub-grade is prevented and therefore the problems in 2 and 3 are overcome.)

4. The firm membrane/sub-grade interface allows moisture to be lost with time from the sub-grade material rather than included in it, and the forcing out of this water can increase the shear strength of the upper layer of the sub-grade (Figure 3.11).

5. Having gained the advantages of the above points by placing the membrane on the sub-grade, one of the most important ways of getting the best performance out of it is to use its separating capability to allow the vibrated compaction of the lower layers of the granular material. If a vibro-roller is used on granular material over soft cohesive soils or peat without a membrane separator, then the stones of the lower layer are forced downwards and the damage to the sub-grade is increased enormously. Also considerable quantities of material can be lost in addition to the wasted cost of the vibrating roller. However, when a membrane is placed at the sub-base/sub-grade interface, the lowest few centimetres of granular material can be compacted and can form a strong base which is contained by the membrane.

6. Once the membrane has been placed and the granular material compacted, then as the road is used, instead of the base deteriorating and breaking away with mud-pumping, the granular material can become more compacted and stronger whilst the sub-grade material can experience a long-term loss of moisture in the upper layers, which increases its strength further. (The use of lightweight vibro-compactors for constructing roads with membrane separators is, in the Author's opinion, one of the most critical aspects of the construction process — especially in certain parts of the world where the use of heavy vehicles and heavy rollers is impossible owing to the low-bearing capacity of the sub-grade soils. In such circumstances a design philosophy has to be adopted whereby the road is not considered to be capable of carrying its final design load until it is completed. On harder clay soils in temperate countries, this tends to be over-looked since the only effect of over-stressing the sub-grade soils is to reduce trafficability. In contrast, over-stressing of a thick very soft tropical coastal clay by use of heavy rollers or lorries can be catastrophic in that it can result in the total loss of the plant by deep shear failure.)

   On very soft soils, it may prove critically important to be able to assess the weight of roller which can be supported over a given thickness of granular material. There are no formal ways of calculating this, but the Author suggests that a consideration of some simple first principles would give the guide shown in Figure 9.6.

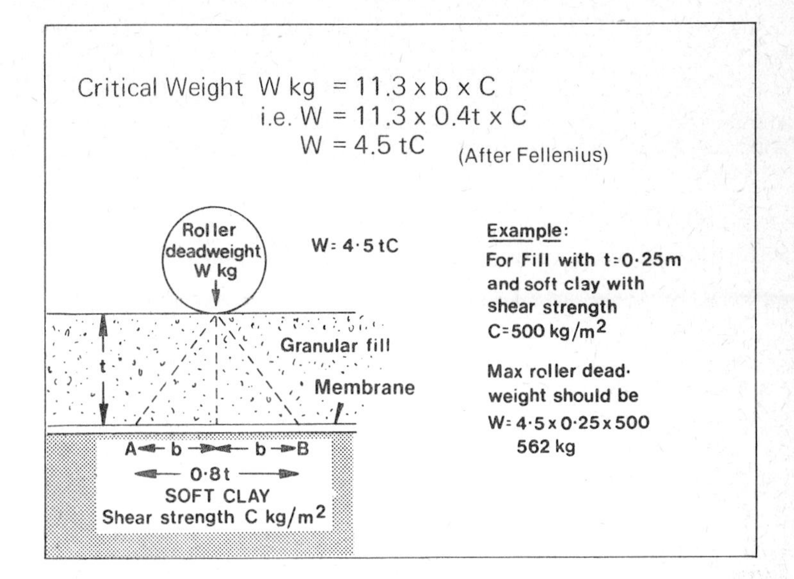

Figure 9.6 Suggested method for assessing the vibro-roller deadweight which can be carried by granular fill over soft ground

The ability of a road to perform work in supporting traffic is reflected in the amount of compaction which can be imparted into the granular material of the road structure.

Figure 9.7B shows the Author's view of the 'strength' of a road where membrane has not been used, and where attempted compaction of the lower layers has resulted in them becoming mixed with soft cohesive sub-grade material. In comparison Figure 9.7A shows the increased strength of the road if the lower layers can be well compacted against a membrane. In an analogy with other physical systems, the work capability of the two road cross-sections may be considered to be proportional to the area contained within the curve of the graph. There is therefore an increased work potential in the road at Figure 9.7A, which may be utilised in the following ways:

(a) The road may be allowed to carry a heavier weight of vehicle than would otherwise be possible.
(b) The road may be considered to have a longer design life for the same type of vehicular traffic, i.e. to have a greater Standard Axle capacity than before.
(c) The type of vehicle and Standard Axle use of the road may be kept the same, but the thickness of road construction may be reduced until the two areas A and B beneath the curves correspond. If the cost of the granular materials saved is greater than the cost of the membrane placed, then there will have been a nett saving in the cost of the road construction *in addition to* the long-term benefits brought about by the presence of the membrane, and also

in addition to the immediate savings of cost in the materials no longer lost at the sub-base/sub-grade interface during the construction process.

Figure 9.8 shows the ideal type of granular material for vibro-compaction work in temporary roads. As a useful 'rule of thumb' the largest stone size present in the aggregate should be smaller than one-third of the thickness of the first aggregate layer to be placed on top of the membrane. Otherwise, any very large single stone could be thrust downwards by overriding traffic and could penetrate the membrane. Although it is advisable to try to avoid the membrane being damaged in this way, it is the Author's experience that, for temporary roads, where roots or large stones penetrate the membrane, providing that the membrane does not rip sideways and provided that it remains wrapped around the penetrating object, then there is little resultant effect on the structure as a whole. The Author believes that membrane tearing is more likely with a woven material than with a non-woven type, but this tendency of some woven materials to develop tears must be weighed against the fact that in general, they have to be stressed to a higher level than non-woven materials for puncturing to occur. On the other hand, there is little doubt that non-woven materials can be extremely strong when taken on a 'weight for weight' basis, and tend to generate their maximum strength when stressed *slowly*.

Figure 9.7 Author's view of road strength variation with depth, with and without a membrane separator

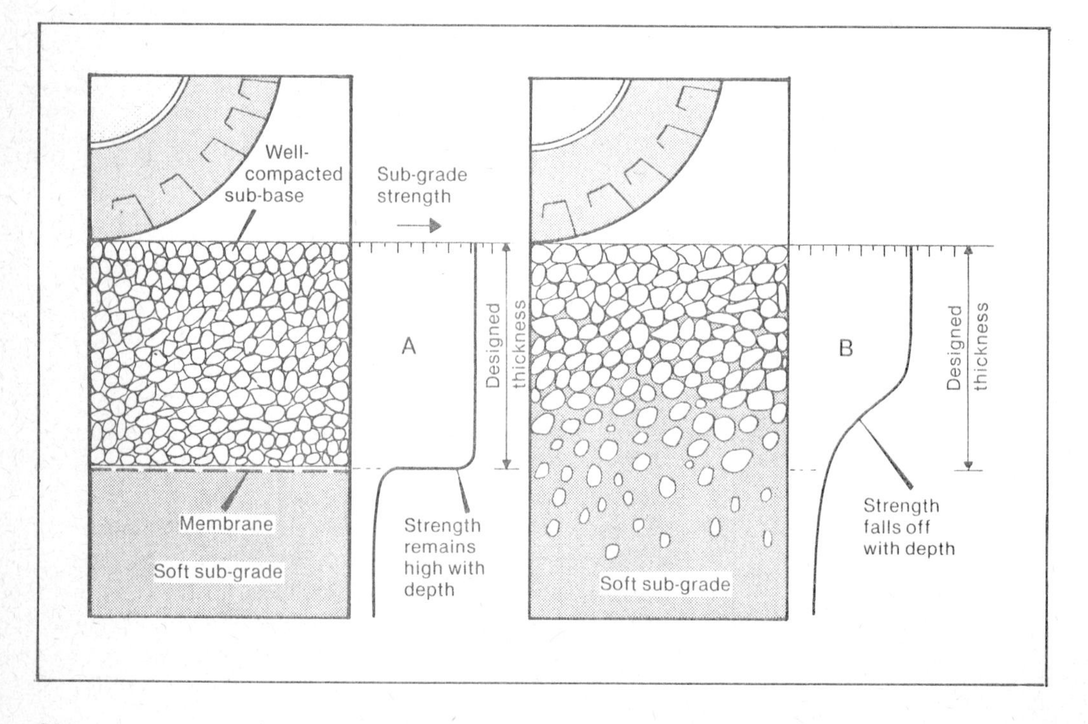

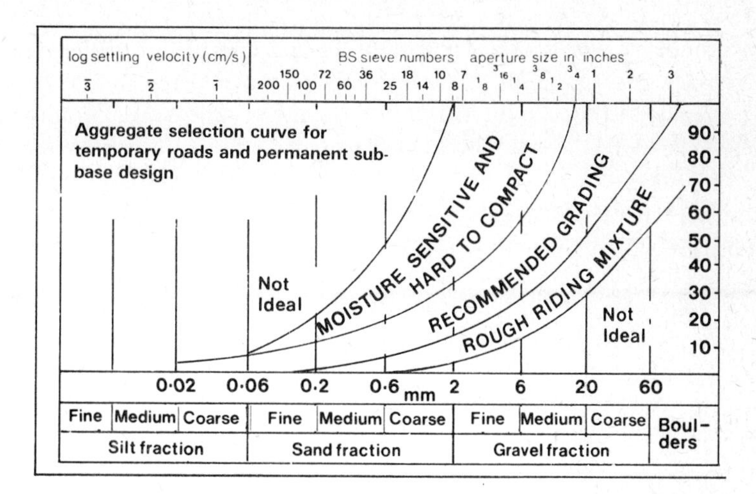

Figure 9.8 Suitable fill grading for compaction on top of membrane separators

This allows time for fibre reorientation and stress redistribution. Where stresses are applied quickly (such as when rocks or stones are dropped on a membrane) there may be a tendency for, say, a spun-bonded material to puncture more easily than a needle punched material, which in turn may puncture more easily than a woven one. However, at the base of a large embankment the woven material would not be able to accommodate as much strain as the needle-punched type and that (in turn) would not be able to extend to the same amount as a melded two-dimensional membrane. Consequently, under rock-drop conditions a woven membrane may be preferable, but once established under an embankment, a non-woven fabric may perform much more satisfactorily without rupturing under high strain conditions. (This latter aspect applies more in consideration of permanent roads than temporary ones.)

Permanent road structures

Membranes are laid at the sub-base/sub-grade interface of permanent roads in order to utilise the same properties outlined in the section on 'temporary roads' above. There is, of course, one important aspect relating to the use of membranes in permanent roads that differs strictly from its use in temporary roads. This is the requirement that in permanent road construction the membrane should not at any time be subject to mechanical stress or abrasion. This contrasts with the case of temporary roadworks where under certain circumstances the membrane is allowed to be stressed and deformed during the laying process, and can even be subject to abrasion and movement during the short life of the road. Indeed, in the extreme case of an access road to a drilling site or some other temporary works where a strictly limited number of vehicular passes are required, then the sub-base thickness might be restricted to such a low figure that the membrane is irrepairably damaged by the time the road ceases to be used. In one sense, providing that no danger were to arise from the use of that road, such a temporary design could be considered to be most cost-effective, since overdesign would be a waste of money. However, in permanent road design where 10, 20 or 50 years life is required, then the concept of low-loading and light-weight vibro-compaction should be strictly adhered to. The membrane should be damaged as little as possible by the placing process of the sub-base, and should not be subject to any mechanical movement from the road traffic. Therefore, the sub-base and base courses should be calculated at an appropriate thickness by a standard method such as that illustrated in Figure 9.9, which is an extract from the TRRL Road Note 29.

It has often been suggested that design thicknesses can be reduced as a result of the introduction of a membrane at the sub-base/sub-grade interface. ICI Fibres produced the diagram shown in Figure 9.10 which was based on their early experience, and which they publish in their manual *Designing with Terram.*

If this diagram is compared with the TRRL work on which it is based, and which is shown in Figure 9.9, then it can be seen that ICI suggests that by using a membrane there may be between 0 % and 3 % saving on the sub-base thickness necessary over soft ground. Similar curves are published by Rhone-

Figure 9.9 Sub-base thickness curves from TRRL Road Note 29

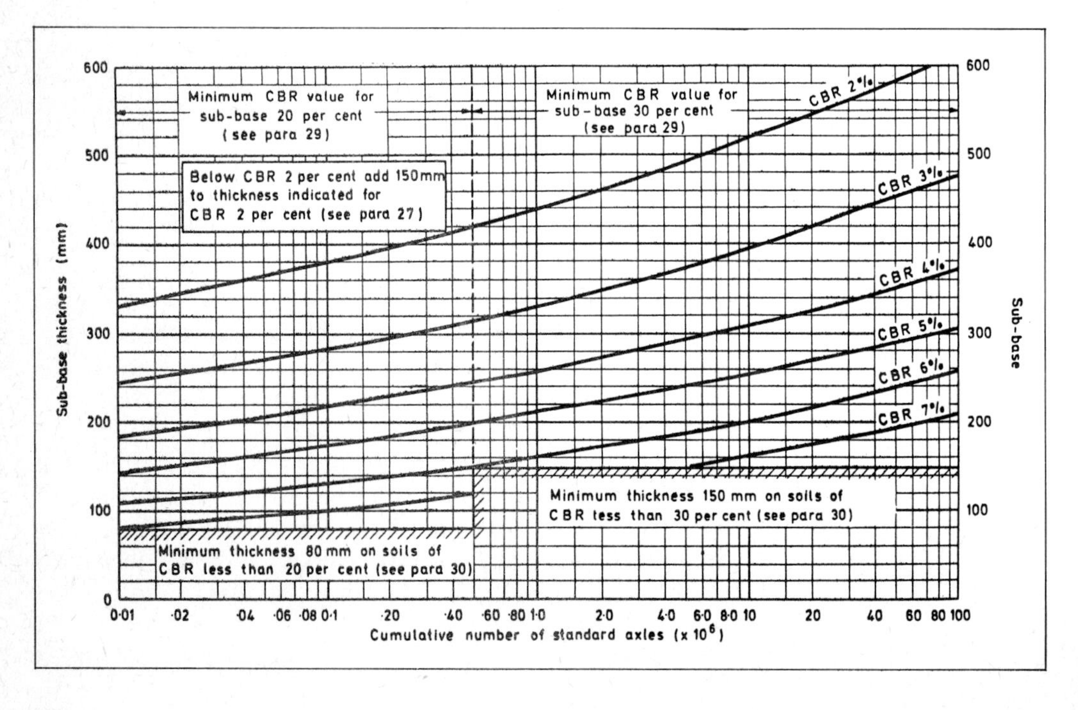

SUB-BASE THICKNESS ASSESSMENT CHARTS

1. Determine Conversion Factor (CF) from vehicle axle load.

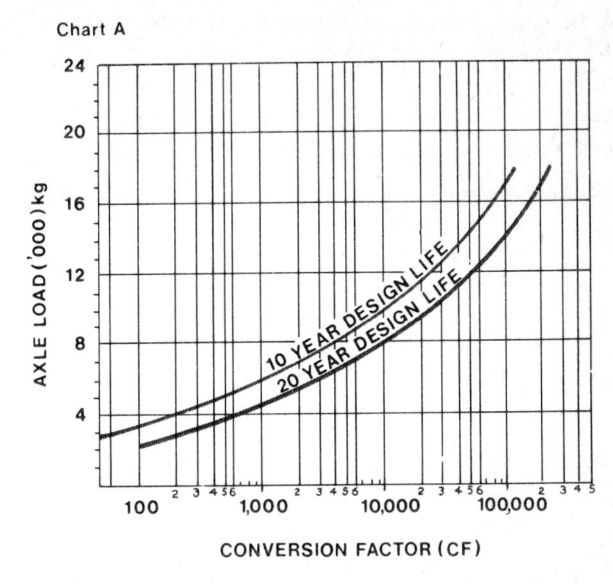

2. Assess number of vehicle passes per day (P).
3. CF x P = Number of standard axles (S).
4. Use S in chart below to find required sub-base thickness.

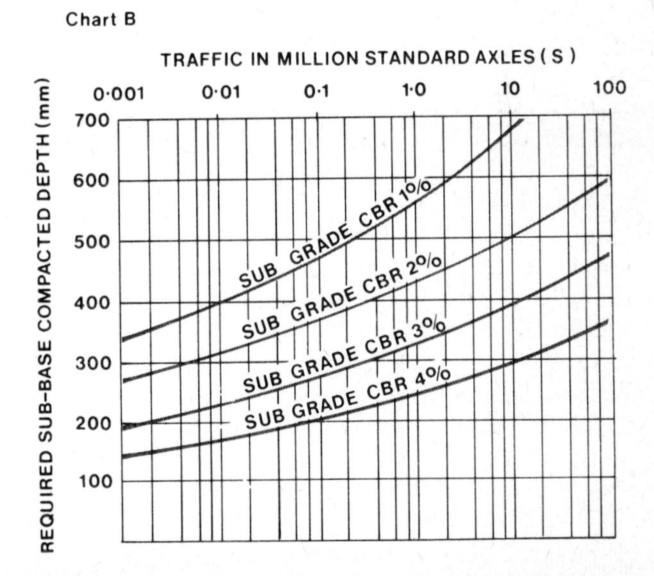

[Based on data from Department of the Environment Road Research Laboratory Road Note 29, 1970 and ICI Field Data]

Figure 9.10 Permanent road design curves published by ICI Fibres in their publication "Designing with Terram"

Poulenc for their Bidim fabric (Figure 9.22) and these indicate greater sub-base reductions. The Author feels that some of these curves as published by commercial firms may often be irrelevant in that, in many applications in permanent road situations where membranes are used, the saving in otherwise lost material at the sub-base/sub-grade interface is the main economic benefit in using a fabric separator. This is not apparent or demonstrated by such design curves. In other cases a membrane is used because the construction of the road would be virtually impossible without it. For instance, in some UK motorway contracts where membranes have been used over soft spots, the excavation alternative would have been of the order of one metre of material. This is the same order of magnitude as the entire sub-base layer itself, and demonstrates a considerable saving beyond that suggested by published curves. Similarly, as shown in Figure 9.11, the use of membranes for constructing permanent roadways over peat and peaty soils enables roads to be constructed in otherwise very difficult situations.

In water-sensitive soils subject to heavy seasonal rainfall and excessive surface saturation, it is possible to use a permeable membrane sprayed with 'cut-back' bitumen to act as a water-resisting encapsulation for a road sub-base. The construction sequence shown in Figure 9.12 results in the sub-base being made from the water-sensitive material originally excavated, *but* being protected from excess moisture by the encapsulating membrane. The soil inside is thus kept relatively strong, and the road structure does not tend to sink since no embankment has been constructed. Thus the whole structure can be very stable in the long term.[30, 31]

Figure 9.13 suggests how membranes might be used to support roads on uniform particle-size sands and to construct capillary breaks to stop salts being carried upwards where they would destroy the road surface.

The same construction principle can be applied to prevent ice lens formation in permafrost conditions in high latitude countries (Figure 9.14) but here the capillary break must be below the frost line and above the water table. Figure 9.15 shows the rate at which water can be lifted by capillary action in fine soils. Note that the most sensitive section is where soils have a particle diameter between 0.01 and 0.1 mm. Consequently, frost heave does not occur in uniform soils with particles larger than 0.01 mm, and graded soils must have at least 3 % smaller than 0.02 mm to exhibit frost heave.

As shown in Figure 9.16, membranes can also be used to protect roads liable to intermittent flood overwash such as is experienced in desert wadis. The chart in Figure 6.7 can be used to assess the required stone needed for 1 : 20/1 : 4 slopes using 'smooth flow' for upstream defence and 'turbulent flow' for

Figure 9.11 Photograph of a permanent road being built on peaty soil in the U.K.

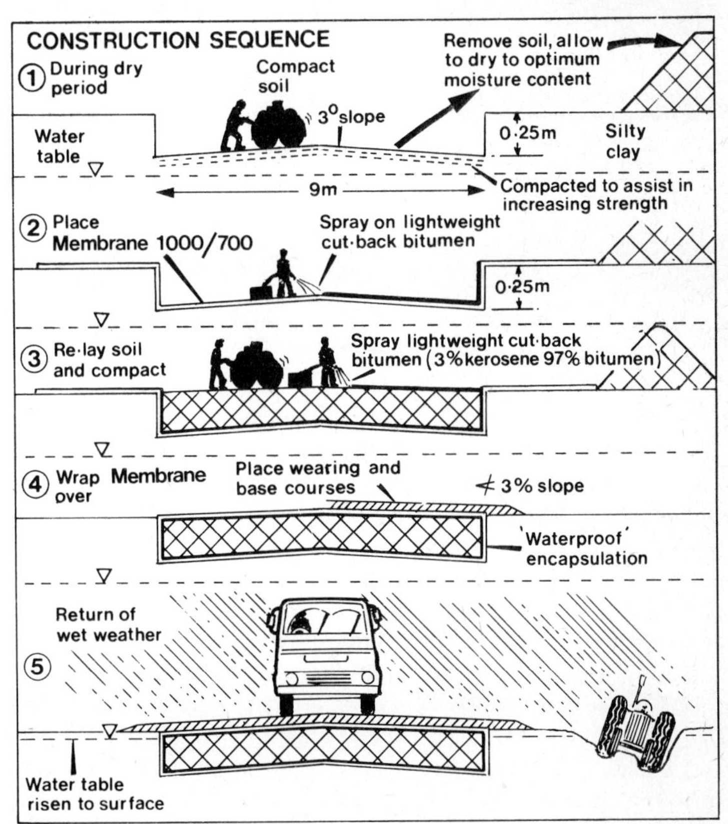

Figure 9.12 Construction of a road sub-base inside a water-resistant encapsulating membrane

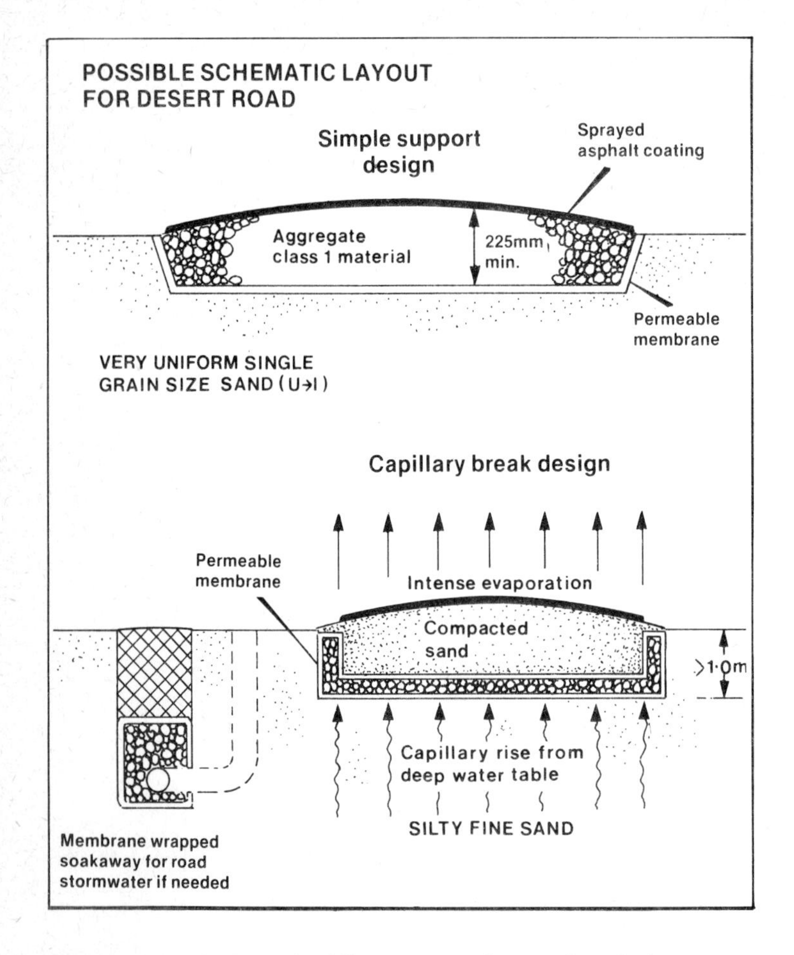

Figure 9.13 Hot arid desert road construction techniques using permeable membrane

downstream stonework. The steeper the angle of slope, the greater will have to be the stone size, taking the chart reading as a minimum for horizontal surfaces.

In tropical countries, subject to continual rainfall, roads must be constructed with good sub-base drainage, or else they will wear rapidly. Figure 9.17 shows how such roads might be constructed using a permeable membrane to protect the drainage integrity of the sub-base/base structure.

In exceptionally soft soils with considerable excessive moisture, permeable membranes have been used to assist in the construction of 'soil-displaced' roads. As shown in Figure 9.18, fill is tipped onto the membrane causing it to displace the soft ground until a firm layer is reached. The membrane holds in the fill effecting some economy, and helping to guard against the possible slip of fill material within the softer soil, thus preventing road damage. This technique can be applied widely from tropical swamps to permafrost summer-thaw soils.

Figures 9.19-9.29 show some typical design techniques and drawing details which are currently in use.

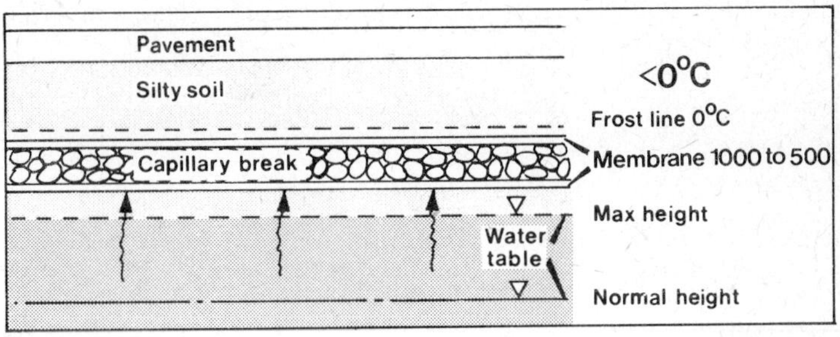

Figure 9.14 Membranes used to construct a capillary break to prevent ice lens formation in permafrost areas

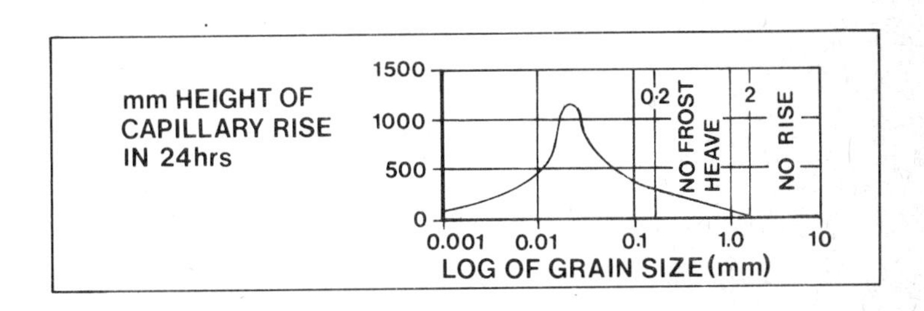

Figure 9.15 Graph showing the rate of rise of capillary water through soils of different particle sizes. (Ref 32)

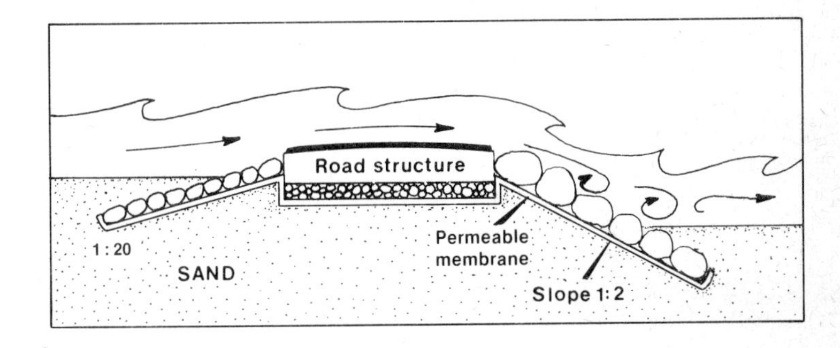

Figure 9.16 Suggested design for wadi crossing road subject to intermittent storm flow overwash

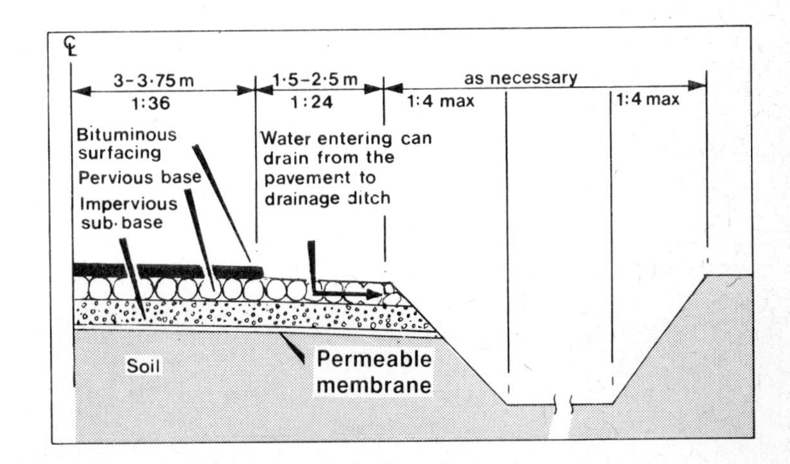

Figure 9.17 Membrane used in the design of a road suitable for tropical countries. (Modified from Ref 33)

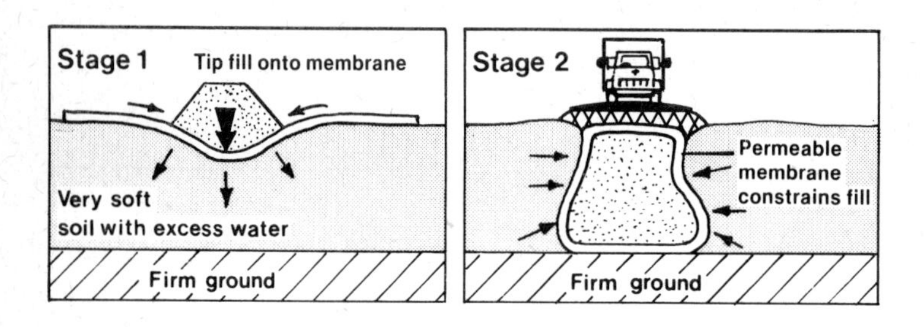

Figure 9.18 The use of membrane to control soil displacement construction technique

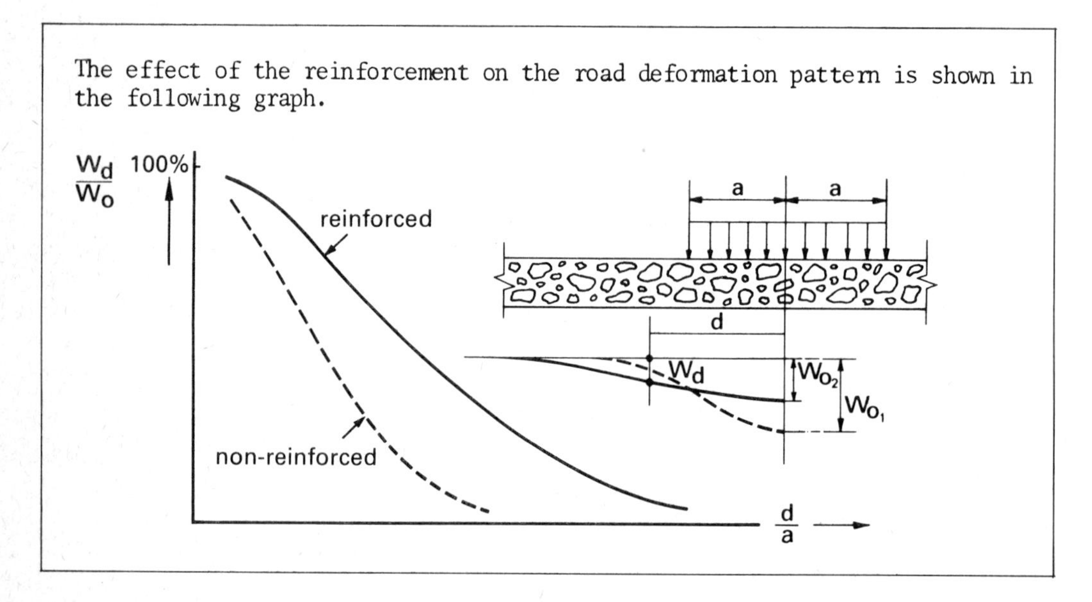

Figure 9.19 Road design curve used by Enka Glanzstoff in their publication on Colbond

Figure 9.20 Paraweb interwoven mat being used directly on the soft sub-grade to form a temporary road. (Courtesy Linear Composites Ltd.)

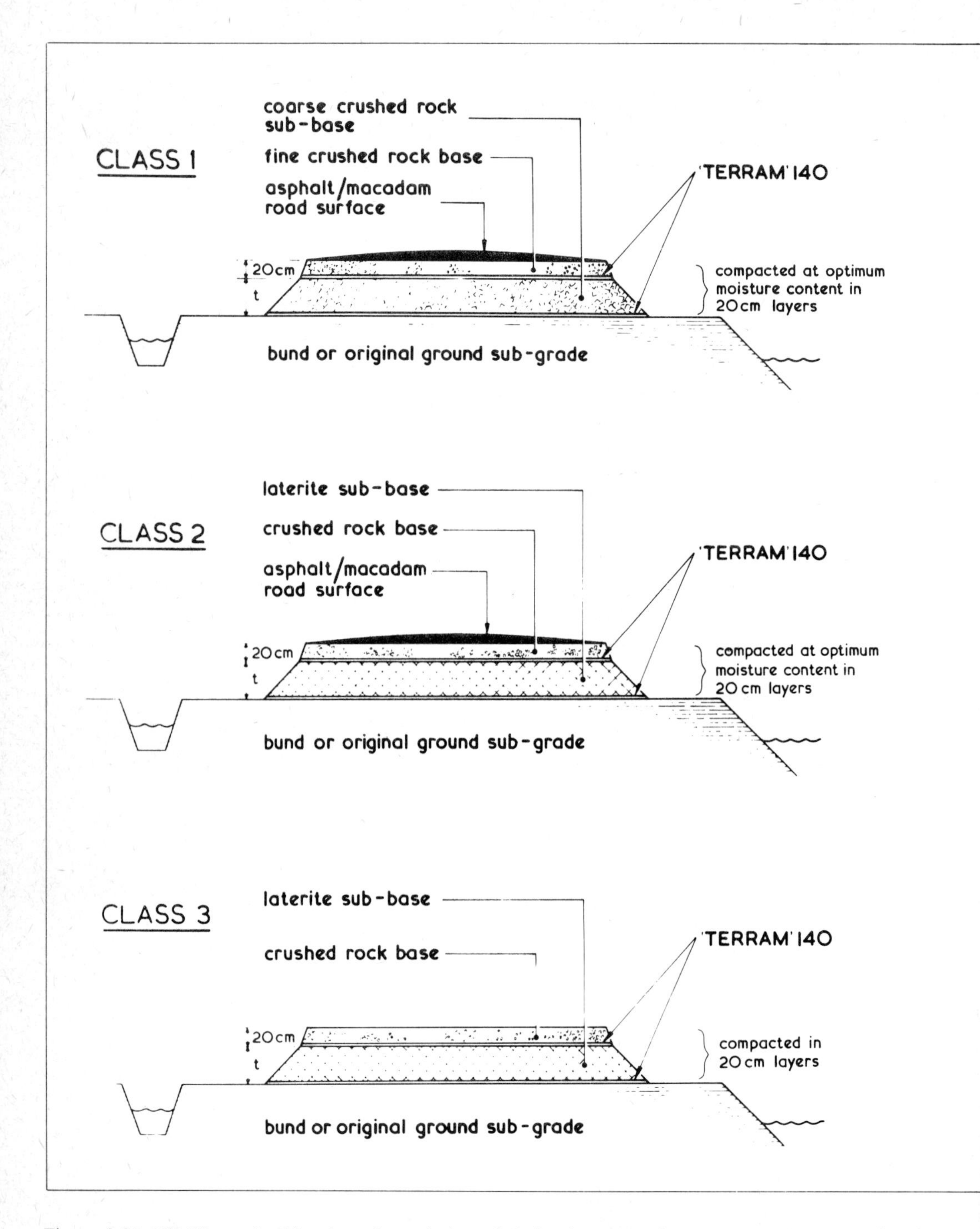

Figure 9.21 ICI Fibres classification of tropical road design incorporating Terram 140 (now replaced by the Terram 700/1000 range)

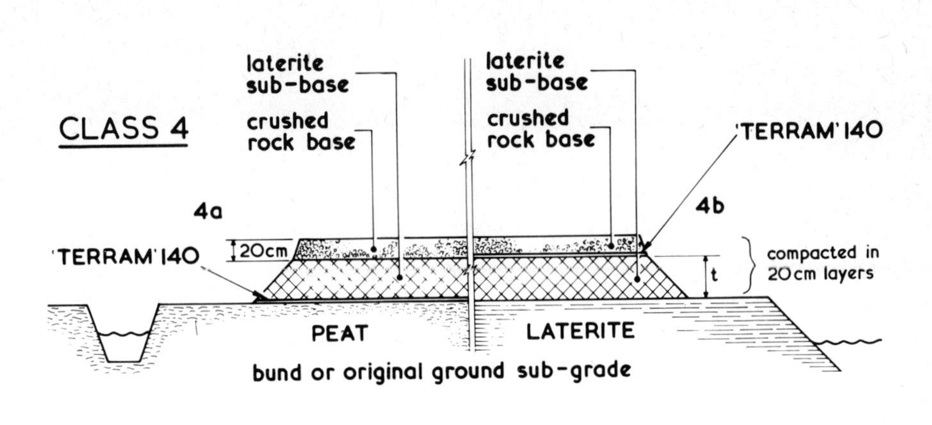
CLASS 4
laterite
sub-base
crushed
rock base
laterite
sub-base
crushed
rock base
'TERRAM' 140
4a
4b
20cm
'TERRAM' 140
t
compacted in
20cm layers
PEAT
LATERITE
bund or original ground sub-grade

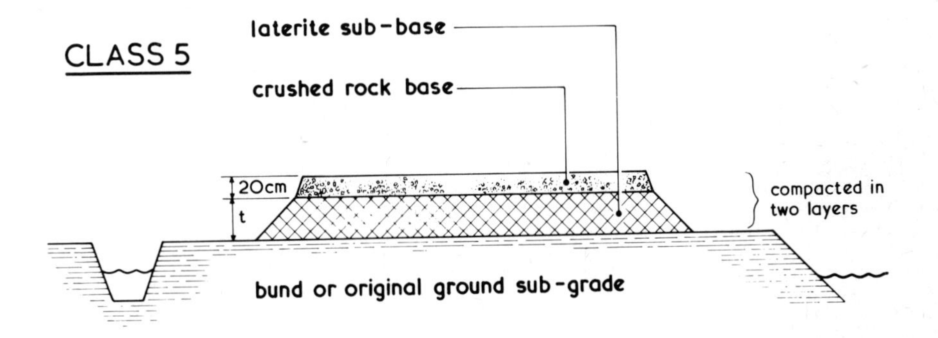
CLASS 5
laterite sub-base
crushed rock base
20cm
t
compacted in
two layers
bund or original ground sub-grade

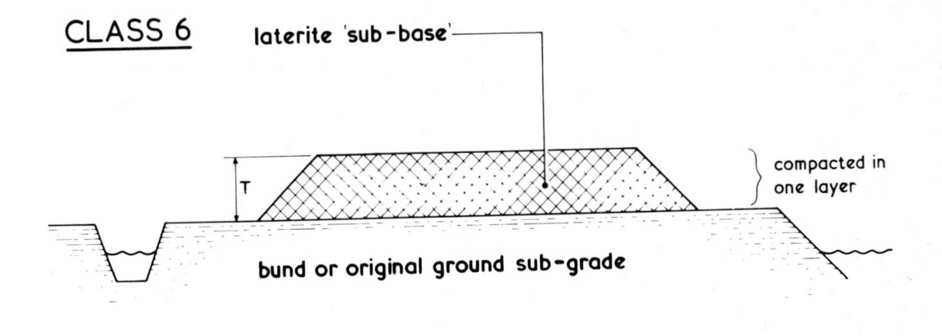
CLASS 6
laterite 'sub-base'
T
compacted in
one layer
bund or original ground sub-grade

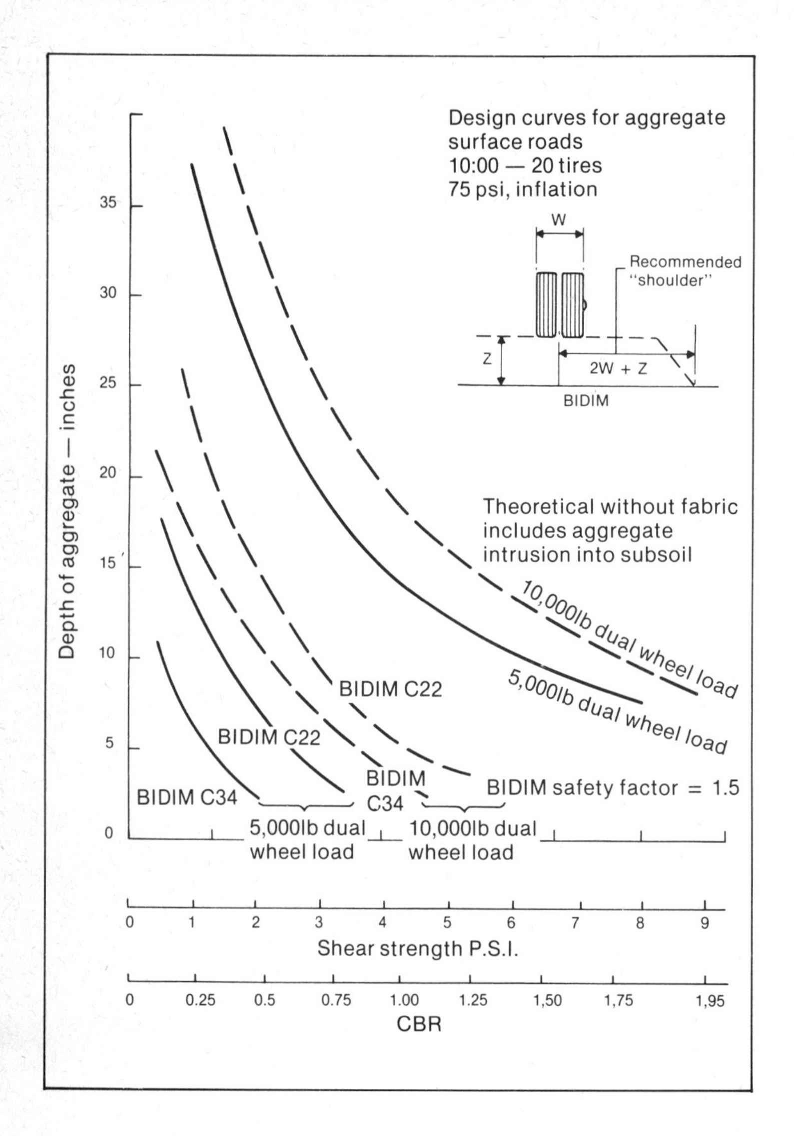

Figure 9.22 Typical design curves for road sub-base construction published by Rhone-Poulenc Textile, and written by G.W. Hamilton in 1977

Figure 9.23 (Opposite) Design for road using fabric separator layer. (Courtesy Carthage Mills)

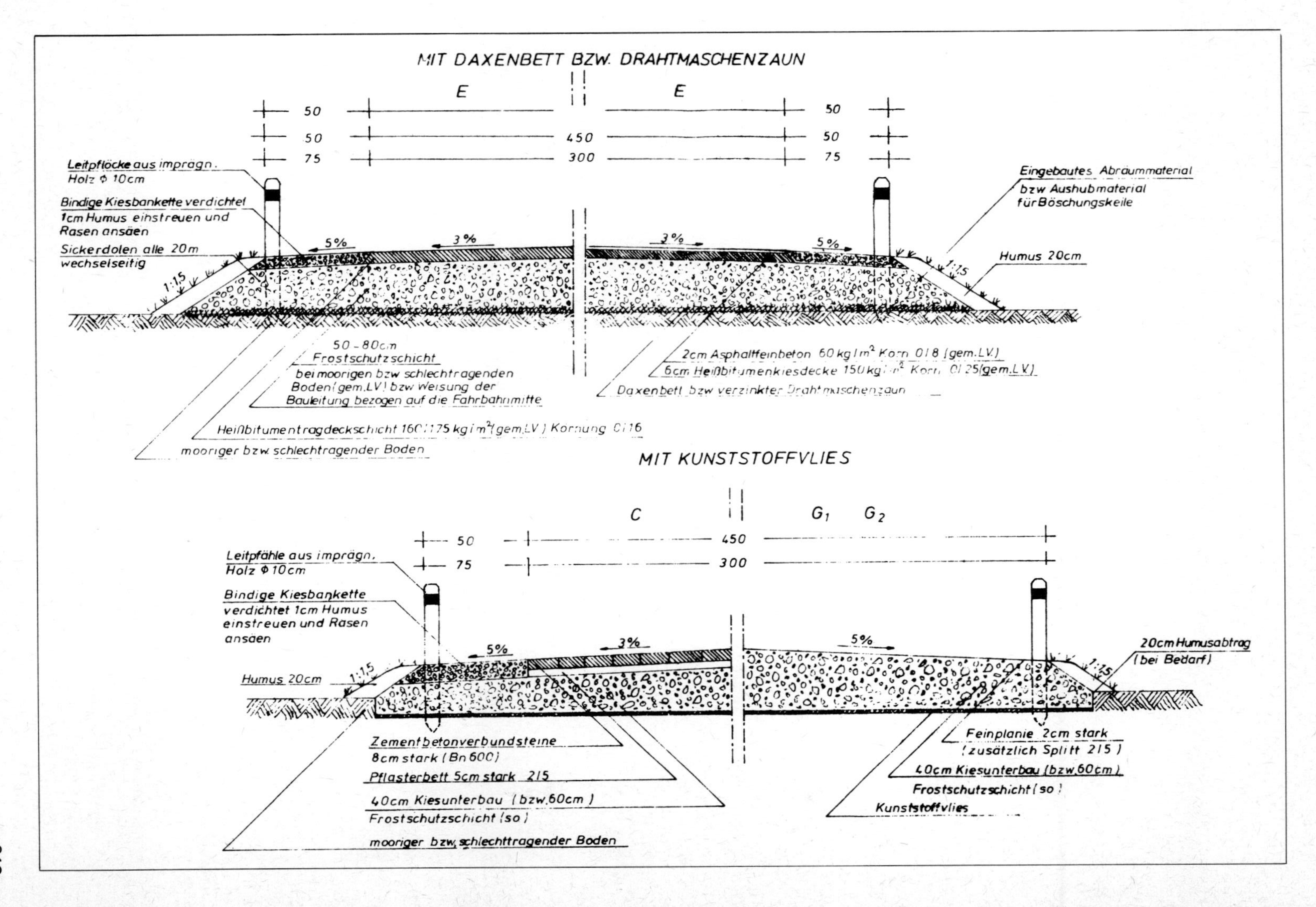
MIT DAXENBETT BZW. DRAHTMASCHENZAUN
E
E
50
50
75
450
300
50
50
75
Leitpflöcke aus imprägn.
Holz ⌀ 10cm
Bindige Kiesbankette verdichtet
1cm Humus einstreuen und
Rasen ansäen
Sickerdolen alle 20m
wechselseitig
1:1,5
5%
3%
3%
5%
Eingebautes Abraummaterial
bzw. Aushubmaterial
für Böschungskeile
Humus 20cm
1:1,5
50 - 80cm
Frostschutzschicht
bei moorigen bzw. schlechtragenden
Boden (gem. LV) bzw. Weisung der
Bauleitung bezogen auf die Fahrbahnmitte
Heißbitumentragdeckschicht 160/175 kg/m² (gem. LV) Körnung 0/16
mooriger bzw. schlechtragender Boden
2cm Asphaltfeinbeton 60 kg/m² Korn 0/8 (gem. LV)
6cm Heißbitumenkiesdecke 150 kg/m² Korn 0/25 (gem. LV)
Daxenbett bzw. verzinkter Drahtmaschenzaun
MIT KUNSTSTOFFVLIES
C
G1 G2
50
75
450
300
Leitpfähle aus imprägn.
Holz ⌀ 10cm
Bindige Kiesbankette
verdichtet 1cm Humus
einstreuen und Rasen
ansäen
Humus 20cm
1:1,5
5%
3%
5%
1:1,5
20cm Humusabtrag
(bei Bedarf)
Zementbetonverbundsteine
8cm stark (Bn 600)
Pflasterbett 5cm stark 2/5
40cm Kiesunterbau (bzw. 60cm)
Frostschutzschicht (so)
mooriger bzw. schlechttragender Boden
Feinplanie 2cm stark
(zusätzlich Splitt 2/5)
40cm Kiesunterbau (bzw. 60cm)
Frostschutzschicht (so)
Kunststoffvlies

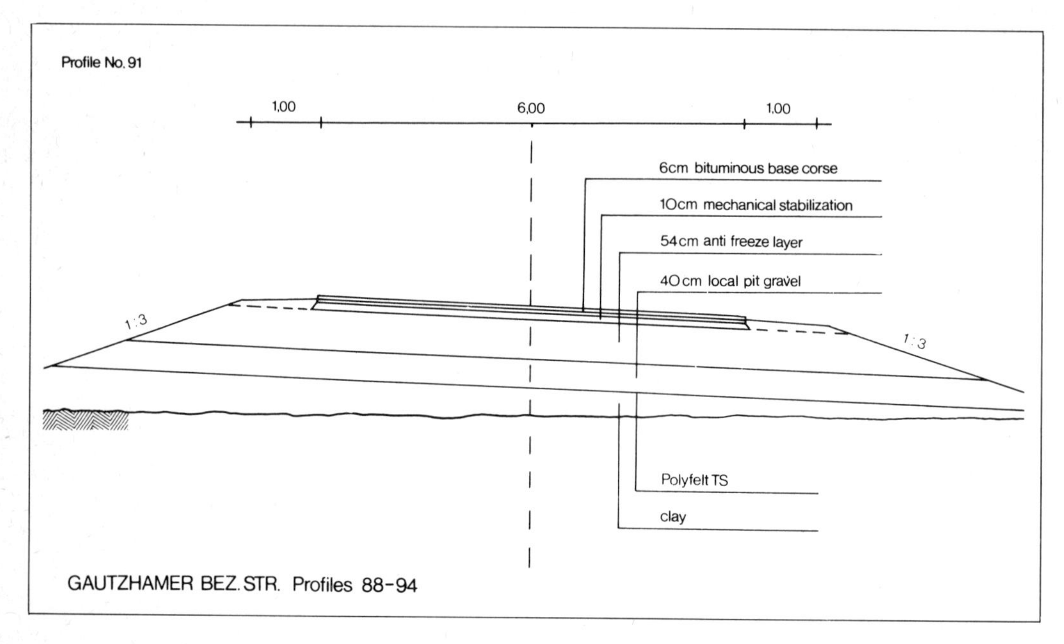

Figure 9.24 (Top) Road embankment cross section using membrane to separate clay and gravel. (Courtesy Chemie Linz AG)

Figure 9.25 (Bottom) Permeable membrane being laid on top of road base, to reinforce the wearing course. (Courtesy ICI Fibres.)

Figure 9.26 (Top) Motorway construction using permeable membrane. (Courtesy ICI Fibres.)

Figure 9.27 (Bottom) Road wearing course being laid by machine running directly on fabric. (Courtesy Low Brothers)

Figure 9.28 (Top) Membrane-based permanent road being constructed over flooded peat soil. (Courtesy ICI Fibres.)

Figure 9.29 (Bottom) Photograph showing that a temporary access road can be taken up after use by lifting the membrane. Especially useful where environmental damage has to be kept to a minimum. (Courtesy Low Brothers)

10

Miscellaneous applications for membranes

There are now so many applications for membranes — both permeable and impermeable — in the fields of Soil Engineering and in Civil Engineering generally, that the Author considers it necessary briefly to show as many as possible. Figures 10.1 to 10.18 are self-explanatory photographs or diagrams illustrating some of the applications of membranes either proposed by manufacturers, or used in recent years in actual construction projects. Diagrams are presented unaltered, from manufacturers' published literature.

The following is a summary list illustrating the extent of applications:

Railway ballast/sub-grade separator.
Dam filters and impermeable barriers.
Roof garden drainage filters.
Artificial seaweed mats for marine sand precipitation and erosion prevention.
Soil reinforcement by webbings in slopes, or below foundation structures.
Wind fences for the precipitation of sand or snow, or for the protection of crops.
Sun blinds or radiation insulators in greenhouse temperature/humidity control systems.
Impermeable linings for reservoirs to prevent water loss.
Impermeable linings for refuse dumps to prevent untreated polluted water running into the ground.
Erosion control fences on steep hillsides.
Water scour prevention at pipe outflow points.

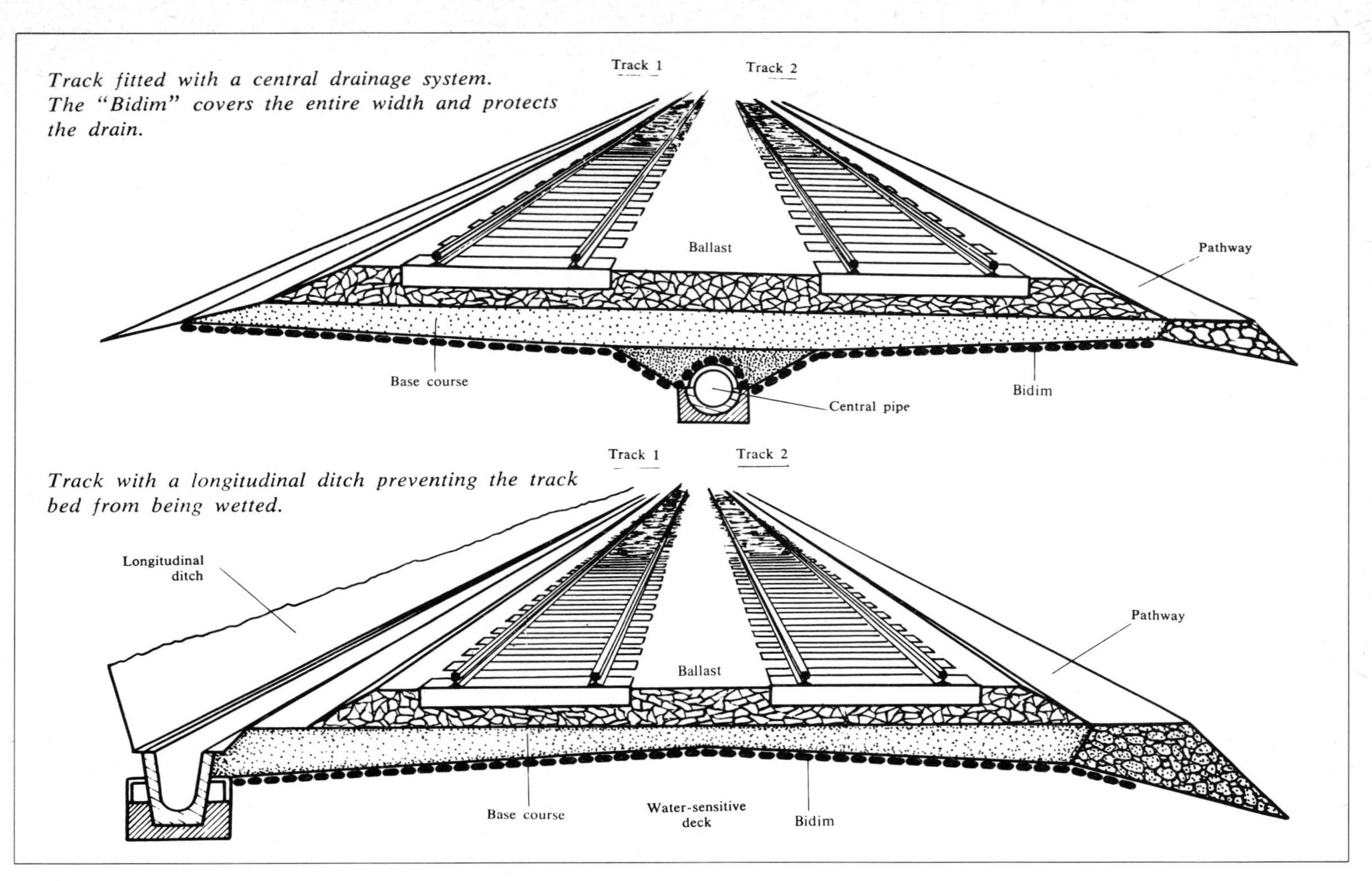

Figure 10.1 Permeable membrane to minimise mud pumping beneath railway lines. The Author's opinion is that a low-permeability membrane usually performs better than a highly permeable one

With the increased dynamic forces resulting from the trend towards higher speeds, there is a need to improve and maintain track to a higher standard. To achieve the design objectives of track stability under high vertical and lateral forces, passenger comfort and ease of maintenance, consideration must be given not only to the geometry and composition of the track and its sub-structure, but also to the earth works below, which must ultimately bear the applied load.

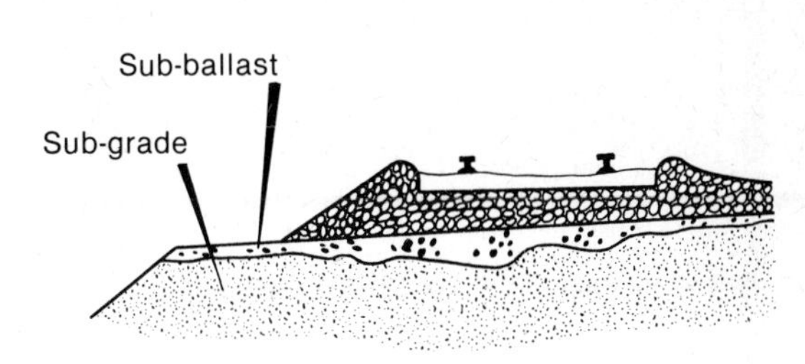

Differential settlement

Problems caused by differential settlement or penetration of the ballast in to the sub-grade, affect the life of the track components, passenger comfort and traffic safety. Netlon laid at the sub-ballast/ sub-grade interface prevents ballast loss, distributes loads uniformly over a greater area, thus reducing stresses and strains in the soil, and protects against tension cracks in the sub-grade.

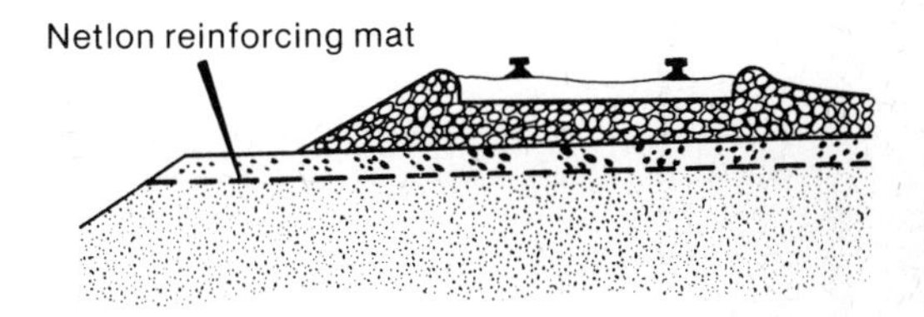

Sub-ballast/sub-grade interface reinforcement

Embankment stability is governed by the ratio of the restoring moment, a function of soil strength, to the disturbing moment, which is the function of the dead and imposed loads. For safety, the ratio must be significantly greater than unity, and the possibility of future increases in live load should not be overlooked. Netlon provides an effective means of increasing the factor of safety, by virtue of the friction induced tensile resistance mobilised under load, without additional earthworks.

The figure below illustrates a similar concept, with the top unreinforced secton of optimum height supported by a reinforced section below. The reinforcement may be continuous, or curtailed a grip length beyond the slip circle, and either straight ended or returned up the slope depending on the relative positions of the slip circle for each side of the bank, and the possibility of slope erosion.

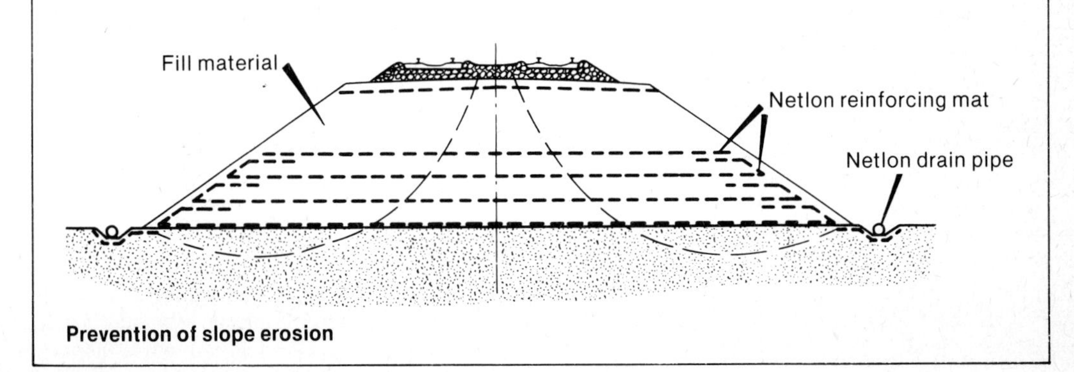

Prevention of slope erosion

Figure 10.2 The use of Netlon mesh for railway sub-grade support and embankment reinforcement

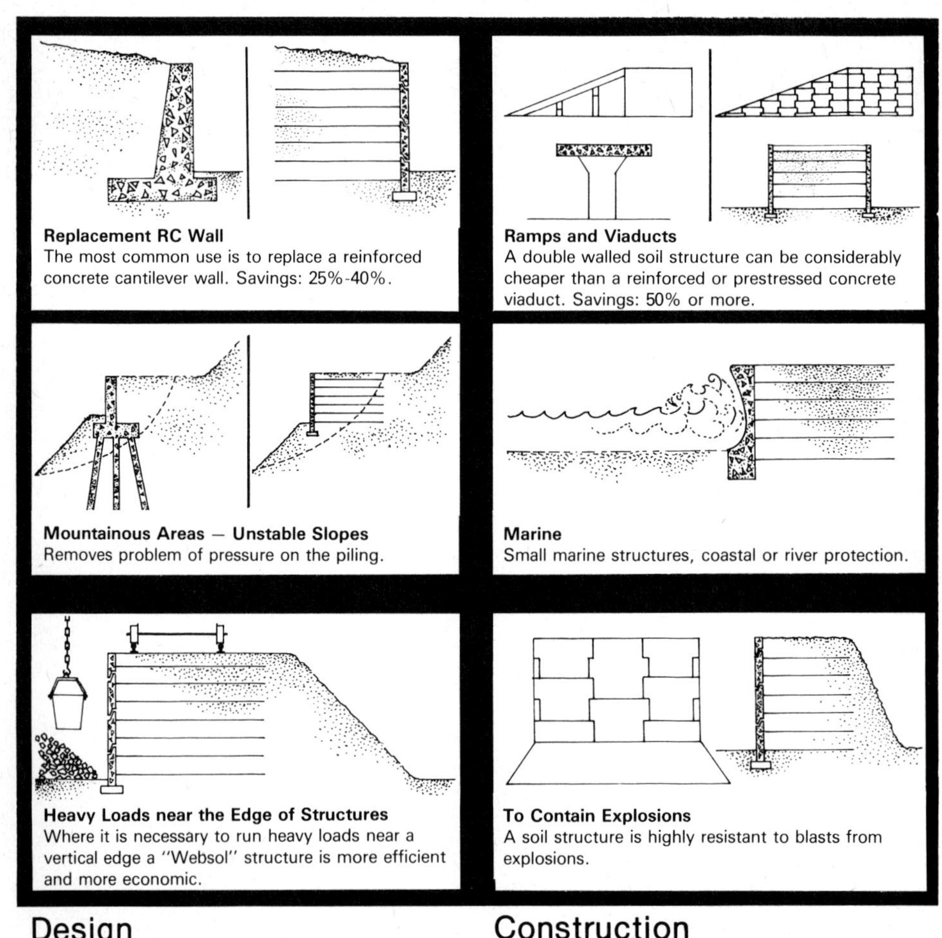

Design

A soil structure is designed to be stable:

(i) Internally
(ii) Externally

(i) **Internally**
This section of the design ensures firstly that there are a sufficient number of anchor reinforcements to carry the forces within the reinforced soil mass as shown by a Rankine analysis, and, secondly that the length of the anchors is sufficient to develop fully the theoretical tensions.

(ii) **Externally**
The soil structure is considered to be a coherent mass, and checks are made to ensure that there is an adequate safety factor (CP2 1951) against sliding, overturning and bearing failures.
Our design section can provide all calculations for a structure, provided that adequate information concerning boreholes and other soil parameters is made available by the Client.

Construction

A small strip foundation approximately 300 mm wide and 150 mm deep on which to place the panels is cast in blinding concrete.
The first row of panels is placed on this bed and adjusted for spacing, line and level. The fill material is brought up to the first layer of anchor reinforcements, and these latter are then placed on the compacted fill and attached to the panels. Fill is placed on top of the anchor reinforcement and construction proceeds in this way until the requisite height of the structure is reached.
In normal circumstances the team required to build a soil structure need consist of no more than a foreman, four labourers, a crane driver and a compactor driver. The only plant required includes a small crane (if space allows work to proceed from the inside of the structure, which is normally the case), a heavy compactor and a light compactor (e.g. Bomag 75S or 90S).
As the progress of the work is controlled by the rate of placing and compacting the fill, it is most important that access is available to supply the fill as and when required.

Figure 10.3 The use of plastic webbings in building soil structures with internal reinforcement. (Courtesy Soil Structures Ltd.)

Plastic Woven Silo Liners

Weldmesh silo photographs supplied by courtesy of A.R.C. Engineering Pty. Limited.

THE PRODUCT

- Made from special woven fabric.
- The toughest, most durable liner produced in Australia.
- Specially designed for lining 1,000 bushel temporary weldmesh silos.
- Reusable.

ADVANTAGES

- **Waterproof yet it Breathes.**
- **Long Life.**
  Lasts longer than any competitive product. Sarlon liners are designed to meet the rigours of the tough Australian climate.
- **Does Not Contaminate** as the fabric is perfectly smooth without fibres and will not taint the grain.
- **Lightweight.**
  Easiest to handle, transport and install as it is both lighter and stronger than hessian.

PRODUCT DETAILS

The Sarlon Silo Liner is supplied 8' 6" wide x 50 feet to suit a 1,000 bushel silo similar to the 1-tier A.R.C. Weldmesh field bin illustrated. The material itself is also available in 50-yard rolls, 72 inches wide.

AVAILABLE FROM

Figure 10.4 The use of reinforcing meshes to make industrial silos

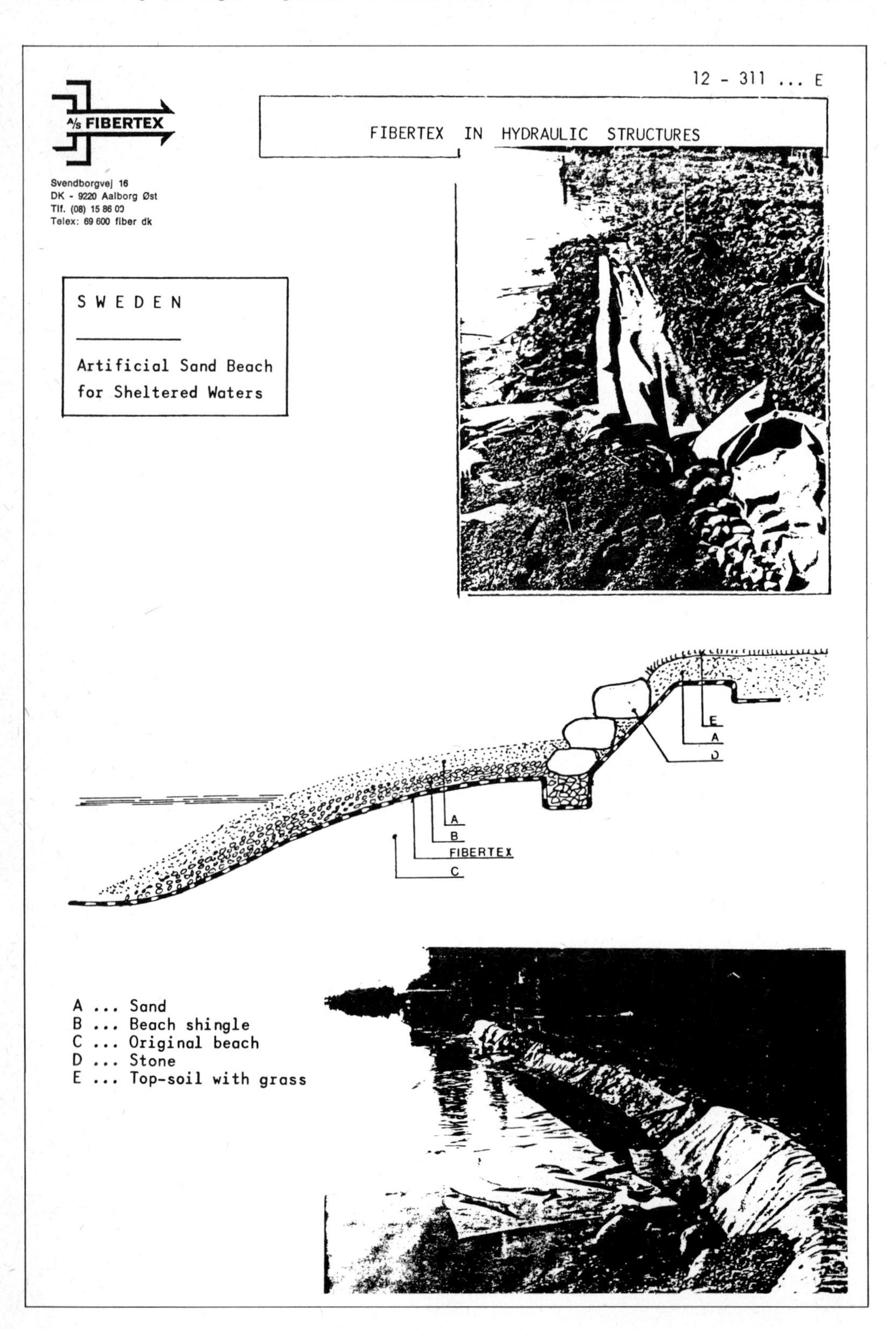
12 - 311 ... E
A/S FIBERTEX
Svendborgvej 16
DK - 9220 Aalborg Øst
Tlf. (08) 15 86 00
Telex: 69 600 fiber dk
FIBERTEX IN HYDRAULIC STRUCTURES
SWEDEN
Artificial Sand Beach for Sheltered Waters
E
A
D
A
B
FIBERTEX
C
A ... Sand
B ... Beach shingle
C ... Original beach
D ... Stone
E ... Top-soil with grass

Sand fencing

Netlon sand fencing provides permanent protection from the heavy maintenance problems caused by drifting sands on roadways, railways and airports.

Reduction of the wind velocity by over 50% causes sand to be deposited in banks angled away from the fence on the leeward side (fig 4).

The net structure in this case is designed for use with finely divided non agglomerating substances (fig 5).

Protection against UV degradation is provided by the use of 2.5% finely divided carbon black pigment.

Size available
Sand fencing is sold in 30 metre rolls with a width of 2 metres.

Fixing
Instruction for installation and fixing will be supplied on request.

Sand fence protecting highway

Net structure for sand fence

FIG 5

Sand bank formed by sand fence

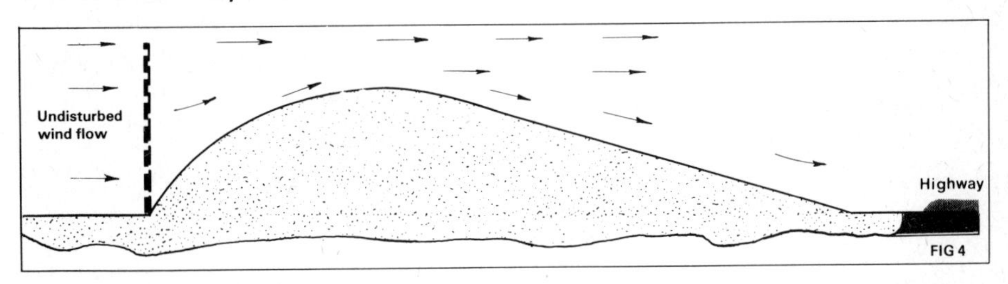

FIG 4

Figure 10.5 (Opposite) The use of a permeable membrane to construct a sand beach artificially over mud or clay

Figure 10.6 (Above) The use of mesh for precipitating sand by slowing wind velocity

Snow fencing

Snow and snow drifts over roads, railways and airports in temperate climates give rise to heavy public expense in equipment, materials and labour for snow clearance. Extensive damage is also caused to road surfaces.

Netlon snow fencing offers a well proven, simple and economic preventive measure against drifting snow. No specialised equipment or skilled labour is required to install the fence in the autumn and dismantle it in the spring. It is light and easy to handle and store.

Netlon snow fencing has been designed with the aid of wind tunnel tests. (fig 3).

The resulting combination of surface roughness, aperture size and the use of both thick and thin strands ensures that drifts are formed away from the areas requiring protection, but without the fence itself becoming choked with snow.

Fig 1 shows how a 50% reduction in wind velocity and the design of the fence causes a gradual formation of drifts to the leeward side of the fence.

Netlon snow fencing is widely used throughout Northern and Eastern Europe to protect roads railways and airports from snow drifts.

Stages in the growth of a drift produced by a vertical fence

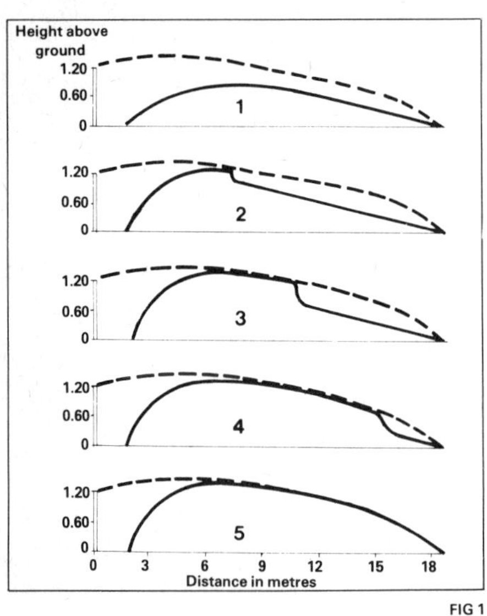

FIG 1

Snow fence erection

Specially designed galvanised steel support posts permit easy erection and dismantling at the beginning and end of the winter. (fig 2).

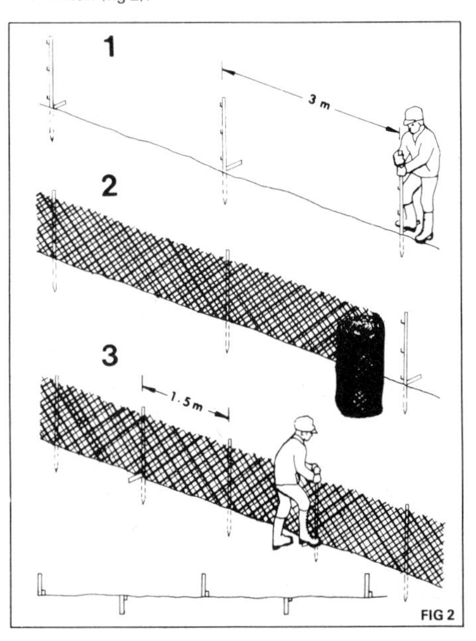

FIG 2

Net structure for snow fence

FIG 3

Figure 10.7 (Left) The use of mesh for precipitating snow by slowing wind velocity

Figure 10.8 (Right) The use of synthetic plastic seaweed attached to a woven ground mat to slow water velocity and cause sand to settle on the sea bed. (Courtesy Nicolou)

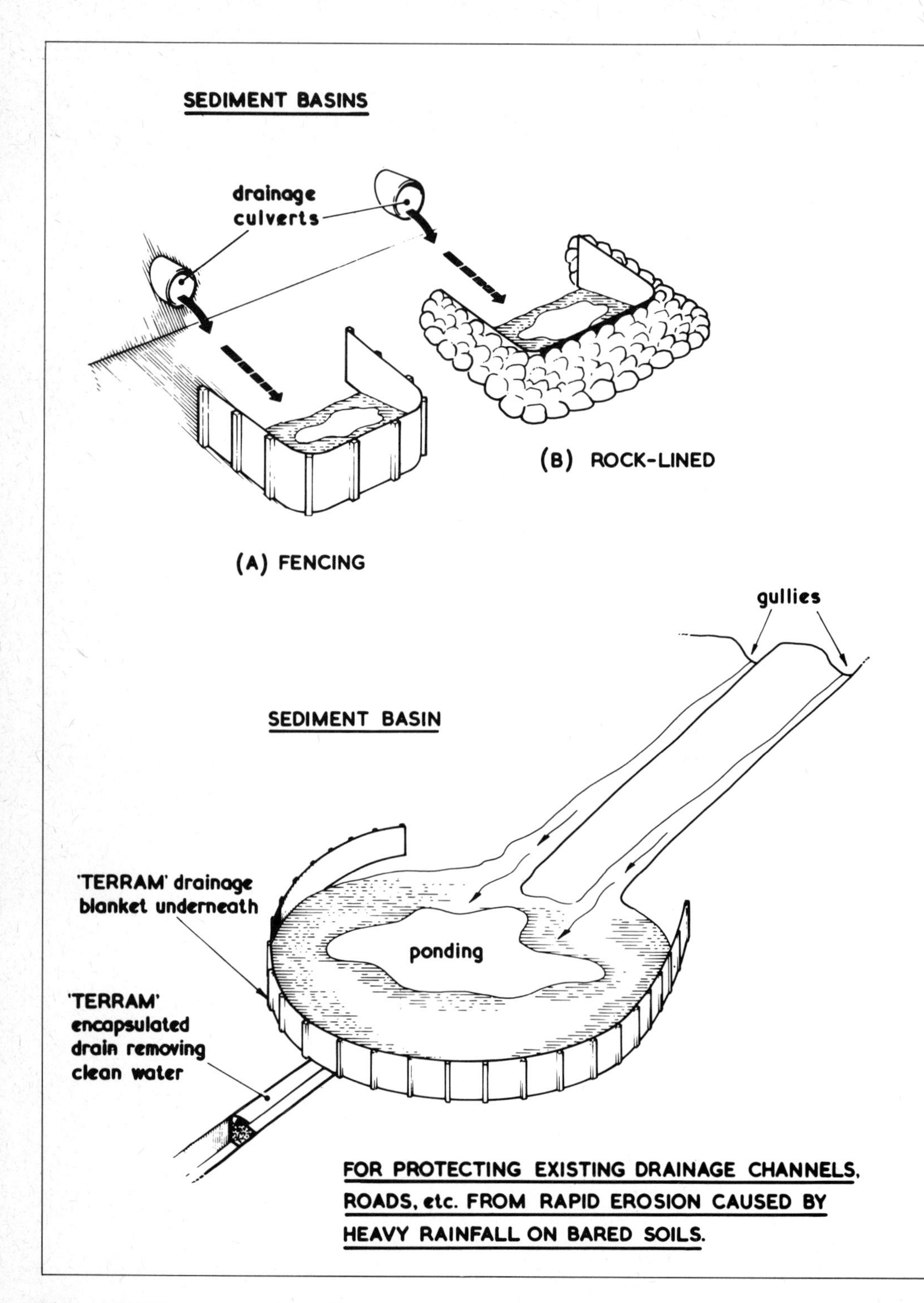

Figure 10.9 The use of a permeable membrane to construct slope erosion control fences. These catch running surface water and filter out the solids

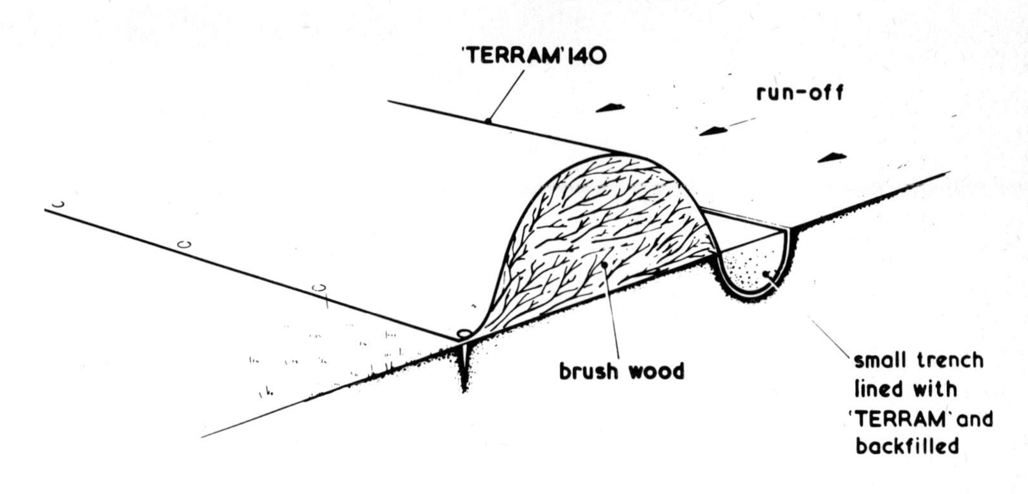
BRUSH BARRIER
'TERRAM'140
run-off
brush wood
small trench
lined with
'TERRAM' and
backfilled

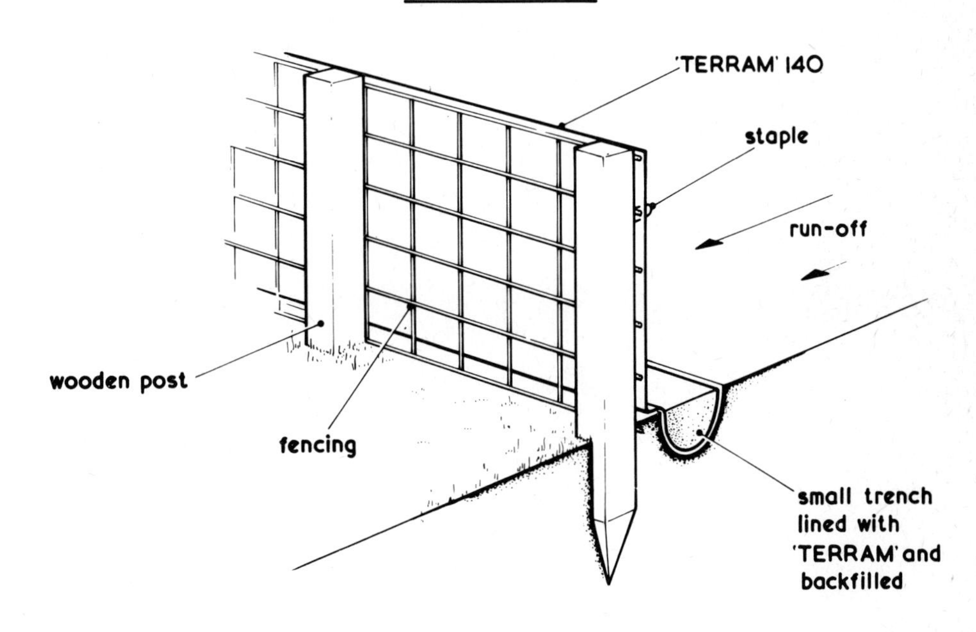
FENCE BARRIER
'TERRAM' 140
staple
run-off
wooden post
fencing
small trench
lined with
'TERRAM' and
backfilled

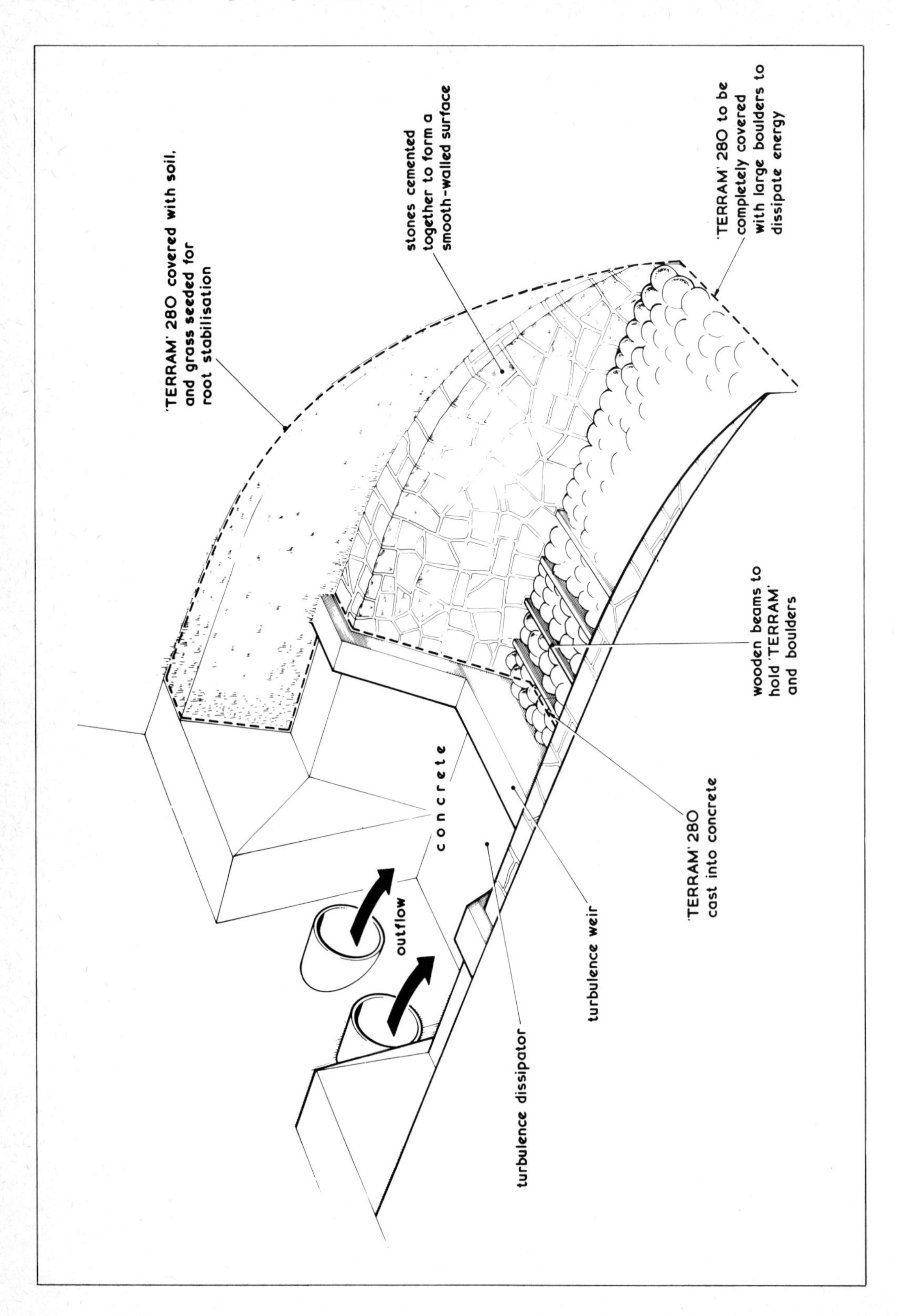
'TERRAM' 280 covered with soil, and grass seeded for root stabilisation
stones cemented together to form a smooth-walled surface
'TERRAM' 280 to be completely covered with large boulders to dissipate energy
wooden beams to hold 'TERRAM' and boulders
'TERRAM' 280 cast into concrete
concrete
outflow
turbulence weir
turbulence dissipator

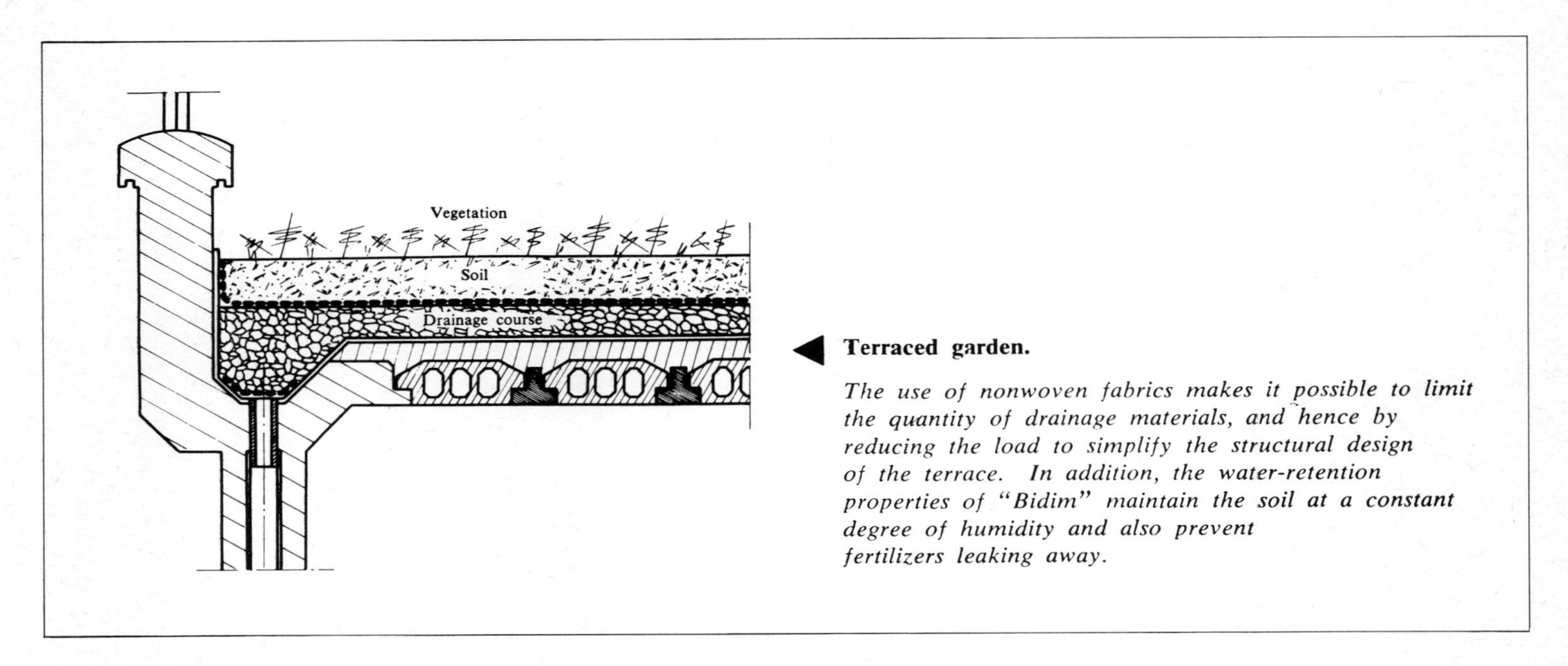

Figure 10.10 (Top) The use of a permeable membrane to protect outflow structures against splash erosion

Figure 10.11 (Bottom) Use of permeable membrane to construct a roof garden soil filter

Figure 10.12 The use of a permeable membrane to protect a foundation material from clogging. (Courtesy Fibertex)

Figure 10.13 The use of Paraweb open mesh for wind fence. Useful for protecting crops; sand precipitation and snow precipitation. (Courtesy Linear Composites Ltd.)

13 - 502 .1. E

FIBERTEX

AS

BITUMEN IMPREGNATED MEMBRANE

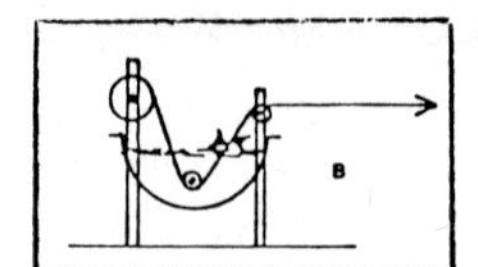

The sketch above shows a method for bitumen impregnating FIBERTEX for use in ponds.

In this tank bitumen of a penetration for instance, 300 is heated to a temperature of about 135 - 150°C with the use of Propane Gas.

A standard 100 metre roll of FIBERTEX type S-170 width 420 cm has been cut to a width of 210 cm.

The fabric is drawn through the tank by hand and carried to the final position on the pond bed. In this way four people can easily handle approximately 5 metre lengths of FIBERTEX.

The 20 cm overlaps are jointed insitu with bitumen.

As is shown on the sketches (plan 1 (d) and cross-section a-a) a drain was placed in advance of placing the membrane in order to de-water the sub-grade.

On top of the impregnated membrane is placed sand/gravel.

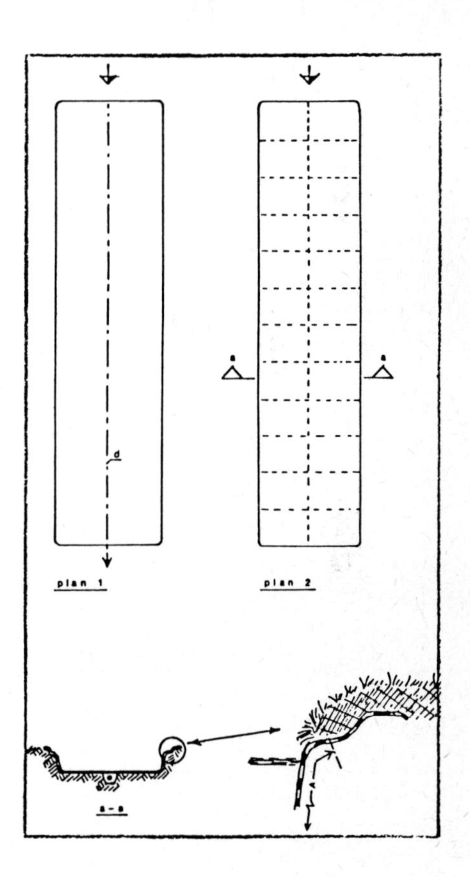

Figure 10.14 The use of bitumen-coated membrane to seal waterproof structures such as small reservoirs

Figure 10.15 (Opposite) RF-12 membrane being used to build a soil-reinforced slope. (Courtesy ICI Fibres.)

Figure 10.16 (Opposite) Permeable membrane being used to line golf bunkers. (Courtesy Low Brothers)

Figure 10.17 Permeable fabric and railway sleepers used to construct a soil-reinforced vertical faced structure. (Courtesy Low Brothers)

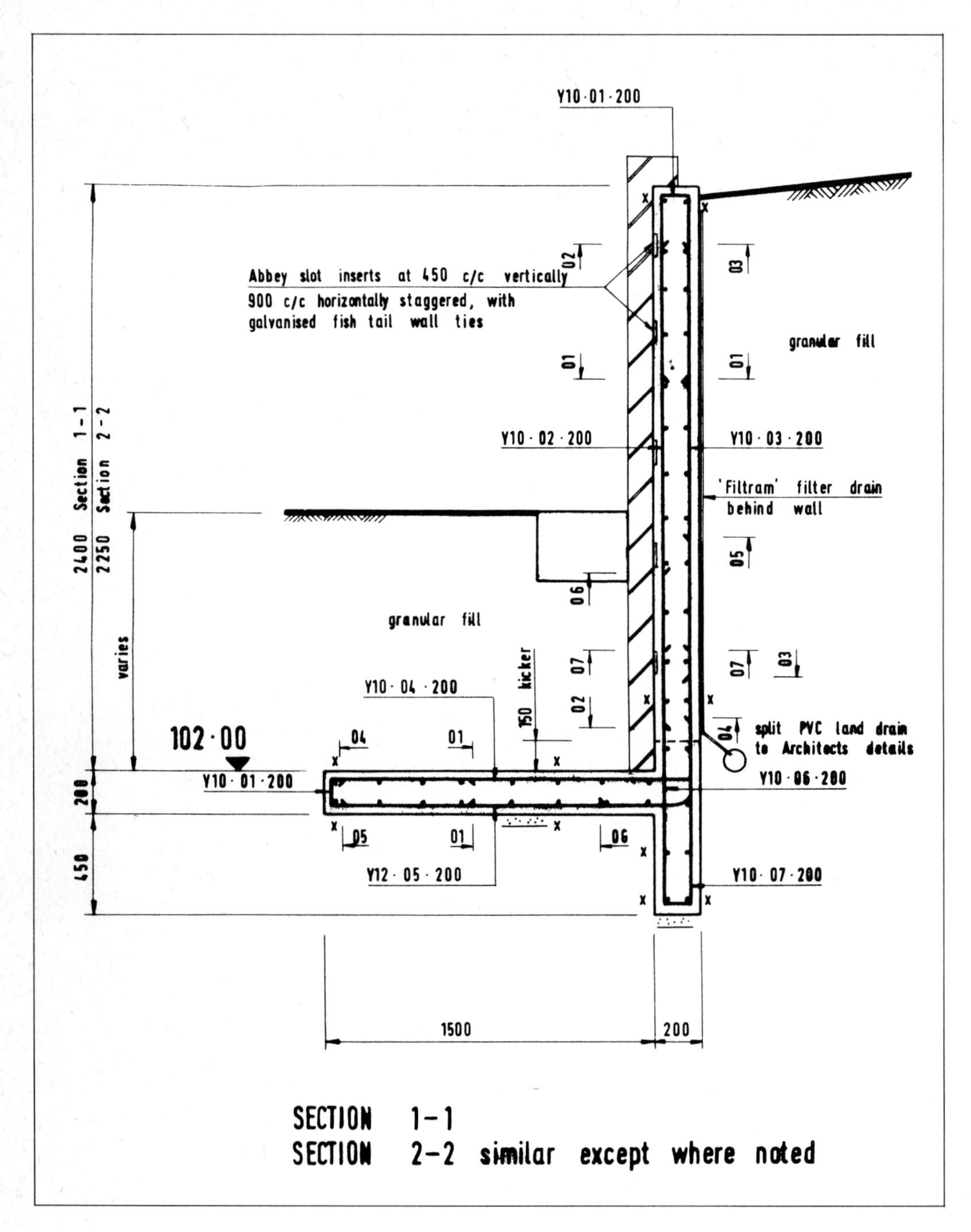

Figure 10.18 Industrial design drawing showing Filtram detailing. (Courtesy ICI Fibres and Andrews Kent and Stone)

Section 3
Laboratory testing

11

Setting up a laboratory for the testing of permeable membranes for soils applications

In the context of the soils uses for membranes described in this book, the term, 'Geotextile' is becoming increasingly used. This term implies that the membrane has been made specifically for use in geotechnical applications. Earlier, it might have been said that some of the membranes used in low-key geotechnical structures were just ordinary textiles adapted to a new purpose. However, the Author believes that even where this was originally the case, most manufacturers have now modified their products to a considerable extent to improve their ground performance capability.

In order to measure the various properties of membranes, a number of standard textile tests have been widely adopted. In addition, a number are still being developed in an attempt to simulate more realistically such site activities as rock dropping and rock abrasion. Some of these tests with brief explanations are listed and illustrated subsequently, but it is important to realise that testing is generally carried out for two purposes.

Testing undertaken prior to ordering

Comparative testing of many different fabrics for a wide range of properties, some of which may be critical to a particular project and other properties critical to a different project. This includes comparisons of published geotextile properties with test results.

Testing undertaken subsequent to ordering

Detailed testing of all properties done occasionally but regularly during delivery (assuming that delivery is over an extensive period), or from random roll samples (assuming that delivery is all at one time).

Regular daily field testing during construction work of a few selected properties such as weight/unit area, tensile strength, and water permeability to ensure consistent product performance to the design specification.

Accurate checking of samples sent from the field which have failed to meet the simple criteria established as being required for satisfactory functioning.

To carry out the above tests (with the exception of field testing), a laboratory is needed, fitted out with the following equipment:

Sink with water supply.
Electric sample cutter.
Tensile testing machine (5000 kg) (For strip tensile, plane strain, grab tensile, wing tear and CBR tests).
Clean water permeability apparatus.
Weighing balance (accurate to 1/100 gramme).
Weighing balance — coarse.
Cone drop test device.
Soil sieve — motor driven, with multi-sieve capability.
Atterberg Liquid Limit and Plastic Limit test apparatus for soil.
Binocular or monocular microscope, preferably with microscope camera.

To carry out the regular daily field tests, a simple field laboratory is necessary to ensure day-to-day quality maintenance. The equipment here must be calibrated against the main laboratory equipment so that the quality can be maintained. Equipment needed is as follows:

Sink with water supply.
Hand sample cutter (preferably dye and weight, but scissors and template will suffice if necessary).
Tensile testing machine (1000 kg) (For strip tensile test).
Weighing balance (accurate to 1/10 gramme).
Cone drop test device.
Hand lens or simple microscope.
Simple soils apparatus.

With regard to the air-conditioning of these laboratories, it would seem sensible that the main laboratory be controlled as follows to allow international comparisons of properties — remembering that quoted figures will often be based on these standard testing conditions:

Main Laboratory: Temperature 20°C ± 5°C
Relative Humidity 65 % ± 5 %

Such a stable environment will allow the confident testing of all fabrics. If fabrics are to be rejected by a Client, then he would be well advised to have such a reliable testing environment before entering into contractual disagreement.

The field laboratory will not need air-conditioning if the fabric being installed on site is made of polythylene, polypropylene or polyester, and temperature is within − 10°C to + 30°C. Other fabrics such as polyamide or viscose can be more sensitive to temperature and/or humidity changes, and for such fabrics, it might be advisable to make some attempt to stabilise the field test environment.

Apparatus and tests

Main laboratory tests and equipment

Electric sample cutter (Figure 11.1)
This may appear to be a luxury, but based upon long experience, is is recommended that this piece of equipment be considered seriously. The press comes down onto sharp cutting dyes that ensure the rapid production of samples that are consistently exactly the same size. It is most important to remember that all textiles must be treated statistically with quantities of 8, 16 or 32 tests to achieve a satisfactory indication of minimum performance. If the fabric is folded prior to placing into the cutting machine, then 8, 16 or 32 samples can be cut at once all at the same time, and from statistically random parts of a fabric sheet.

Strip tensile test (Figures 11.2 and 11.3)
The effect of this test is to measure the contribution of the fibres to overall strength rather than the strength of the fibres themselves. It is extremely useful as an indicator of the 'strength' of a fabric — especially non-wovens — but can not give a proper indication of woven fabric strength at any other angle than parallel to the warp or weft. Consider that if a woven fabric were cut in a strip at 45° to the warp/weft directions, none of the fibres in the top clamp would be held by the bottom clamp. The fabric would simply fall apart. Therefore a cylindrical test (mentioned later) has been developed to try to overcome this aspect, thus effectively becoming a 'plane-strain' test.

Figure 11.2 shows the 50 mm strip test at the initial phase on an Instron testing machine. (Any similar machine with 5000 kg

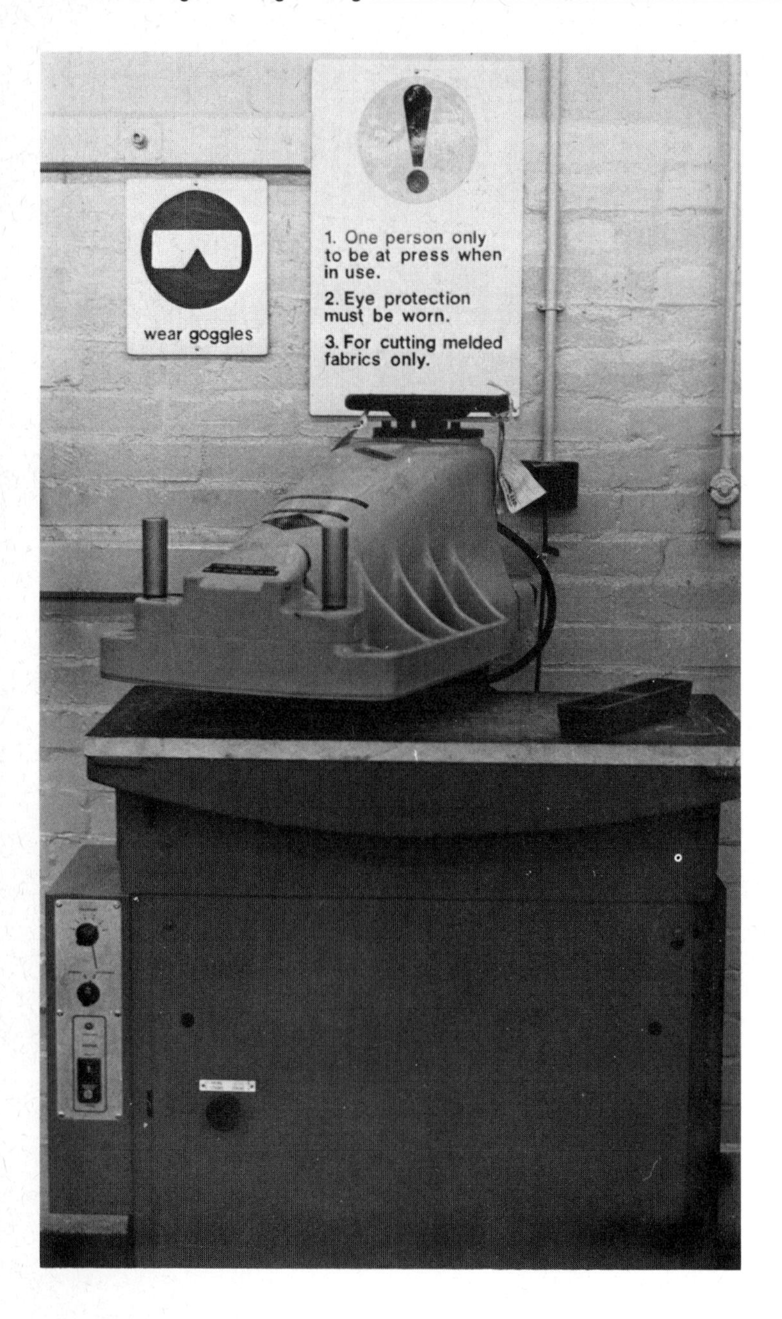

Figure 11.1 Electric sample cutter

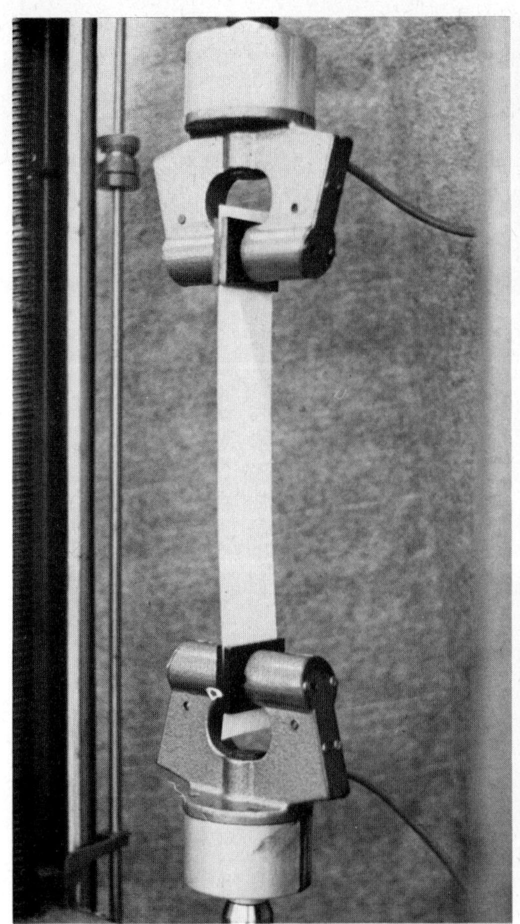

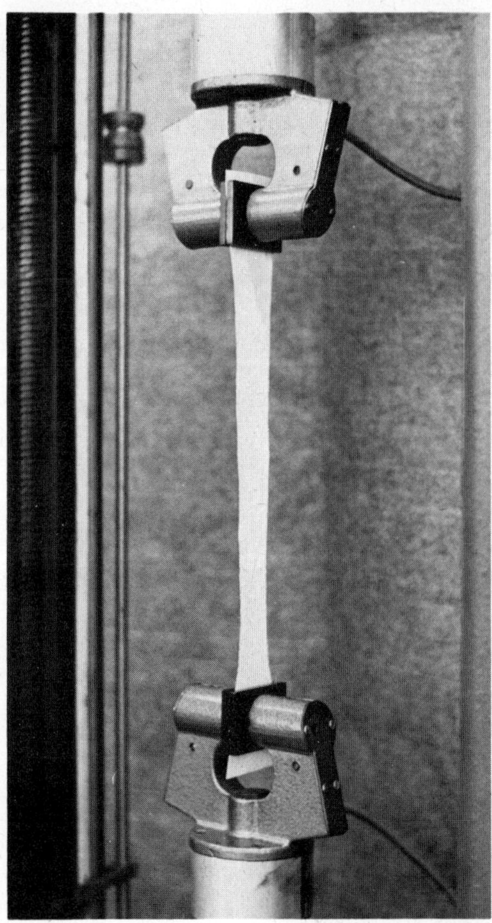

Figure 11.2 (Left)
50mm strip tensile test
— initial stage

Figure 11.3 (Right)
50mm strip tensile test
in progress

capacity would be satisfactory.) Figure 11.3 shows the same strip during the latter stages of the test. Note how the strip has narrowed in width.

Conditions of test
Sample size: 300 mm × 50 mm
Jaw faces: 50 mm wide
Gauge length: 200 mm
Cross-head speed: 200 mm / minute
Chart speed: 100 mm / minute
Results now quoted in Newtons.

% Extension at break is measured, and can vary from 10 % — 140 % so a machine with long travel capability is needed.

Plane strain test (Figures 11.4 and 11.5)
(French Cylinder Test straining in the plane of the fabric)
Standards: DIN 53857; NF G07-001; BS 2576.

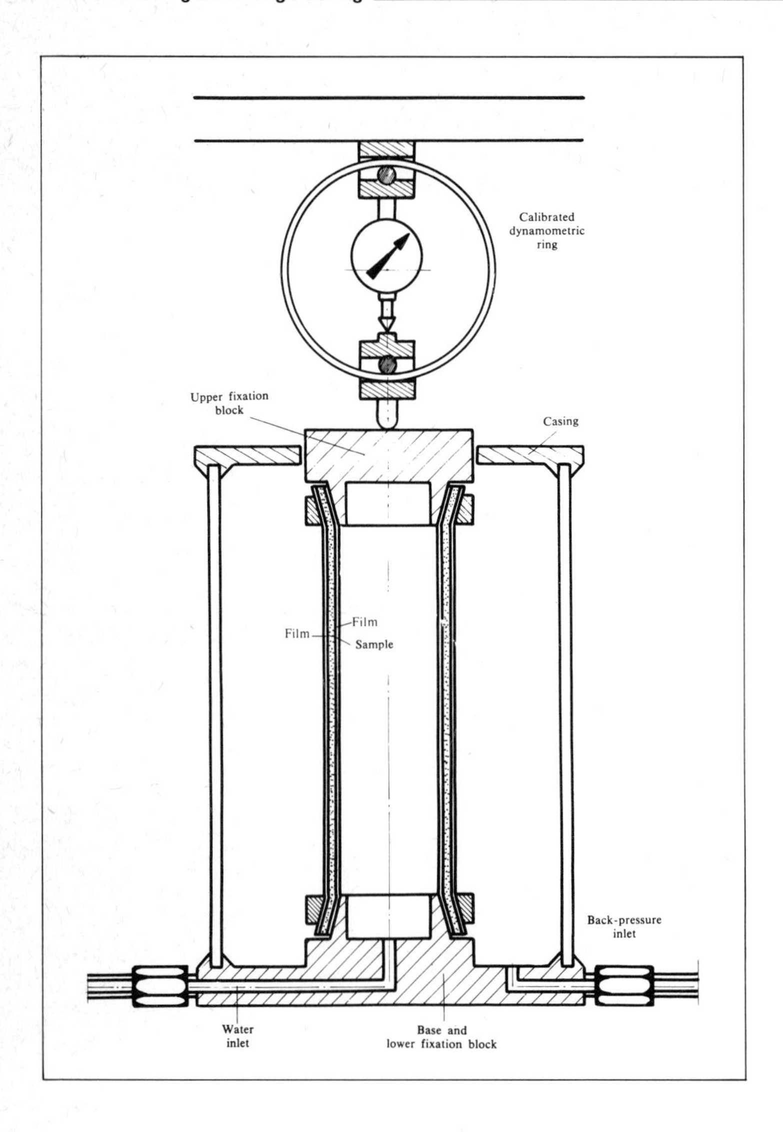

Figure 11.4 Cylindrical test apparatus

It is believed that, although more difficult and slower to set up than a strip test, this plane strain test more realistically represents the tensile behaviour of a sheet of membrane buried in the ground at depth and subject to overall loading.

It is believed that soil restraint in the ground prevents the fabric from narrowing as can be seen in the strip test. The conventional Plane Strain apparatus is shown in Figure 11.4. A cylinder of fabric is made by stitching, and is placed inside two cylindrical rubber membranes. The sleeve is firmly held, together with rubber membranes or films, by a set of cones. The whole is protected from the outside environment by an impervious casing. Water under programmed pressure is admitted through a pipe to the inside of the sleeve, the volume

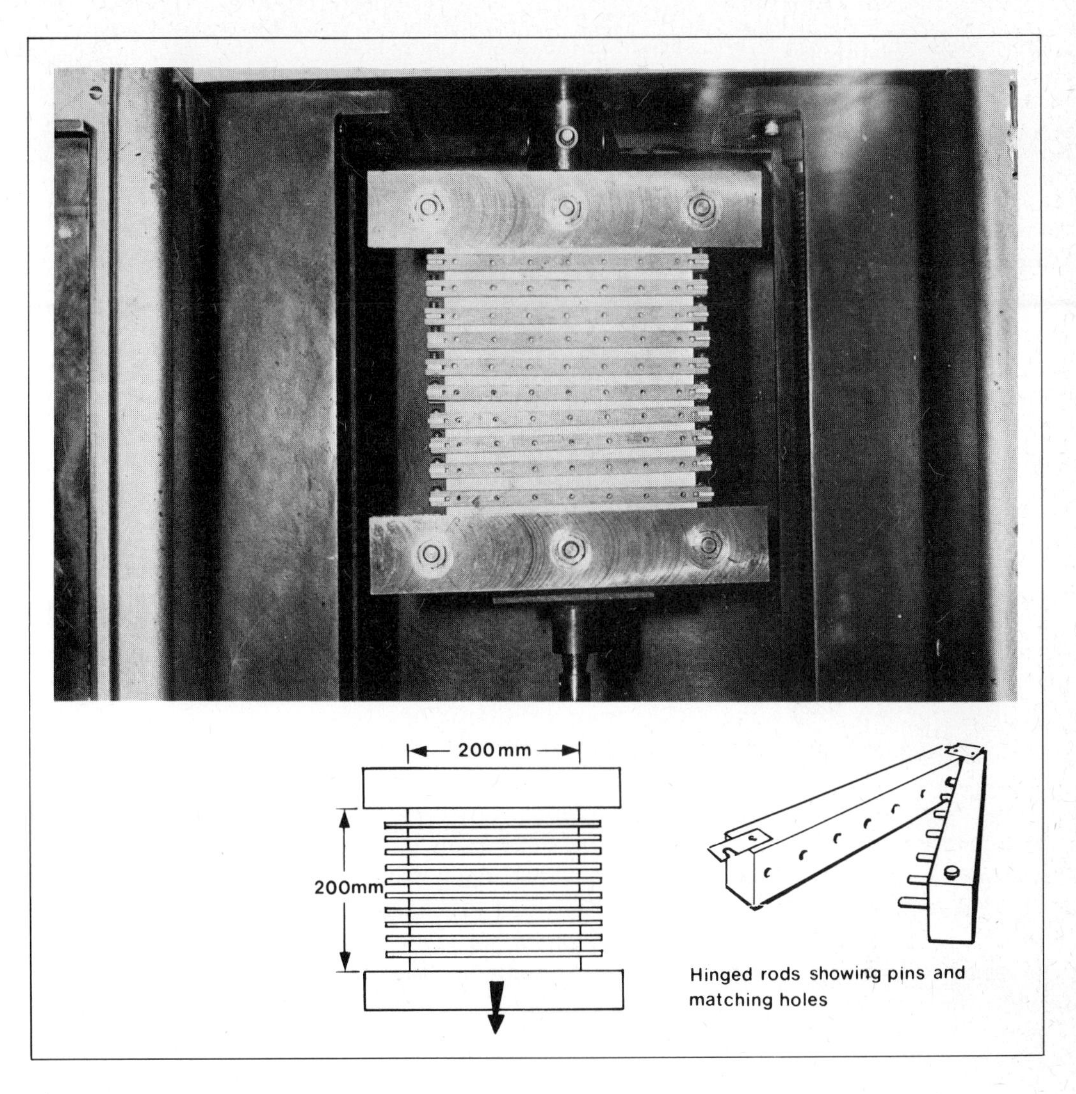

Figure 11.5 Simple plane-strain apparatus used by ICI Fibres

injected being carefully noted. During this stage of the test, the annular space between the sleeve and the casing can be maintained under a constant pressure of air, or it can be filled with water so as to simulate submerged conditions — in which case the external film or membrane is not used.

The bottom of the measuring cell rests on a fixed base, while the top can move vertically, thrusting against a calibrated dynamometric ring system which indicates the reaction when the sleeve is inflated.

Breakage occurs under a certain pressure. By this time the volume of the sleeve has increased by about 50 % in the case of spun polyester non-woven fabric, and by 200 % for spun polypropylene non-woven fabric. The other parameters are kept constant during the test, as is the height of the sample under test, and the data gathered makes it possible to determine percentage strain.

Take:

$\sigma\theta$ Circumferential tensile stress
σ_z Vertical stress
$\epsilon\theta$ Circumferential strain

From this may be calculated:

$\sigma\theta_r$ Breaking strength
E Modulus of deformation
γ Poisson's ratio

The simple plane-strain apparatus shown in Figure 11.5 is now used by ICI Fibres following extensive development. This test uses a 300 mm long × 200 mm wide specimen clamped completely across the top and bottom, in a conventional tensiometer with constant rate of extension, to leave a 200 mm test length. (The use of a 200 mm wide sample itself more realistically represents the field situation than does the conventional 50 mm textile strip.) Ten lightweight hinged wooden rods are spaced at equal distances down the sample and these contain stout pins which pierce the fabric. It is these pins which resist contraction in the width of the sample. The specimen is strained at a rate of 10 %/min and the maximum load achieved is recorded (breaking load) as also is the extension to this load (extensibility). In addition, a measure of modulus is obtained by reference to the load at 5 % extension and the area under the stress/strain curve is a measure of the work necessary to rupture the fabric (rupture energy). Plane strain tensile strength is given in *N*/200 mm and can easily be compared with the 50 mm strip test result by dividing by four.

Grab tensile test (Figures 11.6 and 11.7)
Standards: DIN 53858; ASTM D 1682. (Modified)
If it is required to test the behaviour of a fabric under a more localised concentrated stress, then the grab tensile test (as well as the burst test) is considered to be very useful. The same machine is used for this test as for the strip tensile and plane-strain tests, and is shown in Figure 11.6 where a test is just starting, and in Figure 11.7 illustrating the latter stages of the same test. This is similar to a strip test, but the surrounding fabric is allowed to contribute to the overall strength. The most easily visualised ground situation comparable with this test would be during fill spreading by earth-moving equipment, which can cause the

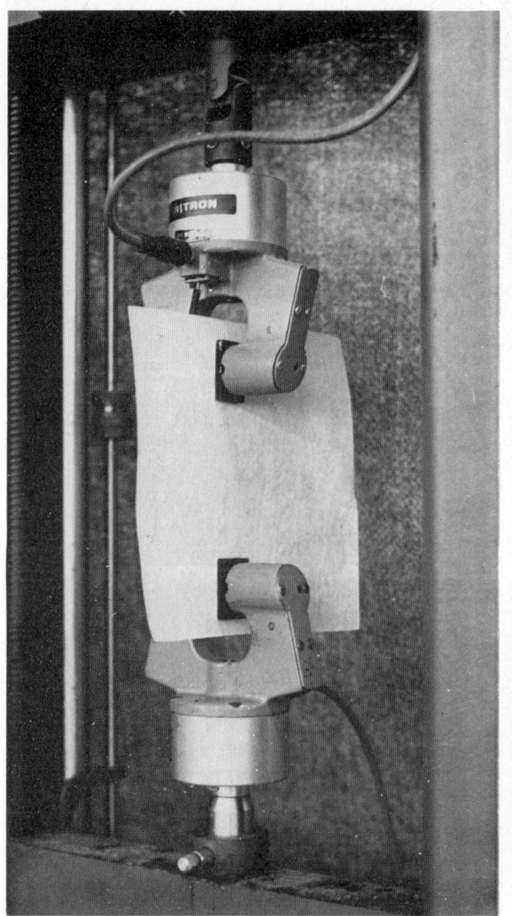

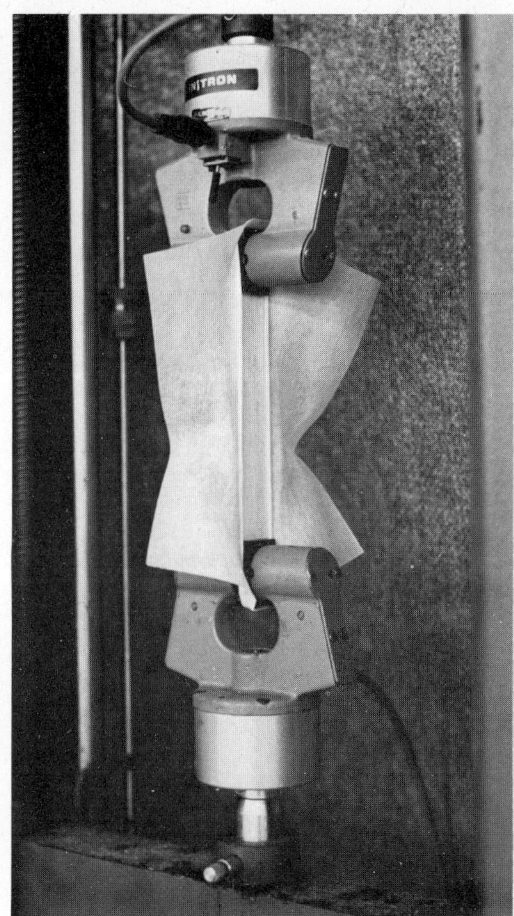

Figure 11.6 (Left) Grab tensile test — initial stage

Figure 11.7 (Right) Grab tensile test in progress

fabric to be stretched by individually-gripping rocks or groups of stones.

Conditions of test

Sample size: 200 mm × 200 mm
Jaw faces: Width 25 mm
Height 38 mm maximum
Gauge length: 100 mm
Cross-head speed: 200 mm / minute
Chart speed: 200 mm / minute
Results are quoted in Newtons

% Extension at break is measured and can vary from 10 % to 140 %.

Wing tear test (Figures 11.8-11.10)
Standards: BS 4303 (1968).
Using the same tensile testing machine as above, this test gives the load required to propagate a tear in a fabric once the tear has

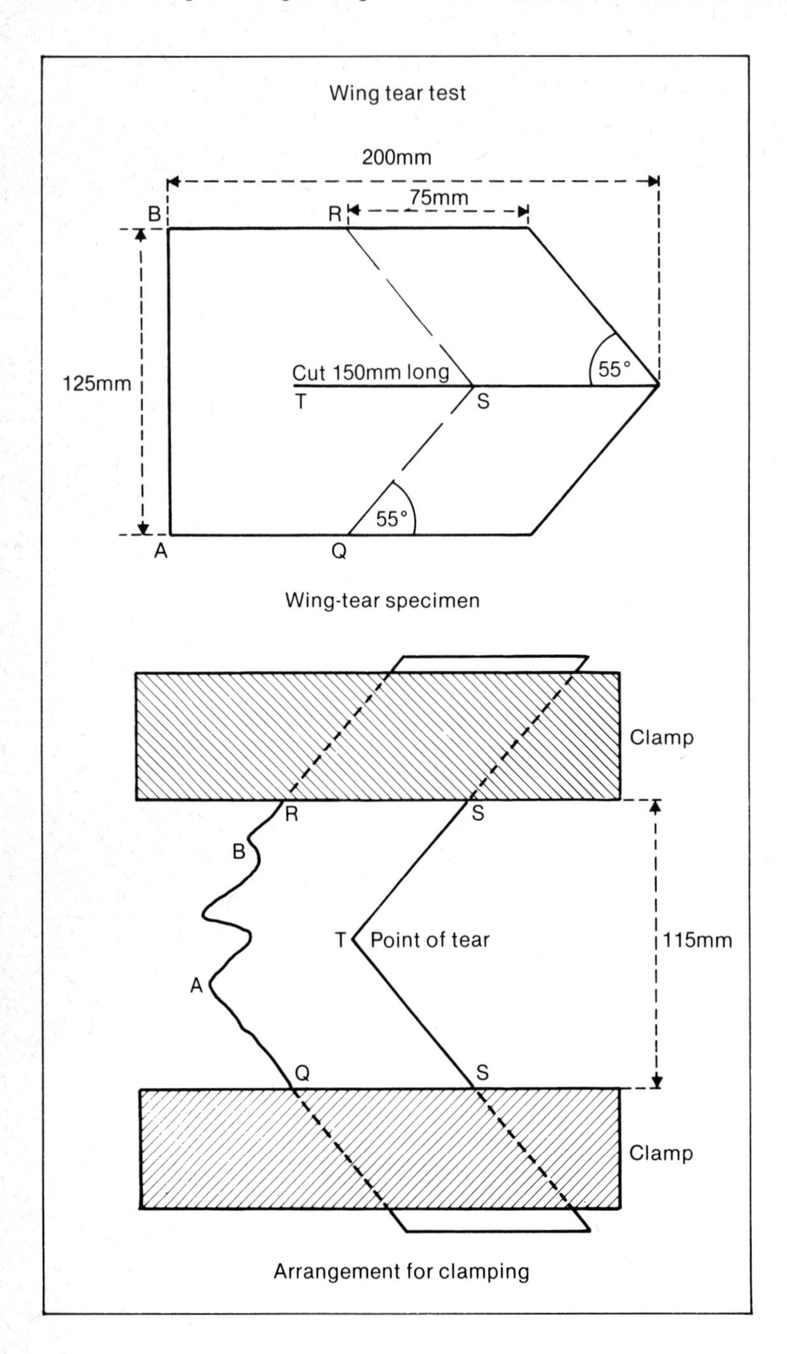

Figure 11.8 Preparation details of wing-tear specimen

Figure 11.9 (Left) Wing tear test — initial stage

Figure 11.10 (Right) Wing tear test in progress

been initiated. For example, if a membrane is pressed onto a sharp boulder by sand fill and is penetrated at one point by the boulder, will the membrane suddenly rip easily and dramatically resulting in a serious loss of sand, or will the fabric tend to cling tightly around the penetrated point? This can be vitally important for hydraulic land-fill reclamation schemes where tidal flow would wash out considerable quantities of fill through large rips.

Conditions of test

Sample size:	200 mm horizontal width prior to cut
	125 mm vertical length prior to cut
	150 mm horizontal cut (See Figure 11.8)
Jaw faces:	25 mm wide minimum
Gauge length:	115 mm
Cross-head speed:	200 mm / minute
Chart speed:	200 mm / minute

The *maximum* load is recorded in Newtons

Figure 11.11 Typical tensile test machine useful for most textile extending purposes. In this Figure, it is set up for a grab tensile test

Figure 11.12 Mullen Burst apparatus

Mullen burst test (Figures 11.12 and 11.13)
Standards: DIN 53861; BS 4768.
In this test, an impermeable rubber membrane is used to blow the fabric into the shape of a small hemisphere. The apparatus is fairly compact as shown in Figure 11.12. At first, as pressure is applied to the rubber, the fabric deforms into a dome shape as shown in Figure 11.13. Eventually, also as shown in that figure, the fabric ruptures. This test gives a measure of the pressure necessary to rupture the fabric, and simulates the effect of a small to medium-sized stone trying to punch through the fabric. Non-woven fabrics usually perform well in this test, demonstrating that they have high extensibility to cope with local deformations thus retaining their critical separation function.

Other tests (Figure 11.14)
The tensile testing machine mentioned earlier, and illustrated in Figure 11.11, can also be used for a tensile hook test and for an effective 'CBR' test. In the former, a hook rips the fabric whilst one end is held in a jaw. (This correlates well with the grab test.) In the 'CBR' test, the tensile machine forces a metal cylinder through a sheet of membrane held taut within a circular clamp. The hook test simulates a membrane being ripped sideways by a

Figure 11.13 Mullen Burst test in progress and after rupture

sharp projection such as a tree root, whilst the 'CBR' test simulates punching by a large stone.

Both of the above tests are still being assessed with regard to their validity for testing membranes, and in view of their controlled speed of strain, they are likely to prove successful. However, there is one new test — the cone drop test — which is being developed and which must be viewed at present with extreme caution. The problem with this test, in the Author's opinion, is that it may produce misleadingly good results for woven fabrics when compared with non-woven types, especially spun bondeds. This, the Author feels, could be due to the rapid rate at which the rupture is applied, which in a woven material simply displaces the warp and weft threads sideways rapidly and is arrested by friction on these threads without breaking them. In a non-woven material this sideways moving facility is not available, and thus thread groups are ruptured individually without being allowed to build up a collective strength. Therefore in the Author's opinion, at the present stage of development, this test is unsuitable for comparing one type of fabric with another in order to assess their field behaviour, and its suggested presence in the main laboratory is only to parallel its recommended presence in the field laboratory where it is used to monitor consistency of quality.

Figures 11.15-11.23 show some of the standard equipment suitable for setting up an effective main membrane testing laboratory. It should be noted that testing of fabrics in isolation is only part of this laboratory's function. Equipment and personnel must be available to test soils in conjunction with particular membranes so that an assessment can be made prior to purchase, of the ideal type of membrane for a particular project. For example, it is doubtful that the same membrane would perform equally well under an embankment as a drainage protector, on a coastal defence as an erosion protector, or beneath a short-life access road to support heavy vehicles. It might be found that the first purpose would best be achieved by a melded or spun-bonded membrane, the second by a needle-punched felt, and the third by a woven fabric. Of course, this is only conjecture, but it highlights the function of the main laboratory in determining the factors to be entered in the Tender Specification for any major purchase.

Figure 11.14 Some newer tests being evaluated at present for usefulness in assessing membrane working qualities

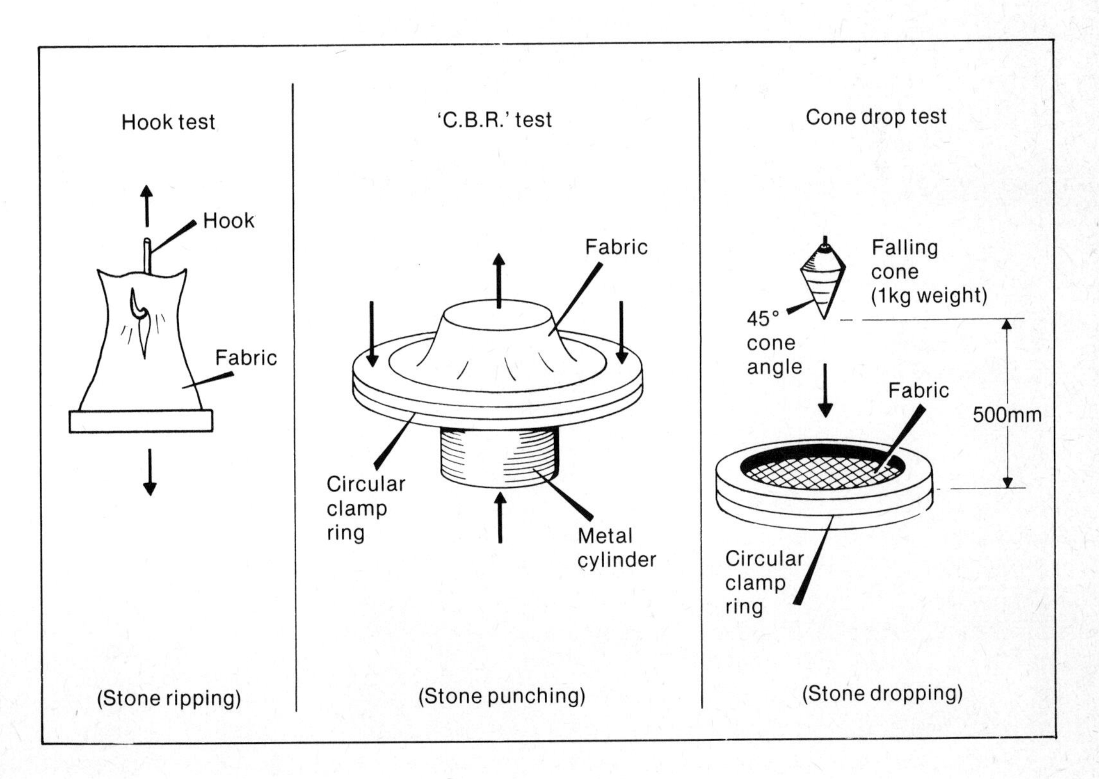

Figure 11.15 Variable head permeability tester for membranes

Figure 11.16 Permeability tester showing fabric support and fabric being placed onto support

Figure 11.17 Motor driven soil sieve

Figure 11.18 Suitable metal sink

BERKEL

800
700
600
500
400
300
200

Figure 11.19 (Opposite, left) Suitable accurate balance

Figure 11.20 (Opposite, right) Suitable coarse balance

Figure 11.21 (Bottom, left) Fine particle settlement apparatus

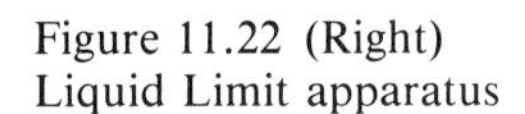

Figure 11.22 (Right) Liquid Limit apparatus

Figure 11.23 Permeameter for testing soil/fabric permeability

Field testing laboratory

Hand sample cutter (Figure 11.24)
This type of simple apparatus is worthwhile having in the field situation since it ensures consistent sample size, and can save much time in preparation work.

Strip tensile tester
A small tensile testing machine (1000 kg) similar to that shown in Figure 11.11 should be used to conduct 50 mm strip tests on fabrics upon receipt, and prior to being used. Its use in the field is simply to check for deviation from the standard fabric ordered, and to ensure that deterioration has not occurred prior to covering.

Weighing balance
The use of this apparatus is perhaps the most important method of keeping a check on the fabric coming onto the site. Of course, if a standard cutter is used as recommended, then each cut sample can be weighed prior to testing. The deviation of this repeated figure gives:

(a) a quality deviation measure — particularly for non-woven fabrics
(b) a check that, by accident, a roll of a different grade has not been enclosed with a large batch.

This is measured in weight per unit area and is most commonly expressed in grammes/m^2 (g.s.m.). Woven fabrics are fixed in warp and weft count, and these can be checked with a hand lens at so many warps or wefts per cm. However, non-woven fabrics are open to marked variation in weight, which is not easily assessed by feel or visual inspection. Therefore, a regular weight check is essential.

Cone drop test device
This device has the following specification and can easily be constructed by any straightforward engineering workshop. It is constructed as shown in Figure 11.14. When the cone is dropped through a fixed height (say 500 mm for fairly strong fabric), the diameter of the hole made (as the cone comes to rest in the fabric) is measured. This is usually measured with a conical piece of wood marked accordingly in mm diameter. The accuracy of such readings is within ± 3mm. The object of this test is to check for the deviation of factors which may not be apparent in the tensile test. For example, in a woven material which is calendered or resin set, the test could reveal a sample which was poorly set by bringing to light the lateral mobility of the warp

Figure 11.24 Hand press for cutting fabric samples using pre-shaped sharp dies

and weft threads. Or, in the case of a non-woven membrane, a roll with poor heat setting would certainly not affect the weight of the product, and might not be so obvious in the tensile tester, yet the cone drop test could pick this up.

A hand lens and simple soils apparatus should be present for keeping a check that the soils types expected from the Site Investigation works are actually proving to be present throughout the works. Extreme variations of soil type could negate the design and should be reported. It could be found that a different fabric be required in one part of the site from that needed in another part.

Other miscellaneous tests

The following tests can be conducted as specially required for particular purposes:

1. Chemical composition.
2. Chemical resistance.
3. Biological resistance.
4. Temperature stability.
5. Sunlight exposure resistance.
6. Thickness — particularly for thick fabrics which may compress.
7. Pore size — measurement of the hole sizes present in the fabric is made by vibrational sieving analysis. 100 g of well defined glass balls (Ballotini) of known diameter are shaken for 5 mins on a sample of the fabric lying in a standard 300 mm coarse sieve. The shaking rate is sufficient to cause the balls to move over the fabric surface but not so severe that the particles start hopping on the fabric. The amounts passing for various fractional sizes are then computed to give the fabric pore size distribution.

(All laboratory test photographs courtesy of ICI Fibres.)

References

Chapter 1 Historical Development of Membrane Utilisation

1. K. Terzaghi and R.B. Peck, *Soil Mechanics in Engineering Practice,* Wiley, 1967.
2. H.R. Cedergren, *Seepage, Drainage, and Flow Nets,* Wiley, 1967.
3. H.A. Agerschou, 'Synthetic Material Filters in Coastal Protection', *Journal American Society of Civil Engineers (Waterways and Harbours Division)* **87**, no. WW1 (Feb 1961), pp. 111-24.
4. Soil Testing Inc., *Report for Carthage Mills Inc.,* Job No. 6595, 1965.
5. C.C. Calhoun, *Development of Design Criteria and Acceptance Specifications for Plastic Filter Cloths.* U.S.W.E.S. Technical Report, Vicksburg, Mississippi, June 1972.
6. H.J. List, 'Investigations of Synthetic Filters Applied in Hydraulic Structures Under the Influence of Non-Stationary Flow', *Mitt. Bl.d. B.A.W.,* **21,** no. 35 (1973), pp. 124-46.
7. H.J.M. Ogink, *Investigations on the Hydraulic Characteristics of Synthetic Fabrics,* Delft Hydraulics Laboratory, Publication no. 146, May 1975.
8. D.B. Sweetland (MSc Thesis), 'The Performance of Non-Woven Fabrics as Drainage Screens in Sub-Drains', University of Strathclyde, 1977.

Chapter 2 Aspects of Soil Mechanics Relevant to Membrane Design

9. A. Casagrande, 'Classification and Identification of Soils', *Proceedings American Society Civil Engineers,* **73** (1947).
10. C.R. Scott, *In Introduction to Soil Mechanics and Foundations,* Applied Science Publishers, 1969.
11. K. Terzaghi and R.B. Peck, *Soil Mechanics in Engineering Practice,* Wiley, 1967.
12. United Kingdom Transport and Road Research Laboratory, 'Further Studies in the Compaction of Soil and the Performance of Compaction Plant', *Technical Paper No. 33,* October 1954. Also consult *Technical Paper No. 17* published by TRRL in 1950.

Chapter 3 Theoretical Functions and Behaviour of Membranes in the Soil Environment

13. A. McGown and D.B. Sweetland, *Fabric Screen Research and Development,* University of Strathclyde Report, Dept. of Civil Engineering, 1973.
14. B.D. Marks, *The Behaviour of Aggregate and Fabric Filters in Sub-drainage Applications,* University of Tennessee Report, 1975.
15. A. McGown, 'The Properties and Uses of Permeable Fabric Membranes', Research Workshop on Materials and Methods for Low-Cost Road Construction, Leura, Australia.

Chapter 4 Types of Membrane Commercially Available

16. D.J. Hoare, 'Permeable Synthetic Fabric Membranes', *Ground Engineering* (July 1978).
17. J. Savage — Chairman, 'Guidelines for the Use of Geo-Fabrics in Civil Engineering', Republic of South Africa Highway Materials Research Committee Draft Paper 1977.
18. J.H. van Leeuwen, 'Stabilising the Soil', *Middle East Construction* (July 1976).

Chapter 5 Drain Filters in One-Way Water Flow Situations

19. F.F. Zitscher, 'Recommendations For the Use of Plastics in Soil and Hydraulic Engineering', *Die Bautechnik,* **52,** 12 (1975), pp. 397-402.
20. *Designing With Terram* (1978), published by ICI Fibres, Terram Section, Pontypool, Gwent, Great Britain NP4 8YD.

21. A. Atterberg, 'Studien auf dem Gebiet der Boden Kunde' (Studies in the Field of Soil Science), *Landu Versuchsanstalt,* **69** (1908).
22. Armco Drainage & Metal Products Inc, Middleton, Ohio, USA, *Handbook of Drainage and Construction Production.*
23. ICI Publication, *City of Gothenburg Playing Fields,* Publication No. S2/A2/C3, April 1974.

Chapter 7 Marine Defences and Filters in Reversing Water Flow Situations

24. E. Hoek, *Rock Slope Stability in Opencast Mining,* Imperial College Publication, London, 1973.
25. J.V. Parcher and R.E. Means, *Soil Mechanics and Foundations,* Merrill Publishing Co., 1968.
26. L.S. Blake, *Civil Engineer's Reference Book,* Newnes & Butterworth, 1975.

Chapter 8 Inland and Marine Land Reclamation

27. ICI Publication, *Hunterston Iron Ore Terminal,* Publication C 12/1, Sept. 1976.
28. 'Fibre Matting Stabilises Rock Fill During Reservoir Construction', *Middle East Construction* (April 1976).

Chapter 9 Permanent and Temporary Roads

29. United Kingdom Transport and Road Research Laboratory, *Road Note 29,* 3rd edition, 1970.
30. C. Lawson and O.G. Ingles, *An Examination of Membrane Encapsulated Soil Layers Under Specific Australian Conditions,* University of New South Wales, Australia, Dept. of Civil Engineering, June 1975.
31. F.H. Schmidt, 'The Use of Terra Firma in Road Engineering', *Highway Engineering in Australia,* **7**, no. 6 (1975).

Glossary:

Terms used in the context of civil engineering membranes

Calendering: A method of stabilising a fabric structure using hot rollers.

Continuous Filament: A synthetic length of yarn produced continuously by plastic extrusion.

Drainage Medium: The material used in a drain (usually crushed stone) through which water passes.

Effective Size (D_n)*:* The diameter of the largest particles in the smallest *n* percent of a soil. Common examples D_{15}, D_{50}, D_{85} are used in the design of granular drains.

Fabric-wrapped Drain: A two-layer drain comprising an inner core of drainage medium and an outer fabric sheath.

Filtration: The placing against a soil of a permeable material containing pores sufficiently fine to allow water to pass out of the soil, but not to permit the passage of soil particles. The permeable material is called a 'filter' or 'filter medium'.

Flow Regime: This term can either describe the steady groundwater conditions at any particular locality, or it can describe the micro-flow conditions adjacent to a soil/filter interface, e.g. laminar, turbulent, radial, planar.

Granular Drain: A traditionally designed drain incorporating a specifically graded aggregate to suit the surrounding soil.

Heterofilament: A synthetic filament comprising a core of one polymer surrounded by a sheath of different material.

Hydraulic Conductivity: The capacity of a drain to remove a certain flow rate of seepage.

Laminar Flow: Fluid follows specific flow paths, obeys Darcy's Law when seeping through porous media.

Melded: A special ICI-patented type of spin-bonding wherein both heterofilaments and monofilaments are heat treated, but only the sheaths of the heterofilaments weld together. This provides less rigidity and greater extensibility in the resultant fabric.

Meld-stitched: A melded product which has been stitched with strong filaments, to produce a high permeability membrane with very high strength characteristics.

Monofilament: A synthetic fibre made from a single polymer.

Needle-punching: A process for producing non-woven fabrics using barbed needles to entangle the fibres.

Non-woven: A fabric produced by methods other than weaving. Fibre structure complex with a random matrix of filaments.

Percentage Frequency of Pores/Unit Area: A measure of the free space within the structure of a non-woven fabric (its porosity).

Percentage Open Area (% OA): A measure of the free space in usually a woven fabric, as a percentage of the total area of the fabric.

Piping: The washing out of fine particles from a soil by water emerging from the soil. Adjacent to drains, the piping of ground soil can cause surface settlement, and the clogging of the drains. Beneath dams, piping can cause the outwash of fine soil, resulting in increasing permeability and increased groundwater flows culminating in catastrophic failure.

Pore, Slot, Aperture: The opening of a screen.

Pore Size Distribution: The range of opening sizes within a screen.

Screen: A porous sheet physically separating two soils.

Self-induced Filtration: The phenomena of creating an upstream filter layer in the parent soil by permitting controlled soil migration through a screen.

Soil Migration: The transportation of soil particles through a screen by seepage flow.

Spun-bonded: A fabric manufacturing process wherein continuous or staple monofilaments are spun, formed into a sheet and then subject to heated pressurised rollers which weld the filaments together at their contact points.

Staple Fibre: Short lengths of filaments normally 100 to 200mm long.

Uniformity Coefficient: The ratio of the D_{60} to D_{10} sizes of the soil, i.e. a measure of the spread of soil sizes.

Warp: Filaments running in the long direction of the weaving machine.

Weft: Filaments running at 90° to the long direction of the weaving machine.

Woven: A fabric produced by weaving comprising filaments at 90° to each other.

O: Symbol representing hole diameter.

D: Symbol representing particle diameter

Appendix I:
Miscellaneous physical data and conversion tables

Conversion tables

AREA AND CAPACITY

Hectares	Ha/acres	Acres
0.41	1	2.47
0.81	2	4.94
1.21	3	7.41
1.62	4	9.88
2.02	5	12.36
2.43	6	14.83
2.83	7	17.30
3.24	8	19.77
3.64	9	22.24
4.05	10	24.71

Cu Metres	Cu m/ cu ft	Cubic feet
0.03	1	35.32
0.06	2	70.63
0.08	3	105.9
0.11	4	141.3
0.14	5	176.6
0.17	6	212.0
0.20	7	247.2
0.23	8	282.5
0.25	9	317.8
0.28	10	353.2

Litres	Litres/ galls	Gallons
4.55	1	0.22
9.09	2	0.44
13.64	3	0.66
18.18	4	0.88
22.73	5	1.10
27.28	6	1.32
31.82	7	1.54
36.37	8	1.76
40.91	9	1.98
45.46	10	2.20

LENGTH

Metres	M/ft	Feet
0.30	1	3.28
0.61	2	6.56
0.91	3	9.84
1.22	4	13.12
1.52	5	16.40
1.83	6	19.68
2.13	7	22.97
2.44	8	26.25
2.74	9	29.53
3.05	10	32.81

Centi- metres	cm/ ins	Inches
2.54	1	0.40
5.08	2	0.80
7.62	3	1.20
10.16	4	1.60
12.70	5	2.00
15.24	6	2.40
17.78	7	2.80
20.32	8	3.20
22.86	9	3.50
25.40	10	3.90

Kilo- metres	Km/ miles	Miles
1.61	1	0.62
3.22	2	1.24
4.83	3	1.86
6.44	4	2.49
8.05	5	3.11
9.66	6	3.73
11.27	7	4.35
12.88	8	4.97
14.48	9	5.59
16.09	10	6.21

WEIGHT

Grammes	G/ozs	Ounces
28.35	1	0.04
56.70	2	0.07
85.05	3	0.11
113.4	4	0.14
141.8	5	0.18
170.1	6	0.21
198.5	7	0.25
226.8	8	0.28
255.2	9	0.32
283.5	10	0.35

Kilo- grammes	Kg/ lbs	Pounds
0.45	1	2.21
0.91	2	4.41
1.36	3	6.61
1.81	4	8.82
2.27	5	11.02
2.72	6	13.23
3.18	7	15.43
3.63	8	17.64
4.08	9	19.84
4.54	10	22.05

Metric tonnes	Tonnes/ tons	Tons
1.02	1	0.98
2.03	2	1.97
3.05	3	2.95
4.06	4	3.94
5.08	5	4.92
6.10	6	5.91
7.11	7	6.89
8.13	8	7.87
9.14	9	8.86
10.16	10	9.84

Useful physical data

1 m = 39.370 ins.
1 ft = 12 ins = 30.48 cm
1 ft^2 = 929.03 cm^2
1 mile = 1.609 Km
1 mile^2 = 2.589 Km^2 = 258.9 ha.
1 acre = 4046.86 m^2 = 0.4046 ha.
1 Km^2 = 247.105 acres
1 hectare (ha) = 10,000 m^2 = 0.01 Km^2
1 ft^3 = 0.0283 m^3
1 cm/sec = 1.9685 ft/min = 1417.3 ft/day
°C = (5/9) × (°F-32)
1 lb = 453.59 gm = 0.45 kg = 0.001 kips (kilo pounds)
1 lb/ft^2 = 4.882 Kg/m^2 = 0.016 ft of water pressure

1 ft water pressure = 2989.07 N/m^2 = 0.43 psi
1 cu ft water weighs 62.4 lbs
1 cu ft water = 6.23 gallons
1 Imp. gallon water weighs 10 lbs
1 cu sec = 1 cu ft water/sec = 6.23 Imp. gall/sec
= 7.48 US gall/sec
1 cu metre = 35.33 cu ft
1 m^3/sec = 35.33 ft^3/sec
1 lb/ft^3 = 0.016 metric tons/m^3 = 16.02 kg/m^3
1 lbf/in^2 = 6894.76 N/m^2
1 Kgf/cm^2 = 98.066 kN/m^2
1 lbf = 4.448 N 1 ton force = 9.964 kN 1 kgf = 9.806 N
1 ton/ft^2 = 107.25 kN/m^2
1 lb/in^2 = 703.1 kg/m^2

Weights & measures

METRIC MEASURES AND EQUIVALENTS

Length

1 millimetre (mm)	=	0.03937 in
1 centimetre (cm)	=	0.3937 in
1 decimetre (dm)	=	3.937 ins
1 metre (m)	=	3.281 ft
1 decametre (dam)	=	10.936 yds
1 hectometre (hm)	=	109.36 yds
1 kilometre (km)	=	0.6213 miles
1 myriametre	=	6.213 miles

Area

1 sq centimetre (cm^2)	=	0.155 sq in
1 sq decimetre (dm^2)	=	15.5 sq ins
1 sq metre (m^2)	=	1.196 sq yds
1 are (a)	=	119.6 sq yds
1 hectare (ha)	=	2.4711 acres
1 sq kilometre (km^2)	=	247.105 acres
1 sq myriametre	=	38.55 sq miles

Capacity

1 centilitre (cl)	=	0.07 gill
1 decilitre (dl)	=	0.176 pint
1 litre (l)	=	1.760 pints
1 decalitre (dal)	=	2.2 gals
1 hectolitre (hl)	=	22 gals

Volume

1 cu centimetre (cm^3)	=	0.061 cu in
1 cu decimetre (dm^3)	=	61.024 cu ins
1 cu metre (m^3)	=	1.308 cu yds

Weight

1 gram (g)	=	15.432 grains
1 dekagram (dag)	=	154.32 grains
1 hectogram (hg)	=	3.527 ozs
1 kilogram (kg)	=	2.2046 lbs
1 myriagram	=	22.046 lbs
1 centner	=	110.231 lbs
1 quintal	=	220.462 lbs
1 tonne (t)	=	0.9842 ton

IMPERIAL MEASURES AND EQUIVALENTS

Length

1 inch (in)	=	2.54 cm
1 foot (ft)	=	0.3048 m
1 yard (yd)	=	0.9144 m
1 mile (ml)	=	1.6093 km

Area

1 sq inch (sq in)	=	6.4516 cm^2
1 sq foot (sq ft)	=	9.2903 dm^2
1 sq yard (sq yd)	=	0.8361 m^2
1 acre	=	4046.86 m^2
1 sq mile (sq ml)	=	259.0 hectares

Capacity

1 pint	=	0.5683 litre
1 gallon	=	4.5461 litres
1 bushel	=	36.369 litres

Volume

1 cu inch (cu in)	=	16.387 cm^3
1 cu foot (cu ft)	=	28.3167 dm^3
1 cu yard (cu yd)	=	0.7646 m^3

Weight

1 ounce (oz)	=	28.350 g
1 pound (lb)	=	0.4536 g
1 cwt	=	50.802 kg
1 ton	=	1.016 tonnes

1 lb/ft^3	=	16.018 kg/m^3
1 ton/yd^3	=	1328.94 kg/m^3
1 chain = 22 yds = 66 ft	=	20.117m
1 lb/ft^2	=	47.880 N/m^2
1 lb/ft^2	=	0.0479 kN/m^2
kN/m^2 × 102	=	kg/m^2
lb/ft^3 × 16.02	=	kg/m^2
lb/ft^3 × 0.157	=	kN/m^3
30 kPa	≒	0.33 ton/ft^2
π	=	3.142857
1 radian	=	57.272 degrees

Equivalent sieve mesh sizes

British Standard 410-1962		US Standard (1924) and ASTM (E11-61) designation			US Tyler (1910)		I.M.M. (1907)		German Standard (DIN 1171-1926)		
Mesh no.	Sieve aperture mm	Mesh no.	ASTM designation microns	Sieve aperture mm	Mesh no.	Sieve aperture mm	Mesh no.	Sieve aperture mm	Mesh per cm	Mesh per cm^2	Sieve aperture mm
					2.5	7.925					
					3	6.680					
		3.5	5,660	5.66	3.5	5.613					
		4	4,760	4.76	4	4.699					
		5	4,000	4.00	5	3.962					
5	3.35	6	3,360	3.36	6	3.327					
6	2.80	7	2,830	2.83	7	2.794					
							5	2.540			
7	2.40	8	2,380	2.38	8	2.362					
8	2.00	10	2,000	2.00	9	1.981					
10	1.68	12	1,680	1.68	10	1.651					
							8	1.574			
									4	16	1.50
12	1.40	14	1,410	1.41	12	1.397					
							10	1.270			
14	1.20	16	1,190	1.19	14	1.168			5	25	1.20
							12	1.056	6	36	1.04
16	1.00	18	1,000	1.00	16	0.991					
18	0.850	20	841	0.841	20	0.833					
							16	0.792	8	64	0.75
22	0.710	25	707	0.707	24	0.701					
							20	0.635			
25	0.600	30	595	0.595	28	0.589			10	100	0.60
									11	121	0.54
30	0.500	35	500	0.500	32	0.495			12	144	0.50
36	0.420	40	420	0.420	35	0.417	30	0.421	14	196	0.43
									16	256	0.38
44	0.355	45	354	0.354	42	0.351					
							40	0.317			
52	0.300	50	297	0.297	48	0.295			20	400	0.30
60	0.250	60	250	0.250	60	0.246	50	0.254	24	576	0.25
72	0.210	70	210	0.210	65	0.208	60	0.211			
									30	900	0.20
85	0.180	80	177	0.177	80	0.175	70	0.180			
100	0.150	100	149	0.149	100	0.147	80	0.157	40	1,600	0.15
							90	0.139			
120	0.125	120	125	0.125	115	0.124	100	0.127			
									50	2,500	0.12
150	0.105	140	105	0.105	150	0.104	120	0.107			
									60	3,600	0.102
170	0.090	170	88	0.088	170	0.089	150	0.084	70	4,900	0.088
200	0.075	200	74	0.074	200	0.074			80	6,400	0.075
240	0.063	230	63	0.063	250	0.061	200	0.063	100	10,000	0.060
300	0.053	270	53	0.053	270	0.053					
350	0.045	325	44	0.044	325	0.043					
		400	37	0.037	400	0.038					

Appendix II:
Further reading

Agarwal, A., Singh, A. and Agarwal, P.M. 1971. A Review of Environmental Deterioration of Synthetic Polymers: Plastics. Pop. Plastic, 16, pp. 1-7.

Belokon, N.F., Tatevosyan, E.L., Filatov, I.S. and Kuklin, O.P., 'Influence of Biocorrosion on Various Properties of Plastics', *Sov. Plastics,* no. 7 (1972), pp. 76-9.

Booth, G.H. and Robb, J.A., 'Bacterial Degradation of Plasticised PVC — Effect on Some Physical Properties.' *J. Appl. Chem.,* **19** (1968), p. 194.

Borisov, B.I., 'Study of the Ageing of PVC Film in Soil', *Zh. Priskl. Khim.,* **43,** no. 5 (1970), pp. 1116-20.

Callely, A.G., Jones, A.M. and Hughes, D.E., 'Microbial Degradation of Styrene Copolymers'. *Proceedings of the Conference on Degradability of Polymers and Plastics,* Institute of Electrical Engineers, London (1973), pp. 13/1-13/3.

Carlene, P.W., *Resistance to Soil Burial of Terylene, Nylon and Fibre A,* ICI Dyestuffs Division, Technical Report 18837, Issued 21.3.46, 1946.

Mondot, M., *Nouveaux procédés et matériaux d'assainissement et de drainage pour terrains en glissement,* Rapport Cerafer — Division protection contre les érosions 38/St-Martin d'Hères, 1971.

Connolly, R.A., 'Soil Burial of Materials and Structures', *The Bell System Technical Journal,* **51** (1972), pp. 1-21.

De Cost, J.B., 'Effect of Soil Burial Exposure on the Properties of Plastics for Wire and Cable', *The Bell System Technical Journal,* **51** (1972), pp. 63-86.

Dudal, R., Tavernier, R. and Osmond, D. *Soil Map of Europa (Scale 1 : 2,500,000)* FAO of the United Nations, Rome, 1966.

Evans, E.M., 'Degradability of Plastics', *Brit. Plast. Fedn.* (Tech. Editor: J.J.P. Standinger) (1973).

FAO, 'Definition of Soil Units for the Soil Map of the World', *Soil Map of the World: FAO/UNESCO Project,* World Soil Resources Office, Land and Water Development Division. FAO of the United Nations, Rome, 1968, p. 72.

FAO, 'Key to Soil Units for the Soil Map of the World', *Soil Map of the World: FAO/UNESCO Project,* World Soil Resources Office, Land and Water Development Division. FAO of the United Nations, Rome, 1970, p. 16.

Fox, R.J., *Resistance of PVC to Attack by Various Agents. A Bibliography covering the Period 1962-1971,* ICI Plastics Division Report, 24.3.72, 1972.

Gaudard, Y., 'Recherche et développement d'un matériau textile dans une application originale', *Revue Textiles chimiques,* no. 9 (Septembre 1970).

Giroud, J.P., 'L'etancheite des retenues d'eau par feuilles deroulées', Conference prononcée le 13 mars, 1973.

Gupta, U.C., *Soil Biochemistry,* Marcel Dekker, Inc., New York, 1967.

Heap, W.M. and Morrell, S.H., 'Microbiological Deterioration of Rubbers and Plastics', *J. Appl. Chem.* **18** (1968), pp. 189-94.

Huber, H. and Joerg, F. 1975. 'Effect of nitrogen oxides on plastics (PET and Polyamides).' Staub — Reinhalt Luft, 35, pp. 184-87 (In German). *Chem. Abs.,* 1975, **83,** No. 132225.

Hueck, H.J. 'The nature of the biodegradation of materials with special reference to polymers'. Proc. Conf. Degradability of Polymers and Plastics, Institute of Electrical Engineers, London, (1973) pp. 11/1-11/5.

Jeuffroy, E.G., 'Conception et construction des chaussee par'. (1967).

Jumikis, A. 'Experimental Studies on Moisture Transfer in a Salty Soil Upon Freezing as a Function of Porosity'. *Engineering Research Bulletin,* **49,** (1969).

Klein, T.H. 'Effect of soil burial exposure on the properties of structural grade reinforced plastic laminates'. *The Bell System Tech. J.,* **51** (1972), pp. 51-62.

Küster, E. and Azadi-Baksh, A., 'Studies on microbial degradation of plastic films'. Proc. Conf. Degradability of Polymers and Plastics, *Institute of Electrical Engineers,* London, (1973), pp. 16/1-16/4.

Kwei, T.K. 'Effect of soil burial exposure on the properties of electrical grade reinforced plastic laminates'. *The Bell System Tech. J.,* **51** (1972), pp. 47-9.

Lambe, W. and Whitman, R., *Soil Mechanics.* John Wiley & Sons Inc. (1969).

Leflaive, M. (L.C.P.C.) Conférence à Paris le 14 novembre 1973. Propriétés mécaniques et hydrauliques des nontissés dans les travaux publics.

Maclachlan, J., Heap, W.M. and Pacitti, J., 'Attack of Bacteria and Fungi on Rubbers and Plastics in the Tropics. A literature Review', *Microbiological deterioration in the tropics,* Soc. Chem. Ind. Monogr. no. 23, (1966), pp. 185-200.

Miner, R.J., 'Effect of Soil Burial Exposure on the Properties of Molded Plastics', *The Bell System Technical Journal,* **51** (1972), pp. 23-42.

Potts, J.E., Clendinning, R.A. and Ackart, W.B., 'The Effect of Chemical Structure on the Biodegradability of Plastics', *Proc. Conf. Degradability of Polymers and Plastics,* Inst. Electr. Engineers, London, pp. 12/1-12/9.

Puig, J., Blivet, J.C., 'Remblai à talus vertical armé avec un textile synthétique', *Bulletin de Liaison laboratoire,* Ponts & Chaussées 64- Mars/Avril.

Rosen, W.J. and Marks, B.D., 'Investigation of Filtration Characteristics of a Nonwoven Fabric Filter', *Transportation Research Record No 532,* Nat. Research Council, Washington, D.C., 1975.

Sakharov, B.L., 'Action of Such Chemical Reagents as Fertilizers, Toxic Chemicals, and a Soil Suspension of Polymeric Materials', *Plast. Massy.* **9** (1975), p. 71. (In Russ.) *Chem. Abs.* **84,** no. 18151 (1976).

Scott, K.A. and Paul, K.T., 'Weathering Resistance of Plastics Composites', *Composites,* **5** no. 5 (1974), pp. 201-8.

Stanley, T.A. and Associates from ICI Plastics Division, *Chemical Resistance of Plastics,* ICI Engng. Res. Brochure on Plastics, ICI Ltd, 1955.

'The chemical resistance of ''Alkathene'' polyethylene', *Technical Service Note A 101,* ICI Plastics Division, 1972.

Tubis, R.I., *The Degradation of Nylon by Iron Rust.* US Dept. of Commerce, Tech. Rep. No. 8-74, Natl. Techn. Inf. Service, 1974.

Zeronian, S.H. *et al.,* 'Reaction of Fabrics made from Synthetic Fibres to Air Contaminated with Nitrogen Dioxide, Ozone, or Sulpher Dioxide', *Proc. Int. Clean Air Congr.,* 2nd (1970), pp. 468-76. *Chem. Abs.* **77,** no. 115793 (1972).

Zeronian, S.H. *et al.,* 'Effect of Sulphur Dioxide on the Chemical and Physical Properties of Nylon 66', *Textile Res. J.,* **43** (1973), pp. 228-37.

Index